STATE BOARD FOR
TECHNICAL AND
COMPREHENSIVE EDUCATION

TECHNICAL COLLEGE OF THE LOWCOUNTRY
LEARNING RESOURCES CENTER
P. O. BOX 1288
100 S. RIBAUT ROAD
BEAUFORT, S. C. 29901

THE BIOLOGY AND MANAGEMENT OF LOBSTERS

Volume II

Panulirus Cygnus, the West Australian Spiny Lobster. Photo by C. Purday.

THE BIOLOGY AND MANAGEMENT OF LOBSTERS

Volume II

Ecology and Management

Edited by

J. STANLEY COBB

Department of Zoology
University of Rhode Island
Kingston, Rhode Island

BRUCE F. PHILLIPS

CSIRO, Division of Fisheries and Oceanography
North Beach, Western Australia
Australia

ACADEMIC PRESS 1980
A Subsidiary of Harcourt Brace Jovanovich, Publishers
New York   London   Toronto   Sydney   San Francisco

COPYRIGHT © 1980, BY ACADEMIC PRESS, INC.
ALL RIGHTS RESERVED.
NO PART OF THIS PUBLICATION MAY BE REPRODUCED OR
TRANSMITTED IN ANY FORM OR BY ANY MEANS, ELECTRONIC
OR MECHANICAL, INCLUDING PHOTOCOPY, RECORDING, OR ANY
INFORMATION STORAGE AND RETRIEVAL SYSTEM, WITHOUT
PERMISSION IN WRITING FROM THE PUBLISHER.

ACADEMIC PRESS, INC.
111 Fifth Avenue, New York, New York 10003

United Kingdom Edition published by
ACADEMIC PRESS, INC. (LONDON) LTD.
24/28 Oval Road, London NW1 7DX

Library of Congress Cataloging in Publication Data
Main entry under title:

The Biology and management of lobsters.

    Includes bibliographies and index.
    CONTENTS: v. 1. Physiology and behavior.--v. 2. Ecology and management.
    1. Lobsters. 2. Lobster fisheries. I. Cobb, J. Stanley. II. Phillips, Bruce F.
QL444.M33B56   595.3'841   79-6803
ISBN 0-12-177402-3 (v. 2)

PRINTED IN THE UNITED STATES OF AMERICA

80 81 82 83   9 8 7 6 5 4 3 2 1

Contents

List of Contributors   ix

Preface   xi

Contents of Volume I   xiii

Part I   ECOLOGY

Introduction
RICHARD F. FORD
- Text   3
- References   7

1 Larval Ecology
B. F. PHILLIPS AND A. N. SASTRY
- I. Introduction   11
- II. Clawed Lobsters   12
- III. Spiny, Slipper, and Coral Lobsters   29
- IV. Conclusions   47
- References   48

2 Ecology of Juvenile and Adult Palinuridae (Spiny Lobsters)
P. KANCIRUK
- I. Introduction   59
- II. Palinurid Habitat   60
- III. Palinurid Reproductive Ecology   69
- IV. Palinurid Fisheries and Ecology   78
- V. Behavioral Ecology   79
- VI. Summary   91
- References   92

3 Ecology of Juvenile and Adult *Homarus*
R. A. COOPER AND J. R. UZMANN

I.	Introduction	97
II.	Distribution	98
III.	Population Structure	115
IV.	Behavior and Activity Rhythms	122
	Appendix	133
	References	139

4 Ecology of Juvenile and Adult *Nephrops*
C. J. CHAPMAN

I.	Introduction	143
II.	Habitat	145
III.	Behavior and Activity Rhythms	151
IV.	Population Structure and Life Cycle	162
V.	Conclusions	174
	References	175

Part II MANAGEMENT

Introduction
D. A. HANCOCK

Text	181
References	188

5 Population Dynamics of Spiny Lobsters
G. R. MORGAN

I.	Introduction	189
II.	Population Parameters	191
III.	Population Dynamics Models	204
IV.	Conclusions	211
	References	213

6 Population Dynamics of Clawed Lobsters
S. B. SAILA AND G. MARCHESSEAULT

I.	Introduction	219
II.	Vital Statistics	224
III.	Population Dynamics—Models	229
	References	239

7 Spiny Lobster Fisheries Management
B. K. BOWEN

I.	Introduction	243
II.	Major Spiny Lobster Fisheries	244

	III.	The Western Australian Fishery	246
	IV.	The Limitation of Fishing Effort	256
	V.	Concluding Comments	261
		References	263

8 The Clawed Lobster Fisheries
R. L. DOW

	I.	Introduction	265
	II.	Methods of Capture	267
	III.	*Nephrops* Fisheries	271
	IV.	*Metanephrops* and *Nephropsis*	276
	V.	*Homarus* Fisheries	277
	VI.	Effects of Sea Surface Temperature Cycles on Landings of *H. americanus*, *H. gammarus*, and *Nephrops*	300
	VII.	Conclusions	304
		Appendix	306
		References	313

9 Perspectives on European Lobster Management
DAVID B. BENNETT

	I.	Introduction	317
	II.	The European Lobster	318
	III.	The Norway Lobster	327
		References	331

10 Aquaculture
JON C. VAN OLST, JAMES M. CARLBERG, AND JOHN T. HUGHES

	I.	Introduction	333
	II.	Culture of Palinurid and Scyllarid Lobsters	335
	III.	Culture of Nephropid Lobsters	337
	IV.	Conclusions	378
		References	378

Index

385

List of Contributors

Numbers in parentheses indicate the pages on which the authors' contributions begin.

David B. Bennett (317), Fairways, Burnham-on-Crouch, Essex CM0 8NR, England

B. K. Bowen (243), Department of Fisheries and Wildlife, Perth, Western Australia 6000

James M. Carlberg (333), Department of Biology, San Diego State University, San Diego, California 92182

C. J. Chapman (143), Marine Laboratory, Aberdeen AB9 8DB, Scotland

R. A. Cooper (97), Manned Undersea Research and Technology, National Marine Fisheries Service, Woods Hole, Massachusetts 02543

R. L. Dow (265), Department of Marine Resources, Augusta, Maine 04333

Richard F. Ford (3), Center for Marine Studies, San Diego State University, San Diego, California 92182

D. A. Hancock (181), Western Australian Marine Research Laboratories, North Beach, Western Australia 6020

John T. Hughes (333), Massachusetts Division of Marine Fisheries, Lobster Hatchery and Research Station, Vineyard Haven, Massachusetts 02568

P. Kanciruk (59), Environmental Sciences Division, Oak Ridge National Laboratory, Oak Ridge, Tennessee 37830

G. Marchesseault (219), New England Regional Fisheries, Management Council, Peabody, Massachusetts 01960

G. R. Morgan* (189), Western Australian Marine Research Laboratories, North Beach, Western Australia 6020

B. F. Phillips (11), CSIRO, Division of Fisheries and Oceanography, North Beach, Western Australia 6020

S. B. Saila (219), Graduate School of Oceanography, University of Rhode Island, Kingston, Rhode Island 02881

*Present Address: Kuwait Institute for Scientific Research, Safat, Kuwait

A. N. Sastry (11), Graduate School of Oceanography, University of Rhode Island, Kingston, Rhode Island 02881

J. R. Uzmann (97), Manned Undersea Research and Technology, National Marine Fisheries Service, Woods Hole, Massachusetts 02543

Jon C. Van Olst* (333), Department of Biology, San Diego State University, San Diego, California 92182

---

*Present Address: California Mariculture Corporation, La Jolla, California 92037

Preface

The animals generally called lobsters fall into several taxonomically distinct groups: the clawed lobsters (Nephropidae), the spiny lobsters (Palinuridae), the slipper lobsters (Scyllaridae), and the coral lobsters (Synaxidae). Despite the taxonomic distinctions, there are many reasons to treat these animals together in a single text. Most notably, they are large, abundant animals that play important roles in the ecosystems in which they are found, and virtually all the abundant species are subject to intense and similarly applied fishing pressure. Much attention has been directed toward management of the valuable lobster stocks and to research related to this management. The study of lobster biology has a long history that well illustrates the intertwining of basic and applied research. All of the contributors to this volume are actively involved in attempts to understand lobster biology more fully and to apply that understanding to aiding management decisions.

In the first of these two volumes an introductory chapter on the general biology of lobsters was included to provide an overview that we hoped would be helpful as a background for the entire work. The remainder of the first volume deals with lobster physiology and behavior. In this volume lobster ecology, population dynamics, management, and culture are discussed in depth.

While the information presented in this book on the clawed lobsters and spiny lobsters is relatively voluminous, it is remarkable that information on the slipper lobster is very sparse, and knowledge of the coral lobster is almost nonexistent. All the authors were specifically asked to include information about the scyllarids, but almost uniformly their response was that there was not enough information to make any sort of statement. We considered but rejected the notion of including a chapter on the ecology of scyllarid juveniles and adults because, although some information exists, it is very fragmented and does not begin to present a satisfactory picture of scyllarid ecology. Despite the lack of information, the Scyllaridae are a widely dispersed group that forms an important fraction of the larger benthic decapods of the world's oceans.

It is axiomatic that sound management practices and appropriate culture techniques must be based on good knowledge of the biology of the animal. In the 70 years since F. H. Herrick wrote his classic monograph concerning one of the lobster species, *The Natural History of the American Lobster,* a great deal more has been learned about the biology of all lobsters. Much of the new knowledge has arisen through the use of new tools. Some of the techniques used in the research reviewed in this volume are illustrative: scuba, submersibles, submarine television, sophisticated bioeconomic models, growth equations appropriate to crustaceans, and advanced technology for culture systems. It is through the use of this new set of research instruments that a firm basis for rational management can be laid. Such management practices are badly needed. Many of the world's lobster fisheries are overfished and overcapitalized, with management strategies based on knowledge a half-century old. However, biological knowledge is not the only prerequisite for sound management. In many countries it appears that the management of heavily fished lobster populations is as much a political as it is a biological and economic problem. The political situation is not dealt with in this book. However, we hope that the biological information combined with reviews of the status of several of the fisheries presented in the following pages will aid in the futher development and refinement of management practices for the many species of lobsters.

We believe that, in an overall sense, more is known about the biology of lobsters, particularly *Homarus americanus*, than any other aquatic invertebrate. Despite this, great holes in our knowledge persist. In this book we have tried not only to compile information but to present an integrated view of basic and applied biology while pointing out areas of exciting current and future research. If lobster research progresses at its current pace, we may soon be able to put together a whole view of the animal from evolutionary history through physiological mechanisms to ecology and behavior. Such a picture will emerge only if both formal and informal communication between workers continues. These volumes are intended to stimulate and foster such communication as well as to provide a baseline of information from which to work.

Many friends and colleagues have helped in the preparation of this book, and to them all we give our sincere thanks. In addition to those noted in the Preface to Volume I the following deserve special mention: P. F. Berry, Oceanographic Research Institute, Durban; N. G. Hall, Department of Fisheries and Wildlife, Western Australia; M. W. Johnson, Scripps Institute of Oceanography, San Diego; G. Newman, Department of Industries, Capetown; and D. Reid, CSIRO Mathematics and Statistics, New South Wales. The author of Chapter 4 thanks B. B. Parrish and J. Mason for comments and the following for permission to quote or reproduce unpublished material: R. J. A. Atkinson, O. Bagge, A. Charuau, H. Eiriksson, A. S. D. Farmer, A. Fernandez-Garcia, M. M. Figueiredo, J. P. Hillis, G. G. Howard, J. A. Kitching, R. Jones, A. C. Simpson, D. J. Symonds, H. J. Thomas, P. Warren, B. I. Dybern, and S. Hay. Both of the editors are deeply grateful for the superb assistance given by P. Barbour, S. Dennerlein, J. Downey, S. Koerner, G. McBride, R. Saunders, and C. Squires.

J. S. Cobb
B. F. Phillips

Contents of Volume I
Physiology and Behavior

General Biology
B. F. Phillips, J. S. Cobb, R. W. George

Part I.   Physiology

Introduction
W. Dall
Molting and Growth
D. E. Aiken
Neurobiology
B. W. Ache and D. L. MacMillan
Reproductive Biology
D. E. Aiken and S. L. Waddy
Nutrition
D. E. Conklin
Diseases
J. E. Stewart

Part II.   Behavior

Introduction
J. S. Cobb
Spiny Lobsters: Patterns of Movement
W. F. Herrnkind
Social Behavior
J. Atema and J. S. Cobb
Index

Part I

ECOLOGY

Introduction

RICHARD F. FORD

From the information described in Chapters 1-4 and in the closely related section concerning fisheries management, it is clear that our knowledge about the ecology and habitat behavior of lobsters is extensive. Research in these areas of lobster biology has increased markedly over the last 10 years.

Both the interest in lobsters and the amount of effort devoted to investigating their ecology and behavior are the result of several factors. Surely one of the most important of these is that lobsters are particularly interesting animals, from the standpoint of both their appearance and what they do. As juveniles and adults they occupy a variety of tropical, temperate, and boreal marine habitats in both shallow and deep water. In many cases they are important carnivores and prey species in the associated benthic communities. They have unusual morphological and behavioral adaptations, modes of life, and population characteristics, both as planktonic larvae and during their benthic existence.

Most species of lobsters inhabiting benthic areas in shallow water are readily accessible for study in nature, in part because of their large size, their relatively high densities, and their tendency to remain in or near established shelters. Most of these species also are tolerant of environmental conditions and handling in laboratory aquarium systems. These characteristics make them excellent subjects for field and laboratory studies.

Another equally important factor is that species of the families Nephropidae (clawed lobsters) and Palinuridae (spiny lobsters) support some of the most valuable fisheries in the world. Because of this, the population ecology and related characteristics of the species involved in these fisheries have been investigated extensively in some cases, using methods of the fisheries biologist.

The catch per unit effort for many lobster fisheries has declined markedly, while the demand for both nephropid and palinurid lobsters and the monetary value of the catch continue to increase. These trends have intensified interest in learning more about the habitat requirements, population ecology, and fisheries-oriented behavior of the species involved, so that effects of the fishery can be evaluated and predicted as a basis for effective management. These trends also have intensified interest in developing fisheries for other species and stocks of lobsters, as for example of *Panulirus penicillatus* in Micronesia (MacDonald, 1971; George, 1972; Ford and Ebert, 1979). Efforts in such fisheries-oriented studies have provided much of the information now available concerning the environmental relationships and behavior of nephropid and palinurid lobsters during their benthic stages.

One major disadvantage of this emphasis on studying exploited lobster species is that we know considerably less about the remaining ones and, in the case of many species, almost nothing at all. This is particularly true in general of the family Scyllaridae (slipper lobsters), the family Synaxidae (coral lobsters), deep water species of the Palinuridae, and many other palinurids which are not of commercial importance or support only minor fisheries. Such gaps in our knowledge are unfortunate, because many of these forms differ from the well-studied species in their larval life histories and benthic habitat requirements, and presumably also in their population characteristics and behavior.

For example, it is significant that there is essentially no discussion of ecological and behavioral characteristics of the Scyllaridae in this volume, other than that concerning larval forms in Chapter 1. This simply reflects the fact that very little is known and still less has been published about these characteristics of slipper lobsters. Similarly, what we know about the ecological characteristics of palinurid lobsters is based on detailed studies of relatively few species, all of which inhabit shallow water areas. Thus, while our knowledge of the ecology of lobsters is substantial, it is not nearly as comprehensive as it should be.

A second major problem resulting from our reliance on studies of exploited lobster populations is emphasized in Chapters 2-4 and in Chapter 7 of Vol. I. For practical reasons, fisheries biologists studying lobsters have tended to obtain ecological and behavioral data that have direct bearing on fishery problems and that can be obtained primarily by sampling from the fishery itself. These fishery data concerning distribution and abundance, size structure of the population, mortality, growth, reproduction, movements, and other fishery-related behavior are valuable not only to the fisheries biologist, but to all those studying lobster populations. Yet, by itself this information does not provide an adequate description of the basic ecological and behavioral characteristics of the lobster species considered. A related problem is that standard trapping and trawling methods used in many lobster fisheries tend to be highly selective against certain life history stages in the population, because of their size and behavior and the characteristics of the benthic habitats sampled. In addition, these fishery sampling methods usually do not provide particularly accurate quantitative estimates of distribution and abundance of the kind needed in detailed studies of habitat requirements and population ecology.

Introduction

Many of the more recent studies described in Chapters 1-4 reflect the fact that those studying the ecology and behavior of lobsters are now relying on a more balanced approach, using contemporary methods of the population ecologist and the behaviorist, as well as traditional approaches of the fisheries biologist. The result has been that for the benthic stages of several species, including *Homarus americanus*, *Panulirus argus*, and *P. cygnus*, we now have an adequate understanding of their habitat requirements, feeding relationships, population ecology, and, in some cases, their environmental tolerances and basic behavioral characteristics.

Ecological and behavioral studies of *H. americanus* by Cooper and Uzmann (Chapter 3), and by Atema, Cobb, and their colleagues (Vol. I, Chapter 8) illustrate this approach. Similar examples from work on palinurid species include studies of *P. cygnus* in Western Australia by Chittleborough (1974, 1975, 1976), Phillips (1975a,b), Phillips and Hall (1978), and Phillips *et al*. (1978); those of *P. homarus* and other species in South Africa by Berry (1970, 1971, 1973); and those of *P. argus* in the United States (Herrnkind and McLean, 1971; Herrnkind *et al.*, 1973, 1975; Cooper *et al.*, 1975; Davis, 1977).

This more basic and balanced approach to ecological and behavioral studies of lobsters also is providing valuable data that have direct applications in fisheries ecology and technology. For example, as Chapman emphasizes in Chapter 4, detailed information about the burrowing behavior and associated activity rhythms of *Nephrops norvegicus* provides a very useful basis for interpreting variations in trawl catches by the fishery. Similarly, studies such as those of Chittleborough (1974, 1975, 1976) concerning homing and dominance behavior and the specific effects of environmental factors on growth and survival of juvenile *Panulirus cygnus* provide valuable information for evaluating and managing the fishery. Such information could not be obtained from the traditional approaches to fisheries biology alone.

Unlike most scientific fields, which have relatively well-defined boundaries, ecology is for practical reasons "an exceptionally multidisciplinary kind of biology" (Deevey, 1964). Both the ecologist and the ecologically oriented behaviorist seek to understand the reasons why, where, and when lobsters occur, their reactions to environmental conditions, and the specific functional capabilities that allow them to respond to remain in a state of dynamic equilibrium with their physical and biological environment. In the process these workers must often rely on information and methods from allied disciplines, including most notably functional morphology and physiology.

For example, as Travis (1954), Passano (1960), Chittleborough (1975), and others have observed, growth plays a dominant role in the life processes of lobsters and other crustaceans, at the level of both the individual and the population. Reproduction, population size structure, susceptibility to predation, and many aspects of behavior and related sensory functions are affected directly or indirectly by the complex molting cycle involved in the growth of lobsters. Many of these life processes and a variety of other ecological factors in turn affect growth (Chittleborough, 1975, 1976). Similarly, there are other important interactions between the physiological processes of reproduction and the ecological characteristics of lobsters.

As Bartholomew (1966) points out, a multidisciplinary perspective and approach on the part of physiologists, ecologists, and behaviorists is not only necessary, but also very important:

> If one attempts to attain an adequate understanding of the relation of an organism to its environment he addresses himself to a problem of such enormous complexity that he must, unfortunately, be reconciled from the outset to obtaining an incomplete answer. The task, which all scientists face, of isolation and simplification of problems is present in particularly acute form to the student of ecologically relevant physiology and behavior. He cannot reduce his problem until only a single variable remains; he cannot restrict his data to a single level of biological integration, nor, as is often the case in most other biological disciplines, even to several adjacent levels. Furthermore, he cannot limit his data gathering to techniques of any one speciality. More than most students, he must recognize that biology is a continuum. Whatever his technique and methods, he must be a naturalist.

The value of this integrative, multidisciplinary approach to lobster research is evident in many of the studies on ecology and behavior reviewed in Chapters 1–4 and in closely related discussions concerning studies of growth and molting processes (Vol. I, Chapter 2), and reproductive physiology (Vol. I, Chapter 4). Also evident is the importance of the modern naturalist's approach, emphasized by Bartholomew and others.

Even when studying benthic populations of lobsters in shallow water, ecologists often are limited in their ability to make long-term observations under natural conditions and to conduct field experiments of the kind common in research on terrestrial animals. As a result, they have come to rely in part on realistic laboratory studies that they design and interpret on the basis of what is observed in nature. Where possible, the more satisfactory approach of conducting closely related laboratory and field studies is employed (Chapman and Rice, 1971; Rice and Chapman, 1971; Chittleborough, 1976).

These problems are still more difficult in conducting ecological and behavioral studies of the planktonic larval stages. As Phillips and Sastry emphasize in Chapter 1, such information on lobster larvae is still very limited and fragmentary. The primary reasons for this are the difficulties and expense of obtaining adequate quantitative samples and our inability to make direct observations of these larval forms in nature.

Recent development of several semiquantitative sampling devices for puerulus larvae of palinurids (Serfling and Ford, 1975; Phillips, 1975a,b; Phillips and Hall, 1978) has made it possible to study this important life history stage effectively for the first time. Similarly, special sampling programs, such as that of Phillips et al. (1978), have been successful in improving our understanding of the environmental factors that affect distribution and abundance of phyllosoma and puerulus larvae.

For obvious reasons, there has been much reliance on laboratory studies to investigate developmental, ecological, and behavioral characteristics of lobster larvae in greater detail. Despite the limitations of such studies, caused by problems of providing an adequate environment and the intrusion of other laboratory artifacts, this approach has been used very effectively in research on nephropid, palinurid, and scyllarid larvae. Studies such as those of Ennis (1975), concerning behavioral responses of *Homarus americanus* larvae to hydrostatic pressure and light, Mitchell (1971) on food preferences and feeding behavior of *Panulirus interruptus* phyllosomes, and Robertson (1968) on the larval development of *Scyllarus americanus* are representative of what can be accomplished by using this approach. Laboratory work conducted with the primary aim of developing methods for commerical aquaculture (Chapter 10) also has improved our knowledge concerning the environmental tolerances and behavior of *H. americanus* larvae (Ford *et al.*, 1975; Dorband *et al.*, 1976).

New techniques also have improved our ability to study benthic populations of lobsters. Underwater television and photography were used by Chapman (1979) to determine densities of *Nephrops norvegicus* in deep water. Diving has been used very effectively in ecological and behavioral studies of many species (Herrnkind *et al.*, 1973, 1975; Davis, 1977). This observational approach also has been extended into deeper water by using submersibles in studies such as those of Chapman (1979) on *N. norvegicus,* and Cooper and Uzmann on the offshore population of *H. americanus* (Chapter 3). Future studies of other lobster species in deep water probably will rely heavily on these techniques.

The following four chapters provide a comprehensive treatment of what we know about the ecology and habitat behavior of lobsters. On one point all of the authors agree—there is much that we do not know, and more research is needed in almost all areas. One major function of this section should be to stimulate interest in studying many of these unsolved problems.

REFERENCES

Bartholomew, G. A. (1966). Interaction of physiology and behavior under natural conditions. *In* "The Galapagos: Proceedings of the Galapagos International Scientific Project of 1964" (R. I. Bowman, ed.), pp. 39-45. Univ. of California Press, Berkeley.

Berry, P. F. (1970). Mating behavior, oviposition and fertilization in the spiny lobster *Panulirus homarus* (Linnaeus). *S. Afr. Oceanogr. Res. Inst., Invest. Rep.* No. 24.

Berry, P. F. (1971). The biology of the spiny lobster *Panulirus homarus* (L.). *S. Afr. Oceanogr. Res. Inst., Invest. Rep.* No. 28, pp. 1-76.

Berry, P. F. (1973). The biology of the spiny lobster *Panulirus delagoae* Barnard, off the coast of Natal. *S. Afr. Oceanogr. Res. Inst., Invest. Rep.* No. 31.

Chapman, C. J. (1979). Some observations on populations of the Norway lobster, *Nephrops norvegicus* (L.) using diving, television and photography. *Rapp. P.-V. Reun., Cons. Perm. Int. Explor. Mer.* **175**, 127-133.

Chapman, C. J., and Rice, A. L. (1971). Some direct observations on the ecology and behavior of the Norway lobster, *Nephrops norvegicus*. *Mar. Biol.* **10**, 321-329.

Chittleborough, R. G. (1974). Home range, homing and dominance in juvenile western rock lobsters. *Aust. J. Mar. Freshwater Res.* **25**, 227-234.

Chittleborough, R. G. (1975). Environmental factors affecting growth and survival of juvenile western rock lobsters, *Panulirus longipes* (Milne-Edwards). *Aust. J. Mar. Freshwater Res.* **26**, 177-196.

Chittleborough, R. G. (1976). Growth of juvenile *Panulirus longipes cygnus* George on coastal reefs compared with those reared under optimal environmental conditions. *Aust. J. Mar. Freshwater Res.* **27**, 279-295.

Cooper, R. A., Ellis, R., and Serfling, S. (1975). Population dynamics, ecology and behavior of spiny lobsters, *Panulirus argus*, of St. John, U.S.V.I. 3. Population estimation and turnover. Results of the Tektite Program, Vol. 2. *Sci. Bull., Nat. Hist. Mus., Los Angeles Cty.* **20**, 23-30.

Davis, G. E. (1977). Effects of recreational harvest on a spiny lobster, *Panulirus argus*, population. *Bull. Mar. Sci.* **27**, 223-236.

Deevey, E. S., Jr. (1964). General and historical ecology. *BioScience* **14**, 33-35.

Dorband, W. R., Van Olst, J. C., Carlberg, J. M., and Ford, R. F. (1976). Effects of chemicals in thermal effluent on *Homarus americanus* maintained in aquaculture systems. *Proc. Annu. Meet.—World Maric. Soc.* **7**, 391-414.

Ennis, G. P. (1975). Behavioral responses to changes in hydrostatic pressure and light during larval development of the lobster, *Homarus americanus*. *J. Fish. Res. Board Can.* **32**, 271-281.

Ford, R. F., and Ebert, T. A. (1979). Population characteristics and fishery potential of spiny lobsters at Enewetak Atoll. *Annual Rep. Fiscal Year 1978, Mid-Pac. Mar. Lab., Enewetak, Marshall Islands* pp. 20-23.

Ford, R. F., Van Olst, J. C., Carlberg, J. M., Dorband, W. R., and Johnson, R. L. (1975). Beneficial use of thermal effluent in lobster culture. *Proc. Annu. Meet.—World Maric. Soc.* **6**, 509-515.

George, R. W. (1972). "South Pacific Islands—Rock Lobster Resources. Report Prepared for the South Pacific Island Fisheries Development Agency," Doc. WS/C7959. Food Agric. Organ. U.N., Rome.

Herrnkind, W. F., and McLean, R. B. (1971). Field studies of homing, mass emigration and orientation in the spiny lobster, *Panulirus argus*. *Ann. N.Y. Acad. Sci.* **188**, 359-377.

Herrnkind, W. F., Kanciruk, P., Halusky, J., and McLean, R. (1973). Descriptive characterization of mass autumnal migrations of spiny lobster, *Panulirus argus*. *Gulf Caribb. Fish Inst., Univ. Miami, Proc.* **25**, 79-98.

Herrnkind, W. F., Vanderwalker, J., and Barr, L. (1975). Population dynamics, ecology and behavior of spiny lobster, *Panulirus argus*, of St. John, Virgin Islands: Habitation and pattern of movements. Results of the Tektite Program, Vol. 2. *Sci. Bull., Nat. Hist. Mus., Los Angeles Cty.* **20**, 31-45.

MacDonald, C. D. (1971). Final report and recommendations to the U.S. Trust Territory Government on the spiny lobster resources of Micronesia. *Mar. Res. Div., Dep. Res. Dev., Saipan, Mariana Islands* pp. 1-82.

Mitchell, J. R. (1971). Food preferences, feeding mechanisms, and related behavior in phyllosoma larvae of the California spiny lobster, *Panulirus interruptus* (Randall). M.S. Thesis, San Diego State University, San Diego, California.

Passano, L. M. (1960). Molting and its control. *In* "The Physiology of Crustacea" (T. H. Waterman, ed.), Vol. 1, pp. 473-537. Academic Press, New York.

Phillips, B. F. (1975a). Effect of nocturnal illumination on catches of the puerulus larvae of the western rock lobster by collectors composed of artificial seaweed. *Aust. J. Mar. Freshwater Res.* **26**, 411–414.

Phillips, B. F. (1975b). The effect of water currents and the intensity of moonlight on catches of the puerulus larval stage of the western rock lobster. *Rep.—CSIRO, Div. Fish. Oceanogr. (Aust.)* **63**, 1–9.

Phillips, B. F., and Hall, N. G. (1978). Catches of puerulus larvae on collectors as a measure of natural settlement of the western rock lobster *Panulirus cygnus* George. *Rep.—CSIRO, Div. Fish. Oceanogr. (Aust.)* **98**, 1–18.

Phillips, B. F., Rimmer, D. W., and Reid, D. D. (1978). Ecological investigations of the late stage phyllosoma and puerulus larvae of the western rock lobster *Panulirus longipes cygnus*. *Mar. Biol.* **45**, 347–357.

Rice, A. L., and Chapman, C. J. (1971). Observations on the burrows and burrowing behaviour of two mud-dwelling decapod crustaceans, *Nephrops norvegicus* and *Goneplax rhomboides*. *Mar. Biol.* **10**, 330–342.

Robertson, P. B. (1968). The complete larval development of the sand lobster, *Scyllarus americanus* (Smith) (Decapoda, Scyllaridae), in the laboratory, with notes on larvae from the plankton. *Bull. Mar. Sci.* **18**, 294–342.

Serfling, S. A., and Ford, R. F. (1975). Ecological studies of the puerulus larval stage of the California spiny lobster, *Panulirus interruptus* (Randall). *Fish. Bull.* **73**, 360–377.

Travis, D. F. (1954). The molting cycle of the spiny lobster, *Panulirus argus* Laetrille. I. Molting and growth in the laboratory-maintained individuals. *Biol. Bull. (Woods Hole, Mass.)* **107**, 433–450.

Chapter 1

Larval Ecology

B. F. PHILLIPS AND A. N. SASTRY

I.	Introduction	11
II.	Clawed Lobsters	12
	A. Hatching	12
	B. Larval Development	13
	C. Distribution and Ecology	15
	D. Culture	18
	E. Factors Affecting Growth and Survival	19
	F. Physiology and Biochemistry	24
	G. Behavior	26
	H. Effects of Pollutants	28
III.	Spiny, Slipper, and Coral Lobsters	29
	A. Hatching	29
	B. Larval Development	29
	C. Distribution and Ecology	36
	D. Culture	39
	E. Factors Affecting Growth and Survival	41
	F. Behavior	45
IV.	Conclusions	47
	References	48

I. INTRODUCTION

The larval life of the clawed lobsters is very different from that of spiny (or rock), slipper, and coral lobsters. The clawed lobsters have a short and simple larval period (3 weeks, 4 stages) while the spiny, slipper, and coral lobsters have a long (3–22 month) and complex larval period, with the larvae possessing a phyllosoma form that goes through many stages. Because of the difference in their development, the two groups are examined separately.

The review of the true, clawed lobsters is confined to the American, *Homarus americanus*, the European, *Homarus gammarus*, and the Norway lobster, *Nephrops norvegicus*. The review of the spiny, slipper, and coral lobsters covers mainly the spiny lobster genera *Panulirus, Palinuris,* and *Jasus*. There is some discussion of the slipper lobsters *Scyllarus, Scyllarides, Thenus,* and *Ibacus,* and also the coral lobster *Palinurellus*. The state of knowledge of the different genera is by no means uniform, and even within the Palinuridae, which are better known, the larvae of the genus *Projasus* as yet are not described. Data on the larval stages of other groups of lobsters discussed in this volume are almost nonexistent. Berry (1969) has suggested that *Metanephrops andamanicus* has a single zoea larva which settles almost immediately after eclosion, a common type of development found in deep water invertebrates. The chapter is written to examine the ecology, physiology, and behavior of the free-swimming larval stages and to compare our present state of knowledge on both groups. Differences in the life cycles, and varying interests of research workers reflect the extent of knowledge about the two groups. Hence, the arrangement of the material of the two sections.

The methods of collection of the larvae are not described; the interested reader is referred to the monograph "Zooplankton sampling", published by UNESCO (1968).

II. CLAWED LOBSTERS

Because of their considerable economic importance, an extensive literature exists on aspects of biology and fishery of the adult clawed lobsters. Although the distribution and survival of pelagic larval stages are important to recruitment to the adult populations, the larvae have received much less attention. The ecology and physiology of clawed lobster larvae, particularly of *Homarus americanus* and *Homarus gammarus,* is discussed, and references made to comparable data on *Nephrops norvegicus*.

A. Hatching

Larvae of *H. americanus* are primarily released into the plankton in May and June, with a peak hatching intensity in late June and early July (Hughes and Matthiessen, 1962). Bradford (1978a) reported that the relationship between the incubation period of *H. gammarus* eggs to hatching and the sum of monthly average temperatures was approximately linear. It was estimated that 11 months are required for incubation at 10.4°C, the mean temperature in the North Irish Sea. This compares with 11.4 months at the annual mean temperatures of about 8.1°C for *H. americanus*. A description of the physical events of eclosion is given for *H. americanus* by Davis (1964) and for *N. norvegicus* by Farmer

(1974). Although the rate of salt uptake increases near the time of hatching, escape from the egg membrane is apparently not simply due to increased pressure within the membrane caused by inflow of water along an osmotic gradient (Pandian, 1970a). Once the hatching is initiated, it takes several weeks before all the larvae are released from the individual female (Wilder, 1954).

Egg hatching of *N. norvegicus* in the laboratory has been observed only at night, when batches of larvae were being released over a period of several days (Farmer, 1974). Ennis (1975b) reported that the larvae of *H. americanus* are released primarily just after dark, but that some larvae continue to be released throughout the day. The rhythm is more pronounced in *H. gammarus* with larval release consistently occurring within a few hours after darkness (Ennis, 1973a). The rhythm is so well defined that a given female will release larvae for a few minutes at similar times every night for at least 3 weeks (Bradford, 1978b). The hatching rhythm persists for several days, even in constant darkness. Although the onset of darkness is the approximate controlling factor, endogenous components in the female must be important. Since the release of larvae involves two steps, the actual escape from egg membranes and the subsequent release by the females (Templeman, 1937), it is uncertain whether the endogenous component acts on the embryos, the adults, or both (Ennis, 1973a; Pandian, 1970a,b). Hatching is arhythmic in continuous light or darkness. The hatching time of eggs in light/dark cycles is not influenced by the photoperiod, but occurs after sunset. The time when females perform their part of the hatching process in light/dark regimes seem to depend on temperature, day length, and previous photoperiodic experience (Bradford, 1978b).

Contrary to popular opinion, hatching in *H. americanus* and *H. gammarus* results in a prelarva and not the pelagic first stage larva. Davis (1964) and K. A. Aiken (personal communication) have found that a prelarva emerges from the egg, but that this prelarva molts to the first stage larva within 24 hours of hatching. Since hatching may occur at any time of day, but release occurs during a relatively brief period at night when the female actively beats her pleopods (Ennis, 1975b), most larvae are able to complete the first molt prior to release. Apparently, larval movements and chemicals released when the egg membranes burst do not stimulate the release of larvae. Moreover, some females were observed to perform larval releasing behavior without actually releasing any larvae (Ennis, 1973a). Farmer (1974) states that in *N. norvegicus* there is a brief prezoeal larval stage which is unable to swim, since the natatory appendages are devoid of setae, and which molts soon after hatching to produce the first free-swimming stage.

B. Larval Development

The larvae of all three species pass through four pelagic stages before settling to the bottom as juveniles. Herrick (1896) and Hadley (1909) described the larval stages of *H. americanus* (Fig. 1), and provided a description of the appendages. A study of the larvae of *H. americanus* has recently been made by Heckman *et al.* (1978), using scanning electron microscopy. The larval stages of *N. norvegicus* (Fig. 2) are similar to those of *Homarus*. They were first described by Sars (1884, 1890) and later by Jorgensen (1925) and Santucci (1926a,b,c, 1927). Details of the rostrum and telson have been given by Andersen (1962) and Kurian (1956).

Generally, the completion of the fourth stage has been considered the termination of larval life. After molting into the fourth stage, the larva swims for a few days before becoming bottom seeking. However, Neil *et al.* (1976) considered the molting of stage III to IV in *H. gammarus* as the metamorphosis between larval and juvenile phases. Templeman (1936) described an intermediate form between the third and fourth stage larvae of *H. americanus*. The intermediate form retains functional appendages on the pereiopods, as well as the long dorsal spine of segments of the abdomen, which are absent on the normal fourth stage. An intermediate form has also been described for the European lobster, *H. gammarus* (Williamson, 1905). The intermediate forms occur when the culture conditions of temperature, salinity, and food are unfavorable.

Gruffydd *et al.* (1975) compared the growth patterns of larvae of *H. americanus* and *H. gammarus*. The larvae of European lobsters are generally larger at all stages, the differences being particularly marked in stage I larvae.

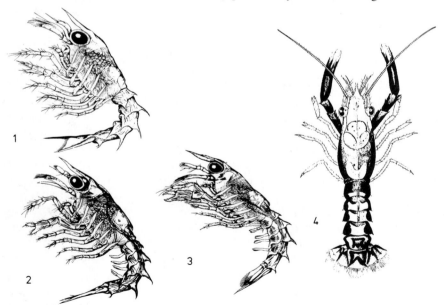

Fig. 1. The four larval (free swimming) stages of *Homarus americanus*. (From Herrick, 1896.)

1. Larval Ecology

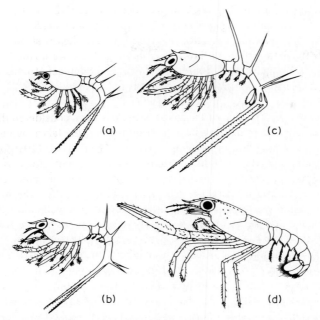

Fig. 2. Larval development of *Nephrops norvegicus* (a) stage I, total length 6.5 mm; (b) stage II, total length 9 mm; (c) stage III, total length 11.6 mm; (d) early post-larval instar, total length 15.6 mm. (From Santucci, 1926a.)

C. Distribution and Ecology

The distribution and abundance of larvae are determined by the distribution of spawning females, the duration of pelagic life, the speed and direction of currents at the depths the larvae are found, the responses of the different larval stages to temperature, salinity, light and hydrostatic pressure, and by stage-specific mortality rates. Templeman (1937), Wilder (1953), and Scarratt (1964, 1968, 1973) studied the distribution of *H. americanus* larvae in Canadian waters in order to determine the relationship between larval abundance and fishery stocks. In contrast, the larvae of *H. gammarus* are rarely taken in plankton samples and hence their distribution and abundance is poorly known (Jackson, 1913; Ennis, 1973a; Nichols and Lawton, 1978).

Herrick (1894) reported that stage I and stage IV larvae of *H. americanus* were frequently taken in surface tows in Vineyard Sound, Massachusetts. Mead and Williams (1903) claimed that the first three larval stages swam near the surface. Templeman (1937) found larvae primarily at the surface during 3 weeks of summer plankton towing off Pictou Island, Nova Scotia, under light conditions ranging from bright sunlight to complete darkness. No larvae were caught when towing near the bottom. Only 333 larvae were caught during his study, and nearly 96% of these were larvae in stage I. Of the rest, ten were in stage II, two were in stage III, and two were in stage IV. The predominance of first-stage larvae was attributed to sampling early in the spawning season. According to Templeman (1937), the predominance of stage IV larvae in Vineyard Sound surface plankton tows taken by Smith (1873) was probably due to collecting later in the breeding season. Sherman and Lewis (1967) collected mostly stage I larvae from June to the middle of August in 257 surface tows made off Maine, with stage IV larvae predominating in the later samples. Their sampling program substantiated previous reports that larval lobsters occur primarily in surface waters. Scarratt (1968) made 100 daytime tows from June 15 to August 3, at 20 stations off Pictou Island. Nearly 2000 larvae were collected, and approximately 95% were stage I individuals. Most of the larvae were taken in the central part of the study area, larval density being especially low further inshore. Presumably most of the larvae were spawned offshore. Wilder (1953) made daytime surface tows in the Northumberland Strait from June to September over 5 years; over 70,000 larvae were taken in 1600 tows. Stage I larvae dominated throughout the season. Scarratt (1964) reported that stage I larvae predominated in collections made in Northumberland Strait between New Brunswick and Prince Edward Island, Canada from 1949 to 1961. Squires (1970) and Squires *et al.* (1971) made plankton tows during July–September in Bay of Islands, Newfoundland to determine the horizontal distribution of larvae over lobster grounds and deeper waters as a function of wind conditions. Light onshore winds appeared to bring the larvae to the surface, and concentrate them toward the shore. The larvae were also present at the surface during the periods of offshore winds, though in lower densities. Only a few larvae were present at the surface when the winds were stronger.

In southern New England, stage IV larvae dominated close to shore, whereas stage I larvae dominated in the offshore samples, suggesting that the larvae move inshore as they develop (Rogers *et al.*, 1968). In contrast, Lund and Stewart (1970) collected few stage I larvae from offshore waters near Long Island Sound between June and August. Within Long Island Sound, the larvae were more abundant and more concentrated toward the western end than the open eastern end, suggesting retention of larvae by currents. Caddy (1979), from an examination of larval survey data on *H. americanus* in Northumberland Strait in Canada for 1948–1963, deduced that despite major water movements in the area, centers of density of successive larval stages maintain position or move upwind, implicating vertical migration of the larvae in position keeping.

Although the studies described above provided some information, the definitive picture of larval distribution has yet to emerge, due to the small number of individuals captured in any single study, the lack of discrete depth sampling programs, and the rarity of night sampling. Advanced stages are especially rare in plankton samples. However, it would seem that most are hatched inshore and remain fairly near their hatching areas. In an analysis of current patterns of Rhode Island Sound and offshore waters (J. S. Cobb and J. R. French, unpublished) it was concluded that larvae released in Narragansett Bay and Rhode Island Sound probably would not disperse from that general area. Thus, it seems that long-distance movements of *H. americanus* larvae are unlikely.

The larvae of *N. norvegicus* are planktonic, and therefore more widespread than the juveniles and adults that inhabit the Mediterranean Sea and the continental shelves of the northeastern Atlantic Ocean. Thomas (1954) gives the relative distribution of the adults and larvae in Scottish waters, from which it is apparent that dispersal is passive. At depths down to 90 m there appeared to be no variation in abundance of the larvae with increasing depth. Jorgensen (1925) reported that larvae were rarely taken near the surface, at least during the day. Fraser (1965) gives the distribution of larvae in Scottish waters throughout the period of 1935-1964. Most larvae were taken well below the surface, although there was a slight increase in numbers in the surface waters during the period dusk to dawn. Fraser concludes that the *N. norvegicus* populations of the Firth of Forth area and the Fladen ground are probably self-sustained, although there was evidence of considerable dispersal of larvae away from the parent stock, due to currents.

Hillis (1968, 1972a,b) has carried out detailed studies of larval distribution of *N. norvegicus* in the Irish Sea. It was found that generally the larvae occupied the same areas as the adults, although there was some evidence that tidal currents in the southeastern part of their range affected the distribution (Hillis, 1972a). It was noticeable that the older larvae tended to be caught in the southern and eastern parts of the area sampled. Hillis (1972b) suggested that this might be due either to differential periods of hatching and growth rates, or to irregular waves of hatching. Larvae were found to be most abundant at a depth of 20-40 and 80 m during daytime sampling. According to Williamson (1956), *N. norvegicus* larvae normally occur at depths greater than 17 m. Santucci (1926b) has reported larvae from 25 to 150 m in the Tyrrhenian Sea.

Hillis (1972b) estimated survival rates of the larvae of *N. norvegicus* in the plankton as stage I to stage II to be 16%; stage II to stage III to be 1.8-19% and stage III to postlarval stage I to be 2.3-26%.

The predators and predation rates on lobster larvae in the field are not known, but probably include all the larger plankton-feeding fishes and possibly comb jellies (ctenophores). Mendall (1934) found that herring gulls (*Larus argentatus*) and common terns (*Sterna hirundo*) were predators of "young lobsters," and Mills (1957) confirmed the role of the common tern as a predator of *H. americanus* larvae, but at a very low level. Mead and Williams (1903) suggested that the mortality rate of pelagic stages of *H. americanus* is very high. It has been speculated that less than 10% reach the bottom stage (Wilder, 1953; Kensler, 1970; Hughes, 1972). However, Lund and Stewart (1970) estimated a survival rate of 52% from stage I to stage IV for larvae collected from Long Island Sound. Scarratt (1973) has estimated the variability in larval survival from year to year in Canadian waters and attempted to relate this to recruitment to adult populations. There is some evidence to suggest a relationship between stage IV larval abundance and subsequent commercial lobster stocks, but sampling errors were too great to provide accurate predictions. Caddy (1979) suggests that, in most years, *H. americanus* larvae produced in the southern Gulf of St. Lawrence at low ambient temperatures early in the season have the best chance of survival.

D. Culture

A historical review of artificial hatching and rearing of *Homarus* lobster larvae is given in a number of papers (Wilder, 1954; Kensler, 1970). Major problems in culturing larvae have been diseases (Fisher *et al.*, 1975, 1976a; Nilson *et al.*, 1975, 1976), and the difficulty of establishing suitable foods (Dannevig, 1936, in Scattergood, 1949; Ehrenbaum, 1903). The larvae are carnivorous, and provided with a diet of live copepods (*Acartia, Eurytemora,* and *Pseudocalanus*) they will complete larval development (Templeman, 1936). However, their diet in nature is not known. Factor (1977, 1978a,b) made a detailed study of the mouthparts and digestive system of *Homarus americanus* larvae. He found that changes which occur in the mouthparts of lobsters during progressive larval stages include an increase in the size of mouthparts and in the number of setae affixed to them. The most striking changes in the mandible during larval stages was the development of the teeth of the cutting edge, in which the delicate setae or spinelike teeth of the first stage transformed into the molars of the fourth stage. A prominent change in the third maxilliped was the development of teeth on the inner medial edge of the ischium. The development of the mandibles and third maxillipeds seem to enable lobsters to deal with the more substantial food they usually encounter in the benthic environment when they enter fourth or fifth (first juvenile) stage. The development of functional teeth on the ischium of the third maxilliped seems to occur at a time when the primary function of these appendages changes from swimming in the first three stages, to feeding in all subsequent stages. The changes in the mouthparts are coordinated with the development of the digestive system, particularly the forgut, the gastric mill, and the filter apparatus. The medial and lateral teeth of the mill develop into substantial structures by the fourth stage, and the filter increases in complexity.

It is now possible to culture *H. americanus* larvae with 70-80% survival rate on a diet of adult *Artemia salina* (Hughes *et al.*, 1974; Serfling *et al.*, 1974; Sastry, 1977). Larvae of *H. gammarus* have also been reared successfully to the fourth stage (Gruffydd *et al.*, 1975; Neil *et al.*, 1976) on a diet of *Artemia*

1. Larval Ecology

nauplii, and attempts are being made in the United Kingdom to achieve large-scale culture (Walne, 1977).

In laboratory experiments, larvae of *N. norvegicus* have been found to feed satisfactorily on mixed live plankton (Farmer, 1972); on *Artemia salina* nauplii (Figueiredo, 1971; Figueiredo and Vilela, 1972; Hillis, 1972b; Farmer, 1972); and on live eggs of the shrimp *Crangon crangon* (Figueiredo and Vilela, 1972). The postlarval instars have been fed successfully on minced *Mytilus* spp. (Figueiredo and Vilela, 1972; Hillis, 1972b); cephalopod flesh (Hillis, 1972b); and on minced cockles (Figueiredo and Vilela, 1972).

Fungal and epibiotic infestations of *H. americanus* larvae under laboratory conditions have been controlled with malachite green, and the infestation of larvae of filamentous bacterium, *Leucothrix mucor*, by bacteriostats (see Chapter 6, Vol. I). Anderson and Conroy (1968) have reported the infestation of *N. norvegicus* larvae by the ciliate *Zoothamnion* sp. The complete body surface of living larvae was covered by the sessile protozoans, and death was thought to have resulted from trauma and interference with respiration. The isopod, *Eurydice pulchra*, has been reported to feed on the larvae of *N. norvegicus* under laboratory conditions (Farmer, 1972). Antibiotics have been used by Figueiredo (1971), Figueiredo and Vilela (1972), and Farmer (1972) to control the growth of bacteria in the laboratory.

At present there are no commercial applications of cultivation using *N. norvegicus*. In view of the very slow growth rate of the juveniles and adults, it is unlikely that this state of affairs will change in the near future.

E. Factors Affecting Growth and Survival

Growth and survival of *Homarus* larvae are affected by temperature (Huntsman, 1924; Templeman, 1936; Hughes and Matthiessen, 1962; Gruffydd et al., 1975), salinity (Templeman, 1936), light levels (Huntsman, 1923; Templeman, 1936), food supply (MacClement, 1917; Templeman, 1936), and degree of crowding (Hughes et al., 1974). Given adequate nutrition and space, temperature seems to exert the dominant regulatory effect on growth (Hughes et al., 1974).

1. Temperature

Larvae of *H. americanus* were reared successfully between 15° and 25°C by Huntsman (1924). At 10°C, larvae were slow to molt and died soon after molting was completed. At 5°C, all larvae died prior to molting. Templeman (1936) determined the relationship between temperature and duration of each of the four larval stages of *H. americanus* over a temperature range of 6°–24°C. At 10°–11°C, the newly hatched larvae required 60 days to reach the fourth stage in full strength sea water. Development to stage IV required about 25 days at 15°C (a

TABLE I

Duration of Molt Period (in days) for Each Life-Cycle Stage of Individually and Communally Reared (20°C and 30 ‰) Lobsters, *Homarus americanus*[a]

	Low density				High density			
	Individual		Communal		Individual		Communal	
Stage	Range	Mean	Range	Mean	Range	Mean	Range	Mean
L1	2-4	2	2-4	2.5	2-3	2.3	1-2	1.5
L2	4	4	2-4	3.1	1-5	3.0	1-6	3.8
L3	3-7	5	3-5	3.7	3-4	3.5	3-6	4.2
L4	7	7	5-11	7.7	7	7.0	6-8	7.0

[a] From Sastry and Zeitlin-Hale, 1977.

temperature typically experienced in the field (Sherman and Lewis, 1967; Scarratt, 1968). At 19°–20°C, the time was reduced to about 15 days. At all temperatures, the duration of stage IV larvae was significantly longer than the duration of the earlier stages. Sastry and Zeitlin-Hale (1977) reported that *H. americanus* larvae complete the four stages in about 18 days in 30 ⁰/₀₀ salinity at 20°C (Table I). Hughes and Matthiessen (1962) reported that the larvae required 13.6 days to complete the fourth stage at 19°–20°C; *H. gammarus* requires 31–32 days to reach the fourth stage at 14 ± 2.0°C (Neil *et al.*, 1976).

According to Poulsen (1946), the larval stages of *N. norvegicus* probably last 2–3 weeks. More accurate estimates of the duration of the larval phase are available only from laboratory experiments (see Table II). Each successive larval stage is longer than the preceding stage. There appears to be a general trend for development to occur more quickly at higher temperatures.

Survival and duration of development of *H. americanus* larvae reared at 10°–20° daily cyclic temperatures differed from those reared at 15°C constant temperature (A. N. Sastry, unpublished). Larvae completed the fourth stage at 10°–20°C and 17.5°–22.5°C daily cyclic temperatures, while those at 12.5°–17.5°C and 15°–25°C daily cycles did not.

Gruffydd *et al.* (1975) investigated the temperature tolerances of larval stages of *H. americanus* and *H. gammarus*. Larvae reared at 20°C were exposed to high temperatures for periods of up to 24 hours. No obvious differences in temperature tolerances between larval stages of the two species were seen. The temperature tolerances were essentially the same for stages III and IV, but there was an indication of slightly higher tolerance by *H. americanus* larvae in the first two larval stages. The upper temperature tolerance limits for 50% survival of *H. americanus* larvae cultured in 30 ⁰/₀₀ salinity at 20°C showed that second-stage

TABLE II

Duration of the Larval Stages of *N. norvegicus* Maintained under Laboratory Conditions

Reference	Temperature (°C)	Duration (days)		
		First zoea	Second zoea	Third zoea
Farmer, 1972	10–18	10	15	—
Figueiredo, 1971	11–14	11	—	—
	13–15	9	—	—
	15–18	5–6	—	—
Figueiredo and Vilela, 1972	7–10	14–15	14–15	—
	11–14	10–11	10–11	—
	13–17	5–6	5–6	—
Hillis, 1972b	9–14	—	13	16
	17–19	—	7–9	8–12

larvae have much lower tolerance limits than any other stage (Sastry and Vargo, 1977).

According to Figueiredo (1971), survival of *N. norvegicus* larvae in the laboratory from the first to the second zoeal stage was 7.7–13.2%, and to the first postlarval instar 0–7.3%. The optimum temperature for the incubation of the eggs and survival of the larvae was found to be 11°–14°C.

A. N. Sastry and S. L. Vargo (unpublished) determined the relationship between dissolved oxygen levels and temperature tolerances of *H. americanus* larval stages. At saturated oxygen levels, the second stage was least tolerant of high temperatures (30°C). The relationship between temperature tolerance and dissolved oxygen levels is different for each larval stage.

2. Salinity

Gompel and Legendre (1927) reported that *H. gammarus* larvae were tolerant of specific gravities between 1.017 and 1.033. Templeman (1936) attempted to rear *H. americanus* larvae over a salinity range of 11.4–31.8 °/oo, at approximately 15°–17.5°C. No larvae reached the third stage at 16.4–17 °/oo. Thus larvae of both species appear intolerant of salinities below about 17 °/oo. At higher salinities at least some larvae reached stage IV successfully. The survival of larvae to the fourth stage was highest in 30–31.8 °/oo salinity. At salinities above 20 °/oo, the time required for *H. americanus* larvae to reach the fourth stage was not significantly affected by salinity (Templeman, 1936). Scarratt and Raine (1967) found that newly hatched *H. americanus* larvae actively avoided water with a salinity of 21.4 °/oo (see Section G).

3. Temperature and Salinity

Sastry and Vargo (1977) determined the combined effects of temperature and salinity on survival and duration of larval stages of *H. americanus*. Larvae developed to the first juvenile stage (stage V) in salinities between 20 and 35 °/oo at 15°C, and between 15 and 35 °/oo at 20°C; i.e., raising the temperature extended the range of tolerated salinities. At 15°C, survival to the first juvenile stage was highest in 35 °/oo, and at 20°C maximum survival occurred between 20 and 30 °/oo. The time required to reach the first juvenile stage was not significantly affected by salinity at any given temperature.

4. Light

Templeman (1936) reported higher survival and shorter development time of *H. americanus* larvae when they were reared in complete darkness rather than under normal laboratory lighting. The stage IV larvae reared in darkness were larger than those reared in the light. Huntsman (1923) reported that *H. americanus* larvae died after several days of direct exposure to light. Possibly the bright light interferes with ingestion of food or encourages the growth of harmful

TABLE III

Percent Survival of Communally (C) and Individually (I) Reared (20°C and 30‰) Lobsters *Homarus americanus* of Each Life-Cycle Stage[a]

Stage	Replicate groups ($n = 20$)		High-density % survival to next stage		Significance[b]	Replicate groups ($n = 20$)		Low-density % survival to next stage		Significance[b]
	C	I	C	I		C	I	C	I	
L1–L2	6	3	67.5	77.6	NS	6	3	66.7	74.0	NS
L2–L3	6	3	30.8	82.0	$P \leq .01$	6	2	29.2	91.0	$P \leq .01$
L3–L4	6	2	46.0	82.0	$P \leq .01$	6	2	42.5	94.0	$P \leq .01$
L4–J1	3	1	62.5	100.0	$P \leq .01$	3	1	60.2	100.0	$P \leq .01$

[a] From Sastry and Zeitlin-Hale, 1977.
[b] Student's *t* test. NS, not significant.

microorganisms (Templeman, 1936). However, light levels of approximately 1 to 310 lux had no observable effect on the duration of the fourth larval stage (Cobb, 1970).

5. Crowding

Cannibalism has been observed among larval lobsters under crowded conditions. Sastry and Zeitlin-Hale (1977) determined the survival rates of individually and communally reared larval stages of *H. americanus*. Survival of larvae held individually was significantly higher than that in corresponding groups of communally reared animals (Table III). Asynchronous molting in groups of communally held larvae apparently contributes to the higher mortality rates. Mortality rates were always higher during the period of molting. Among communally held larvae, the mortality rate was highest in the second stage and then gradually decreased over the later stages.

6. Substrate

Substrate type appeared to have an effect on duration of the fourth larval stage of *H. americanus* (Cobb, 1968), but further work (Cobb, 1970) showed that the increased duration on all substrates except gravel was due to social interactions between the larvae. Individuals held singly molted an average of two days earlier than those held in pairs, despite substrate type. For the paired larvae, presence of the first individual appeared somehow to delay the molting of the second animal.

F. Physiology and Biochemistry

1. Respiration and Excretion

The metabolic-temperature responses of successive larval stages have been determined for *H. americanus* cultured at constant 20°C and at 15°-25°C daily cyclic temperatures (A. N. Sastry and J. P. Laczak, unpublished data). The metabolic rate of all the larval stages cultured at 20°C was depressed at test temperatures between 25° and 30°C. Second and fourth stage larvae compensated their metabolic rate over a temperature range of 15° to 25°C.

Logan and Epifano (1978) determined the energy balance for the larvae and juveniles of the American lobster *H. americanus* cultured at 20°C in 28 °/oo on a diet of adult brine shrimps, *Artemia salina*. The measured ingestion, egestion, excretion of ammonia, routine (starved) and fed metabolism, and growth and production of exuvia. Energy loss from excretion and molting was found to be negligible. Efficiency of digestion was 82%, and efficiency of assimilation was 81% for both the larval and juvenile stages. The assimilation efficiency did not increase for lobster larvae with successive stages. Gross growth efficiency (rate of growth/consumption) increased from stage I (18%) to stage IV (44%) and generally decreased with increasing size to stage XI, (seventh juvenile stage-18%). The calorigenic effect (difference between fed and routine metabolism) was about 10% of ingestion for both larval and juvenile lobsters. The portion of metabolic energy lost to routine metabolism and molting remained fairly constant. The portion chanelled into growth increased during larval life, then decreased and remained fairly constant during the post-larval period. A high protein diet appears to be necessary for lobster growth. A study of the respiration rates, ammonia excretion rates and biochemical composition of *Artemia*-fed larval and first postlarval stages of *H. americanus* by Capuzzo and Lancaster (1979a,b) examines energy utilization both during larval development and metamorphosis to postlarval stages. The weight-specific respiration rates of fed and starved lobsters, and the ammonia excretion rates of fed lobsters increased with each larval stage (I–IV) and decreased with the first postlarval stage (V). The rate of change in metabolic rates was greater than the rate of change of body size of the larval stages indicating an increased energy demand of the later larval stages. There was no significant difference in the O:N ratio for the first three larval stages, but a reduction was observed in stage IV and V lobsters, reflecting an increased dependence on protein catabolism for energy.

Although protein is the principal energy source, lipids and/or carbohydrates are also utilized to some extent, and this is reflected in the significant decrease in lipid content and increases in ash and chitin in the last larval (IV) and first postlarval (V) stages.

2. Endocrinology

The ontogeny of the lobster endocrine system has not been studied, so the endocrine influence on development and growth is not known. Bilateral eyestalk ablation of larval stages I through IV significantly accelerates molting beginning with stage II (A. N. Sastry and J. P. Laczak, unpublished data). Rao *et al.* (1973) showed that bilateral eyestalk ablation of fourth-stage larvae of *H. americanus* accelerates proecdysal preparation and causes precocious initiation of ecdysis. Injection of either ecdysone or 20-hydroxyecdysone accelerates proecdysal preparation, leading to precocious initiation of ecdysis. Lobsters injected with 20-hydroxyecdysone died during an early stage of ecdysis (E_1). Ecdysone-injected lobsters either completed ecdysis or died in the later stages of ecdysis (E_2 or E_3).

3. Biochemistry

Barlow (1969) made an electrophoretic analysis of homogenates of larvae in a series of developmental stages of *H. americanus* from stage I to adulthood. No pronounced changes in enzyme patterns were observed during larval development. Qualitative and quantitative changes in fatty acid methyl esters have been examined for all the larval stages of *H. americanus* cultured at 10°–20°C daily cyclic and 20°C constant temperatures (A. N. Sastry, unpublished data). There is a tendency for greater unsaturation in certain fatty acids (e.g., 16:0) at cyclic temperatures. Certain chain length fatty acids (e.g., 18:2 and 18:3) abundant at cyclic temperatures were present in only trace amounts at constant temperatures.

G. Behavior

Homarus larvae swim upward by increasing the beating of swimming appendages and by occasionally using short bursts of tail flexes. The exopodite apparatus provides the lift and propulsive power for swimming (Neil *et al.*, 1976). Swimming is accomplished by metachronous beating of the exopodites. The branched setae on each exopodite flagellum are articulated to allow movements from a fully extended position during the power stroke to a fully folded position during the return stroke. Frequently, stage I and II larvae swim upward and forward with the abdomen slightly curved and folded under the body. The larvae also travel by gliding and by swimming in upward and backward directions. Larvae move by rapid tail flexes when they encounter solid objects, suggesting an escape response. The third-stage larvae use the same modes of locomotion, but they most often swim close to the bottom. Orientation to gravity is not well developed in the larvae (Neil *et al.*, 1976). A functional statocyst does not appear until stage IV. According to Foxon (1934) *N. norvegicus* larvae "orientate themselves with the telson and abdomen towards the light, but with the head and thorax in a plane at right angles to them. In this posture all the specimens moved towards the light telson first."

Macmillan *et al.* (1976) made a quantitative analysis of exopodite beating in the larvae of *H. gammarus*. There is a tight coupling between ipsilateral appendages and a loose coupling between contralateral exopodites. This loose coupling seems to result in a gliding motion (Macmillan *et al.*, 1976). The exopodite beating patterns are the same for the first three larval stages. However, the preferred frequencies are different for each larval stage. Changes in beat frequency occur discontinuously with age and are associated with larval molts (Laverack *et al.*, 1976). In the third stage a heightened rate of beat occurs. In the absence of a proportional increase of exopodite dimension in the third stage, a faster beating rate provides the extra lift necessary for keeping the larvae off the bottom, compensating for the increase in body weight between stage II and III. The exopodites become nonfunctional in the fourth stage, when the endopodites become functional in walking.

Herrick (1894) and Hadley (1908) reported that young *H. americanus* are positively phototactic. However, first- and second-stage larvae are said to swim away from the surface in very bright sunlight (Templeman and Tibbo, 1945). It was indicated that larvae gradually disperse to greater depths during the night and return to the surface during the daylight period. Older larvae seem less responsive to light than the first stage (Ennis, 1973a). *H. gammarus* orients to light in larval stages I to IV (Rice, 1967).

1. Larval Ecology

Scarratt and Raine (1967) demonstrated that the phototactic response of *H. americanus* larvae can be overridden by the presence of low salinity water, and determined the approximate salinity which would produce an avoidance response. Lobster larvae swam just at the surface in the 31.7 ⁰/oo sea water, actively avoided 21.4 ⁰/oo, but tolerated 26.7 ⁰/oo. The larvae were much more active in 26.7 ⁰/oo than they were in 31.7 ⁰/oo control. These observations were limited to newly hatched larvae.

Ennis (1973a, 1975a) examined the swimming responses of *H. gammarus* and *H. americanus* larval stages to changes in hydrostatic pressure and light. Most of the experiments with *H. gammarus* were conducted with stage I larvae, due to high mortality rates in the culture system. When larvae were subjected to sudden decrease in hydrostatic pressure, they responded by swimming into the upper one-quarter of the water column in which they were held. When the pressure was reduced to 1 atm, the larvae returned to lower levels. The direction rather than the magnitude of change in pressure seems to elicit the response. The larvae were sensitive to gradual pressure changes as well. A rapid increase in the number of larvae swimming in the upper part of the water column was brought about by the slightest pressure increase. A decrease in number of surface swimmers was brought about by the smallest decrease in pressure. Sensitivity to pressure changes was retained through all larval stages, but the third and fourth stages were much less responsive than stages I and II.

The larvae were much less responsive to changes in light intensity than to changes in pressure. No clear phototactic or photopathic behavior was observed in a vertical plane. Larvae of the two species showed only slight differences in their behavioral responses, and these differences cannot account for the relatively rare occurrence of *H. gammarus* larvae in plankton collections.

The actual swimming behavior and distribution of lobster larvae in nature is now well known. Ennis (1973b, 1975a) released laboratory-reared larvae of *H. gammarus* and *H. americanus* at different depths in the sea and followed their movements. Stage I and II larvae of *H. gammarus* released at the surface remained there, and larvae released at depths of 10, 20, and 30 feet tended to swim toward the surface. Older larvae were less responsive than younger individuals and swam upward more slowly when they did respond. Stage III larvae tended to move downward when released in the upper 25 ft. Surprisingly, stage IV larvae always moved to the surface, even when released near the bottom. D. J. Scarratt (personal communication) has observed that late stage IV larvae invariably sank when released near the surface under the observation of a diver. They sometimes paused or reacted vigorously while passing through the thermocline, but subsequently continued their descent to about 60 cm above the bottom (20 m depth). The single stage V (juvenile) individual studied by Ennis showed erratic behavior when released at the surface, ending up at a depth of 18 ft after 10 min, but it took shelter under a rock when released just above the ocean bottom. Scarratt's observations of stage V and VI lobsters corroborate Ennis' observation.

Stage I, II, and III larvae of *H. americanus* exhibited limited downward movement when released at the surface, and generally moved upward when released beneath the surface (Ennis, 1975a). Newly molted stage IV larvae generally remained at the surface when released there and swam to the surface when released at the bottom. When older stage IV larvae were released at the surface they remained there, but when released at the bottom they remained at the bottom and found shelters. Newly molted stage V (juveniles) generally remained where they were released, both at surface and at the bottom. The older stage V juveniles swam to the bottom when released at the surface. Those released at the bottom remained there and found shelters.

H. Effects of Pollutants

Lobster larvae are sensitive to a number of organic and heavy metal pollutants. Wells and Sprague (1976) determined the effects of crude oil on the feeding and development of *H. americanus* larvae. These experiments were done under static conditions, so that oil concentrations declined substantially during each experiment. Initial oil concentrations of 0.24 ppm and above increased the duration of development. Exposure of 0.19 ppm depressed the ingestion rates. Fifty percent of the larvae held at an initial concentration of 1 ppm died in approximately 1 week.

The effects of chlorine and chloramine on mortality and respiration rates of *H. americanus* larvae were determined by Cappuzo *et al.* (1976). Chloramine was more toxic than free chlorine, based on mortality observed 48 hr after a maximum 60-min exposure to toxicant. LC_{50} values at 25°C were 16.30 mg of applied chlorine per liter and 2.02 mg of applied chloramine per liter. Respiration rates were depressed, with chloramine exerting the more pronounced effect. After transfer to uncontaminated water, the respiration rates failed to return to control values during a 48-hr observation period.

Sprague and McLeese (1968) found that pulp mill effluent had an acute lethal threshold of about 100,000 ppm for *H. americanus* larvae. The 48-hr LC_{50} values for *H. gammarus* larvae were 0.02 mg/liter for copper and about 0.07 mg/liter for mercury (Conner, 1972). McLeese (1974b) tested the effects of Fenitrothion (organophosphate insecticide) between 0.001 and 10 ppb concentrations. Normal behavior of larvae was observed below 0.01 ppb concentrations. At concentrations greater than 0.01 ppb, the chromatophores expanded to red coloration. At concentrations above 1 ppb, larvae swam abnormally and sank to the bottom. At 1 ppb, 50% mortality occurred within 4 days. At 10 ppb, 50% mortality occurred in approximately 2 days. A second group of larvae had a slightly lower resistance to the toxicant. However, Scarratt (1969) found that the bleached mill effluent did not have an immediate or direct effect on the distribution or abundance of *H. americanus* larvae examined in an area off Pictou, Nova Scotia.

III. SPINY, SLIPPER, AND CORAL LOBSTERS

The Palinuridae (spiny lobsters), the Synaxidae (coral lobsters with the single genus *Palinurellus,*) and the Scyllaridae (slipper lobsters) are the only decapod families possessing phyllosoma larval stages in their life cycles. Phyllosoma larvae have pelagic adaptations, in that the majority of species are true long-lived larvae, characteristic of the open ocean. The body is extremely dorsoventrally flattened, leaflike, transparent, and with highly setose appendages. This larval form has been described appropriately, the name "phyllosoma" having been derived from the Greek *phyllos* (a leaf) and *soma* (body). A variable number of stages, depending on the worker investigating a species, is found in the phyllosoma larval development.

This section concentrates mainly on the Palinuridae, particularly the major extant genera *Palinurus, Jasus,* and *Panulirus,* in which some 48 species have been described. Some references to the Scyllaridae and Synaxidae are made where appropriate.

A. Hatching

The eggs are attached to the female setae on the endopodites of the pleopods, on the ventral side of the abdomen. Most hatching occurs at night and has been recorded to take place over a period of 3 to 5 days in *Jasus lalandii* (Silberbauer, 1971).

B. Larval Development

By comparison with the 20- to 30-day larval life of *Homarus,* the larval life of the spiny lobster is a lengthy 4–22 months (Berry, 1974a; Johnson, 1960a; Lazarus, 1967; Chittleborough and Thomas, 1969; Lesser, 1978). Hence the larvae can be dispersed over a wide area. Sims (1966) estimates the larval life of the synaxid, *Palinurellus gundlachi gundlachi* as up to 10 months. Scyllaridae usually have a shorter larval life of 30 days to 9 months (Robertson, 1968a).

1. Prephyllosoma Stages

Larval stages have been described that occur prior to the phyllosoma stages in some species. These stages have been referred to as "naupliosoma," "prenaupliosoma," and "prephyllosoma." All are referred to here as naupliosoma. The naupliosoma stage lasts from only a few minutes to a few hours before molting to the stage I phyllosoma.

The naupliosoma stage has been found in *Jasus lalandii* (Gilchrist, 1916; Silberbauer, 1971), in *Jasus verreauxi* (Lesser, 1974), in *Jasus edwardsii* (Archey, 1916; Batham, 1967; Lesser, 1974), in *Ibacus novemdentatus* reported by Harada (1958) as *I. ciliatus,* in *Scyllarides aequinoctialis* (Robertson, 1968b,

1969c), and in *Scyllarides herklotsi* (Crosnier, 1972). Sheard (1949) reported a naupliosoma in *Panulirus cygnus,* but George (1962) referred to the hatched larvae as "early phyllosoma stage I," stating that "The 'naupliosoma' is not here regarded as a distinct stage separated by a moult." Robertson (1968b, 1969a) cautioned that the naupliosoma may have escaped detection in some cases since it is of short duration and hatching often occurs at night. The body of the naupliosoma is in a cramped condition, the appendages in a more or less folded position. Naupliosoma larvae hatch and remain quiescent for about 10-15 min until the locomotory antennae extend fully and are thus able to propel the larva vertically to the surface in a series of rapid, jerky movements. At this time both cephalothorax and abdomen are curved, and their appendages gradually straighten into the normal position as the larvae rise to the surface of the water (Fig. 3). At this stage the body still contains some of the food yolk and the larvae are not transparent. Rapid beating or swimming movements of the biramous setose antennae maintain the naupliosoma at or near the surface.

Von Bonde (1936) initiated the term "prenaupliosoma" for an even earlier stage in *J. lalandii* that molts to a naupliosoma after about 8 hr. However, Silberhauer (1971a) found that although prematurely hatched larvae resembling the "prenaupliosoma" of Von Bonde were observed, none of these survived for longer than 5 hr, and stated that "A prenaupliosoma larva, therefore, is not accepted as a stage in the life history of *Jasus* although the above findings do suggest that early embryonic forms may be prematurely released when the water temperature is raised above 15°C." Feliciano (1956) described a nonswimming prenaupliosoma stage in the embryonic development of *Panulirus argus.* However, these forms have been found in the field and Sims (1965) described prenaupliosoma for *P. argus* from plankton samples. He suggested that perhaps

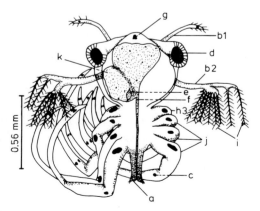

Fig. 3. Newly hatched naupliosoma larva of *Jasus lalandii* (Dorsal view). a, abdomen; b, antenna; c, chromatophore; d, eye; e, heart; f, intestine; g, median eye; h, maxilliped; i, natatory setae; j, pereiopod; k, yolk. (From Silberbauer, 1971.)

1. Larval Ecology

hatching in the naupliosoma stage is the norm for certain species. Baisre (1964) records prenaupliosoma and naupliosoma stages in the slipper lobster *Ibacus ciliatus*. Deshmukh (1968) noticed a form distinct from the naupliosoma and phyllosoma while examining freshly hatched larvae of *Panulirus dasypus* (= *homarus*), and proposed the name "prephyllosoma." Prasad and Tampi (1959) also observed this form in *P. burgeri* (= *homarus*). The prephyllosoma was distinctly free-living and lasted about 3 hr after which it molted into the first-stage phyllosoma larva. However, every larva that hatched out from the egg was not necessarily a prephyllosoma.

There has been controversy as to whether any of the prephyllosoma stages (prenaupliosoma, naupliosoma, prephyllosoma) are a normal part of the natural free-living development, or whether they are the result of premature rupture of the egg. Harada (1958) mentions that when eggs in the advanced stage of *Scyllarides squamosus* were artificially shaken in the laboratory, the larvae hatched in the form of naupliosomas, but were unable to swim. Sims (1965) found that eggs of *P. argus* normally hatch as first-stage phyllosomas in the laboratory, as did Crawford and DeSmidt (1922), Lebour (1950), and Lewis (1951). Sims (1965), however, was able to produce a prenaupliosoma stage in *P. argus* and *Scyllarus americanus* from eggs in late development by placing the eggs in seawater of low salinity.

The prenaupliosoma stage probably represents an embryonic form occurring as a result of premature rupturing of the eggs in most species, although the finding of prenaupliosoma of *P. argus* in the plankton by Sims (1965) shows this is not universal.

2. Phyllosoma Stages

Molting into the phyllosoma stage in the aquarium has been observed in *J. lalandii* by Silberbauer (1971). After 8-12 hr a naupliosoma larva descends to the bottom of the tank, and after 5-20 min it molts into the first phyllosoma stage.

The body of the first phyllosoma larvae is flat and leaflike, about 1.8 mm long (Fig. 4). It has long legs and large, protuberant eyes on long, unstalked peduncles. Since there is no pigment or calcium in the skeleton, the body is colorless and transparent. The swimming setae of the antennae of the naupliosoma stage have been thrown off. Locomotion of the phyllosoma is brought about by the exopodites of the first two walking limbs, which bear long featherlike setae. Robertson (1968a), who worked with *S. americanus,* was the first to study the complete larval history of any phyllosoma species by rearing larvae in the laboratory through all phases of development to the postlarval stage. However, a certain amount of caution is necessary in accepting a series of stages as determined under artificial conditions as normal. Differences between plankton and laboratory-reared specimens have been observed by Saisho (1966a), Ong (1967), and Robertson (1968a, 1969c).

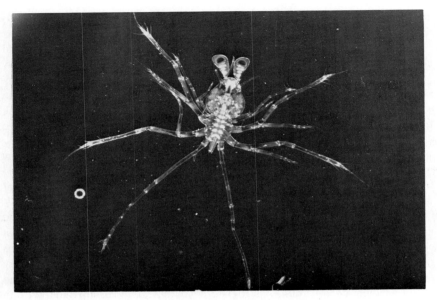

Fig. 4. Stage I phyllosoma larvae of *Panulirus cygnus*. (Photo by C. E. Purday.)

The larva grows continuously, but increases in size by a series of molts, concurrently undergoing small, progressive changes in its development. A series of 7–13 arbitrary stages is used to describe the development of the phyllosoma, from the time that the larva hatch, when it is just visible to the human eye, until it reaches a length of about 35 mm. Johnson (1968) and Robertson (1969b) point out that the number of phyllosoma stages depends on the range allowed in features used. Eleven stages are described in the larval development of the spiny lobster, *Panulirus inflatus,* yet Johnson and Knight (1966) suggest that 25 or more ecdyses may take place in attainment of the final phyllosoma stage. Each stage does not necessarily represent a single instar (see Saisho, 1966a; Inoue, 1978). Several molts may occur between stages. The features used in separating the stages apply only to each particular stage as the sequence in which other features appear may vary. Nevertheless, the characters used in separating the stages do give a generalized account of the sequence of developmental changes. The following sequence for the phyllosoma larvae stages of *P. cygnus* is typical:

Stage I	Eye stalk unsegmented
Stage II	Eye stalk segmented
Stage III	Exopod of third pereiopod setose
Stage IV	Fourth pereiopod as long or longer than abdomen
Stage V	Exopod of fourth pereiopod setose and antennule comprising three segments
Stage VI	Antennule comprising four segments and uropods not bifid

Stage VII	Uropods bifid and distal pair of pleopods not bifid
Stage VIII	Distal pair of pleopods bifid, gills not present
Stage IX	Gills present, lateral spines on uropods present

A fuller description of the stages of *P. cygnus* and a discussion of the relationship between stage and instar number will be found in Braine *et al.* (1979).

Table IV lists the number of phyllosoma larval stages and length of larval life for some species on which data are available.

Adult species to which phyllosoma larvae belong are known in relatively few cases. On several occasions, investigators have assigned a series of larvae to the most common adult species occurring in a particular geographic region. Such questionable identifications are often accepted and perpetuated by subsequent investigators. Specific relationship between phyllosoma larvae and an adult form can only be made by culturing the larvae from eggs of known parentage, or by observing metamorphosis of final stage larvae to identifiable postlarvae.

A few very large "giant" phyllosoma larvae have been reported. Johnson (1951) described a scyllarid specimen collected in the Pacific with a body length of 70 mm, and Sims and Brown (1968) reported a similar specimen from waters north of Bermuda. Robertson (1968c) caught a phyllosoma 65 mm long in the Caribbean Sea, and Prasad and Tampi (1965) reported four larvae 67–80 mm long from the Indian Ocean. Some of these specimens bear a close resemblance to one another, but as yet all are unidentified.

3. Puerulus or Nisto Stage

A further planktonic or nektonic stage occurs in both palinurid and scyllarid lobsters and has been variously identified as "puerulus," "postlarvae," "natant," "nisto," "pseudibacus," and "reptant larvae" (see Lyons, 1970). In this chapter, this stage of the palinurids is identified by the term puerulus, and the scyllarids by the term nisto. Although not a true larval stage, it is normally caught along with the phyllosoma larvae and as a consequence is discussed in this chapter.

The last phyllosoma stage (Fig. 5), which is usually about 35 mm long, metamorphoses into this transitional puerulus or nisto stage (Fig. 6). Gurney (1942) describes this metamorphosis as "The most profound transformation at a single molt known among Decapoda." The most obvious character separating the puerulus or nisto from later juvenile stages is that an absence of calcium and of pigment except in the eyes and odd spots on the exoskeleton causes them to be almost colorless and transparent until just prior to ecdysis. In addition, the puerulus or nisto stage has generally a smooth, rather than spiny or granular exoskeleton, and proportionately large pleopods equipped with long setae. The puerulus of Palinurids is approximately 8–10 mm in carapace length.

Evidence concerning the ecology of the puerulus and nisto stage is fragmentary. This partly results from the small numbers that have been caught by

TABLE IV

Number of Phyllosoma Larval Stages and Length of Larval Life for Some Species of Scyllaridae and Palinuridae

	No. of stages	Reference	Length of larval life	Reference
Family Palinuridae				
Panulirus argus	11	Lewis 1951	>6 Months	Lewis, 1951; Sims and Ingle, 1966
Panulirus interruptus	11	Johnson, 1956	7.75 Months	Johnson, 1971
Panulirus penicillatus	10	Prasad and Tampi, 1959	Probably 7–8 months	Johnson, 1971
Palinurellus gundlachi	12	Sims, 1966	10 Months	Sims, 1966
Panulirus cygnus	9	Brain *et al.*, 1979	9–11 Months	Chittleborough and Thomas, 1969
Panulirus homarus	9	Berry, 1974a	4–6 Months	Berry 1974a
Jasus lalandii	13	Lazarus, 1967	9–10 Months	Lazarus, 1967
Jasus edwardsii	11	Lesser, 1978	12–22 Months	Lesser, 1978
Panulirus japonicus	11	Inoue, 1978		
Family Scyllaridae				
Scyllarus americanus	6–7	Robertson, 1968a	32–40 Days	Robertson, 1968a
Scyllarus depressus	9–10	Robertson, 1971	25 Days	Robertson, 1968b, 1971
Scyllarus chacei	6–7	Robertson, 1968b	About 6 weeks	Robertson, 1968b
Scyllarus arctus	9	Stephensen, 1923; Santucci 1925; Kurian, 1956	Probably 3–4 months	Robertson, 1968b
Scyllarus bicuspidatus	>10	Saisho, 1966b; Robertson, 1968a	Probably 3–4 months	Robertson, 1968b
Scyllarus planorbis	8	Robertson, 1979	54 Days	Robertson, 1979
Scyllarides aequinoctialis	11	Robertson, 1969a	Perhaps 8 or 9 months	Robertson, 1969a
Scyllarides nodifer	12	Sims, 1965; Robertson, 1969c	9 Months	Sims, 1965; Robertson, 1969a

1. Larval Ecology

plankton sampling (e.g., 73 puerulus of *Panulirus inflatus gracilis* by Johnson, 1971; 13 of *Panulirus ornatus*, 14 of *Panulirus homarus rubellus*, and 4 of *Projasus parkeri* by Berry, 1974a; 2 of *Panulirus interruptus*, by Serfling and Ford, 1975; 12 of *Jasus edwardsii* by Lesser, 1978; and 301 of *Panulirus cygnus* by Phillips et al., 1978). Another difficulty arises from the changes in behavior that occur during this stage, in association with the transition from a planktonic or nektonic to a settled existence. Serfling and Ford (1975) have suggested a duration of 2–3 months for the puerulus stage of *P. interruptus*. It subsequently settles and molts into a young juvenile stage, in which the adult morphology has been assumed. Most workers have assumed that the puerulus stage comprises a single molt. However, Ting (1973) states that in *P. argus* the puerulus taken in inshore waters usually undergo two molts within 5–10 days in captivity before molting into the first juvenile stage.

The puerulus stage of *Jasus edwardsii* has been collected from under boulders in considerable numbers, at Castlepoint in New Zealand (Booth, 1979). This is

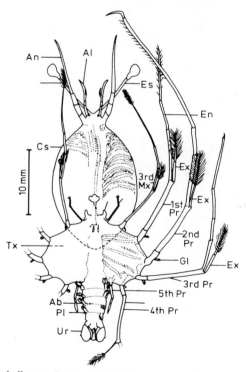

Fig. 5. Stage IX phyllosoma larvae of *Panulirus cygnus*. Ab, abdomen; Al, antennule; An, antenna; Cs, cephalic shield; Es, eyestalk; En, endopod; Ex, exopod; Gl, gill; Mx, maxilliped; Pl, pleopod; Pr, pereiopod; Tx, thorax; Ur, uropod.

Fig. 6. Puerulus stage of *Panulirus cygnus*. (Photo by A. C. Heron.)

the only species in which the puerulus stage has been found in any numbers after natural settlement.

C. Distribution and Ecology

Several workers, including Johnson (1960a,b), Sims and Ingle (1966), Chittleborough and Thomas (1969), and Berry (1974a), have shown that early phyllosoma stages generally occur in highest densities along the coast and near islands, in areas known to be inhabited by the adult populations. The later phyllosoma stages generally occur at distances farther from shore.

Although the long pelagic larval life gives a high potential for dispersal, studies of *P. interruptus* off the coast of California (Johnson, 1960a,b) and *P. cygnus* (George and Cawthorn, 1962; Chittleborough and Thomas, 1969) suggest that the larvae are not entirely randomly dispersed (George and Main, 1967). The larvae can move great distances offshore (the larvae of *P. cygnus* have been found over 1500 km offshore in the Indian Ocean (Phillips *et al.*, 1979). However, this is not exceptional; Johnson (1974) reported the larva of *P. penicillatus*, *P. gracilis* and *Scyllarus astori* in the Pacific Ocean at distances of 1800–2000 nautical miles from their likely point of origin. However, the larvae are rarely found beyond the latitudinal geographic limits of the adult population (see George and Main, 1967).

1. Larval Ecology

The mechanisms transporting the larvae are not fully understood. Unfortunately, an adequate description of the water currents of the ocean, to a level that will permit satisfactory assessment of larval transport, is not available for most areas. The hydrology of many of the areas in which the larvae are found is extremely complex (see Lazarus, 1967; Ingle *et al.*, 1963; Johnson, 1971, 1974; Phillips *et al.*, 1978) and characterized by eddies and countercurrents.

The most comprehensive analysis of the effects of water movements on phyllosoma larvae have been made by Johnson (1960a, 1971, 1974). Johnson (1960a, 1971) postulated that after release, the larvae of *P. interruptus* are carried southward by the California current. During this time many individuals are carried to areas unfavorable for survival. The surviving larvae, now in intermediate stages, are then carried northward again by the Davidson current. Recruitment in this and other species, is suggested to be dependent on groups of individuals that are retained relatively near the coast by counterclockwise currents, retarding eddies, or larvae transported into nearshore areas at the time when they are ready to metamorphose (Johnson, 1976). Johnson (1971) also suggested that soon after hatching, a large part of the larval population of *P. interruptus* (although generally highly planktonic) seeks the bottom water where currents are minimal and thus stay within or near the adult area. He states that these hypotheses have to be further tested by sampling with appropriate gear in the immediate coastal areas.

It was hypothesized that the early phyllosoma larvae of *P. cygnus* were probably transported offshore by surface wind drift (Chittleborough and Thomas, 1969). This hypothesis is supported by the results of studies of wind driven surface transport in the southeastern Indian Ocean by Cresswell (1972), and of the vertical movements of phyllosoma larvae (Rimmer and Phillips, 1979). Studies of the distribution of the phyllosoma larvae of *P. cygnus* in the southeastern Indian Ocean (Phillips *et al.*, 1979) showed that these larvae undergo an extensive migration. Thus, by the time the larvae have developed to the midstages, the majority have been transported well offshore and distributed over an extensive geographical area. As a result of the changes in vertical migration behavior that occur in mid- and late-stage larvae, producing a generally deeper vertical distribution, these phyllosoma stages become more subject to the oceanic circulation features underlying the immediate surface layer, and the effect of surface transport is greatly reduced. Thus, the coastward mass transport of the oceanic waters in the upper 300 m becomes the dominant factor influencing the larval migration, and consequently the larvae are transported back to the coast of Western Australia.

The numbers of puerulus that have been caught by plankton sampling are small, but the general indication is that they are to be found in greatest numbers in nearshore waters (Harada, 1957; Ritz, 1972b; Lesser, 1978). Sweat (1968) reported the absence of late-stage phyllosoma larvae of *P. argus* in inshore areas of Florida and hypothesized that metamorphosis from the last phyllosome took place offshore, followed by an onshore migration of puerulus. Studies of the late phyllosoma larvae and puerulus stage of *P. cygnus* (Phillips *et al.*, 1978) support this hypothesis. Phillips *et al.* found that at the end of the planktonic period the phyllosoma larvae of *P. cygnus* are concentrated just beyond the edge of the continental shelf off Western Australia, between August and December each year. Mixing of oceanic waters with the waters of the continental shelf only occurs on the outer third of the shelf, and the nearshore waters on the shelf are largely independent of the offshore water mass. Because of this lack of mixing of shelf and oceanic waters, the majority of late-stage phyllosoma larvae of *P. cygnus* are not carried onto the continental shelf but are held in eddy flows and carried southward down the West Australian coast beyond the shelf by the water currents. The puerulus of *P. cygnus* molts from the last phyllosoma stage in the waters beyond the continental shelf and completes the larval cycle by swimming approximately 40 km over the shelf across the prevailing currents before settling in the shallow inshore reef areas. The puerulus never comes to the surface while in the waters on the shelf, but in the inshore reef areas it rises to the surface at night, just prior to settlement (Phillips and Olsen, 1975). The puerulus stage, unlike the phyllosoma larvae, is therefore an active, free-swimming stage capable of swimming considerable distances.

The overall direction of the phyllosoma larvae during the larval movements seems to be fortuitous, and there is no suggestion that the larvae actually seek to travel in a particular direction. Under this assumption the presence of larvae in a water mass has been used by oceanographers as indications of the origins of that particular water mass (see Murano, 1957; Johnson and Brinton, 1963). Larvae released in shallow inshore areas are typically described as being carried offshore in the particular water mass in which they are released. Thus, while the mechanisms of return of the larvae back to the coast are not fully understood, their return can be taken as indicating the presence of water movements. The data on the early phyllosoma larvae of *P. cygnus* show that these larvae are transported offshore by the action of surface wind drift, passing over the top and moving contrary to the direction of the described major circulation of the area. This suggests that caution should be used in the interpretation of the movements of the larvae in relation to water circulation features, although the basic truth of the statement is unchanged.

The larvae which drift farthest from the coast are probably lost, but the maximum distance offshore from which the larvae can return successfully is not known for any species. Those larvae that do not return may be the source of recruitment to other areas. Studies of the larvae of *P. argus* in South Florida waters by Ingle *et al.* (1963) and Sims and Ingle (1966) suggested that most of the recruitment of *P. argus* and the other rock lobsters of the area was from other regions of the Caribbean. Similarly, R. W. George (personal communication) has suggested an interconnection of larval distributions for the Australian and New Zealand populations of *Jasus verreauxii,* and Berry (1974b) has suggested that larval exchanges of *P. homarus homarus* may occur between Madagascar and southeast Africa. Studies in progress by alloenzyme analysis, using polyacrylamide gel electrophoresis on local populations of *P. argus* in the Florida area may identify genetically distant populations and thus provide an answer to one of these questions (Menzies *et al.,* 1977; Menzies and Kerrigan, 1978).

D. Culture

Until recently all attempts to rear the larvae of Palinurids in the laboratory from hatching to puerulus have been unsuccessful, although several workers have kept phyllosomas alive for periods of up to several months (see Provenzano, 1968). Greater success has been achieved with Scyllarid larvae. The complete larval development of *S. americanus* was reported by Robertson (1968a), and Takahashi and Saisho (1979) have now reared *Ibacus ciliatus ciliatus* and *Ibacus novemdentatus* through their full larval period.

A major obstacle in rearing the phyllosoma larvae is that the nutritional requirements of the larvae change as the larvae develop, and different sizes and types of food are required by the different larval stages. Determining the most suitable food is particularly difficult because food particles are not normally observed in the gut of phyllosoma larvae caught in the plankton. The larvae of *S. americanus* (Robertson, 1968a) and the early larval stages of *P. inflatus* (Johnson and Knight, 1966), *Panulirus japonicus* (Saisho, 1962; Inoue, 1965), and *Panulirus longipes* (Saisho and Nakahara, 1960) were successfully fed on *Artemia* nauplii. Inoue (1978) reported the rearing of the phyllosoma larvae of *P. japonicus* up to the last stage of 29.6 mm in body length, fed on the nauplii and adults of *Artemia salina,* adults of *Sagitta* spp., and fish fry of several species. It took 253 days after hatching to reach this stage. However, examination of the illustration of this last stage shows that it does not possess gills, and so it is possible that it may not be the final phyllosoma stage of this species. Ritz and Thomas (1973) reared the first three stages of the scyllarid *Ibacus peronii* on live euphausids.

Batham (1967) reported that the early larvae of *J. edwardsii* showed no feeding response to *Artemia* nauplii, but they accepted fish muscle tissue and fed readily on adult capitellid polycheates. It was concluded from these observations and from the absence of an expodite on the third maxilliped that the larvae of different palinurid species probably have different feeding habits and that the larvae of *J. edwardsii* may possibly be benthic scavengers. It has also been reported that the phyllosoma larvae of *J. lalandii* would not eat *Artemia* nauplii (Silberbauer, 1971).

Mitchell (1971) described the feeding mechanisms and behavior of the early

Fig. 7. Phyllosoma of *Scyllarus* spp. riding on medusa of *Aurelia aurita* Linne. Taken in the field over deep water off the coast of Bimini. (Photo by J. G. Halusky.)

phyllosoma larvae of *P. interruptus*. She found that, of the freshly collected wild plankton offered to the larvae, they preferred chaetognaths, fish larvae, hydromedusae, and ctenophores when these were offered as individuals or in a mixture of prey species, whether alive, damaged, or freshly killed. Observations on the feeding functions of the mouthparts and thoracic appendages indicated that the phyllosoma larvae were best suited for capturing and eating large soft bodied forms such as medusae. The phyllosoma larvae of a number of Scyllaridae (but not Palinuridae) have been found clinging to certain hydrozoan and scyphozoan medusae by several workers (including Shojima, 1963; Thomas, 1963; Herrnkind *et al.*, 1976) (Fig. 7). They do not appear to be merely chance occurrences, since some 20% of about 500 medusae examined by Herrnkind *et al.* had phyllosomes attached. Even when displaced from their riding position on the aboral surface and directed into the oral tentacles, the phyllosomes were immune to the effects of the nematocyst-rich underbody, and they quickly maneuvered themselves to their original aboral riding position after such displacement (P. Kanciruk, personal communication). In addition, Sims and Brown (1968) reported a specimen of a scyllarid phyllosoma that contained undigested nematocysts in fecal material extruded from the anus. These observations suggest that the food of the phyllosoma larvae may be relatively liquid in form, which explains the absence of food particles in the gut.

Feeding in the puerulus stage has not been reported.

E. Factors Affecting Growth and Survival

1. Temperature and Salinity

Belman and Childress (1973) measured the oxygen consumption and swimming behavior of the early phyllosoma larvae of *P. interruptus* under different temperature regimes in the laboratory. They found that between 12.5° and 24.6°C (the range tested) temperature markedly affected respiration rate of the larvae. Although activity levels, measured as swimming behavior, remained relatively constant between 17.4° and 24.6°C, at 12.5°C activity ceased. An optimal thermal range with an upper level slightly below 24.6°C was suggested. However, larvae may show varying optimal thermal ranges since decreased mortality of *P. interruptus* larvae was found in larvae held at 25°C in the laboratory (Dexter, 1972).

Lazarus (1967) found that the presence or absence of phyllosoma larvae of *J. lalandii* was related to the specific salinity and temperature conditions of particular water masses, but he was not able to establish a correlation between these factors and larval abundance. However, he did observe that the stages of the phyllosomes were directly related to temperature and salinity. This relationship is probably a consequence primarily of drift and currents, with larvae hatching close to shore in a body of water and then being carried farther offshore in the same body of water.

Similarly, Chittleborough and Thomas (1969) speculated that *P. cygnus* larvae were released into waters with a salinity range of 35.4 to 36.0 ⁰/oo, while other species occurred only in waters having a salinity of 35.0 ⁰/oo or less. In addition, they noted that a wedge of low salinity water extended down the coast during the winter and that the late phyllosoma larvae did not appear to move back toward the coast until this wedge of water disappeared. No relationship between the densities of phyllosoma larvae of *P. cygnus* and surface water temperatures and salinity were found for the early or late phyllosoma stages (Ritz, 1972b; Phillips et al., 1978).

No relationships were found between the densities of the puerulus stage of *P. cygnus* and surface water temperature and salinity (Phillips et al., 1978). A correlation between the peak period of settlement of the puerulus of *P. interruptus* and maximum annual water temperature has been found (Serfling and Ford, 1975). However, this may be a reflection of "summer" and "winter" water movements of the area, returning late phyllosoma larvae to the area during the summer period. Both Serfling and Ford (1975) and Chittleborough and Phillips (1975) found that little or no settlement of the puerulus stage of *P. interruptus* and *P. cygnus*, respectively, occurred during the "winter," the peak of puerulus settlement being confined to the "spring" and "summer" of each area.

2. Plankton Biomass

Ritz (1972b) reported that the densities of early stage phyllosoma larvae of *P. cygnus* appeared to be independent of plankton biomass, but found a correlation between plankton biomass and the densities of late stage phyllosoma larvae. No relationship was found between the density of the puerulus of *P. cygnus* and plankton biomass (Phillips et al., 1978).

3. Currents

The phyllosoma larvae are assumed to be largely at the mercy of water currents because of their morphology and poor swimming ability, although no direct observations are available to confirm this.

Phillips (1975a) examined data on the water currents flowing into a settlement area and the catches of the puerulus of *P. cygnus* on collectors. No relationships were found between the volume of water flowing into the area and the catches on the collectors, either as a cyclic flow related to moon phase, or from night to night at the time of settlement. The puerulus entered the area and settled on the collectors during the night. On the nights on which settlement occurred, the puerulus may have made use of the incoming water currents to aid them in swimming into the area.

The effect of currents on the larvae is further discussed in Section C under larval distribution.

1. Larval Ecology

4. Light

Many workers, including Ritz (1972a), have reported that early phyllosoma larvae show a strong photopositive reaction to dim light. Ritz (1972b) found that the early phyllosoma larvae of *P. cygnus* were concentrated at the surface at night, even under bright moonlight conditions. Recent studies have shown that the early stage larvae of *P. cygnus* were unable to concentrate at the surface on nights when the weather was rough. Turbulence caused mixing in the upper waters and the larvae became randomly distributed through the mixed layer. The later stages were strong enough swimmers to overcome the effects of turbulence and achieved their preferred depth distribution regardless of the weather (Rimmer and Phillips, 1979).

Some other workers have recorded phyllosoma larvae at much greater depths. Prasad and Tampi (1965) found that the concentrations of larvae at 500 and 600 m were greater than at the surface, 150, 250, or 400 m. No scyllarids were found below 2000 m, but a few palinurid larvae were collected from as great a depth as 3500 m. Gurney (1936) also reported data demonstrating phyllosoma larvae to depths of 2480–2580 m. The majority of specimens from very deep water were all late stages. The identity of these specimens is unknown, but it is possible that they are the larvae of deep water species of some of the rarer genera. The depth to which some *Panulirus* larvae descend by day or night in the Gulf of Mexico has been found to be restricted to the waters above the thermocline, if one is present (Austin, 1972).

Russell (1925) was the first author to describe diurnal changes in the vertical distribution of spiny lobster phyllosoma larvae. As for other planktonic crustaceans, light intensity is probably the most important single factor in influencing the depth distribution of the larvae. Recent studies have found that the phyllosoma larvae of *P. cygnus* undergo a daily vertical migration, rising toward the surface of the water at night and then descending to lower depths during the day. Light was the key factor in influencing the depth distribution of the larvae. The newly hatched larvae were usually found within the upper 60 m of the water column, while later stages were usually found at depths of 80–120 m during the daylight. The deeper daytime depth distribution of late stages was partly due to the greater depth of light penetration in the clearer oceanic waters where they are found. The later stages were at the surface only on dark nights and avoided the surface on moonlight nights when lunar illuminance exceeded 5% of full moonlight [densities peaking between 10 and 30 m below the surface (see Fig. 8)]. Early stages were found at the surface at night, regardless of the level of moonlight (Rimmer and Phillips, 1979). Rimmer and Phillips hypothesized that the phyllosoma larvae of *P. cygnus* follow an isolume band of optimal illuminance which is in the order of 50–250 microeinsteins/m^2/sec for early stages, 20–200 μEm^{-2}sec^{-1} for midstages, and 50–500 μEm^{-2}sec^{-1} for late stages.

Phillips (1972) found that the puerulus stage of the western rock lobster *P.*

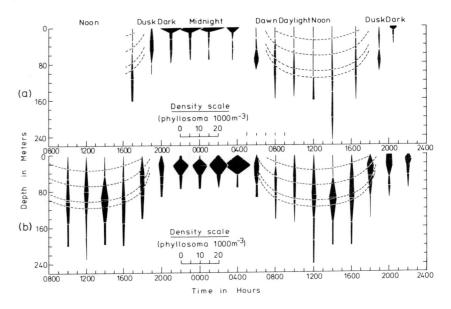

Fig. 8. Diurnal vertical migrations of the late stage phyllosoma larvae of *Panulirus cygnus* (a) at new moon; (b) near full moon. (From Rimmer and Phillips, 1979.)

cygnus could be captured using collectors composed of artificial seaweed. Settlement followed a lunar periodicity, with catches being largely confined to the new moon period. Settlement of the puerulus of *P. argus* has also been found to follow the same pattern (Sweat, 1968; Little, 1977), with most recruitment confined to the new and first quarter moon phases. Serfling and Ford (1975) found no lunar cycle in the settlement pattern of *P. interruptus,* although these data were sparse. Further work (Phillips, 1975a) showed that almost all settlement of the puerulus of *P. cygnus* on the collectors occurs at night, near the time of new moon. The greatest concentration takes place during the period of no moon, but ceases when the moonlight intensity rises above a threshold value of about 10% of full moonlight. It has been suggested that moonlight intensity normally prevents puerulus from entering into the area, hence the settlement on the collectors at nights other than near the time of new moon (Phillips, 1975a). However, since settlement does not commence until after the start of the dark phase, although the light levels in the last four nights of the last quarter of the moon are extremely low, light intensity alone does not fully describe the pattern of puerulus settlement. The manner in which the lunar phase acts on the puerulus to prevent settlement is unknown, as is the reason why moonlight can prevent settlement. However, illumination of the collectors increases their catching effectiveness (Phillips, 1975b). No vertical movements have yet been described for the puerulus stage of any species.

5. Predation

Neither the extent or the causes of larval mortality are well understood for any species. The occasional occurence of phyllosoma larvae in the food of some fish is probably incidental. Typical examples are Chittleborough and Thomas (1969), who noted that fishermen off the coast of Australia had reported finding phyllosoma larvae in the stomachs of mackerel and tuna. Baisre (1964) reported 121 phyllosoma larvae of *P. argus* from the stomachs of skipjack and blackfin tuna off the coast of Cuba. However, two examples of apparently direct predation have been reported. Heydorn (1969) found 150 phyllosomas and 120 puerulus in the stomachs of two longfin tuna (*Thunnus alalunga*), and another specimen had eaten 20 phyllosoma and 50 puerulus. George and Griffin (1972) reported that a single Tommy Ruff (*Arripis geogianus*) had its stomach stuffed full with 35 nisto of a species of *Scyllarus*. Puerulus of *J. edwardsii* were found by Lesser (1978) in the stomachs of dogfish (*Squalus* sp.) and red gurnard (*Cheilidonichthys kumu*). Lesser has pointed out the significance of these observations; i.e., these fish species are benthic or near-benthic feeders, although all the other species mentioned are pelagic feeders.

F. Behavior

The phyllosoma larvae are not well adapted for swimming, but can perform some movements. The first two phyllosoma stages of *P. japonicus* have been seen to be less efficient swimmers than the third and later stages, which had acquired a featherlike exopod on the third thoracic leg. Swimming ability improved further as a result of developments in the later stages. Swimming speeds (direction not described) of the late-stage phyllosoma were 2–3 cm/sec, and those of the early stage phyllosoma approximately one-third this rate. First-stage phyllosoma in aquaria rose vertically at between 0.5 and 0.6 cm/sec (Saisho, 1966b). Early *Panulirus* phyllosoma larvae in the laboratory are reported to have swum the length of a 3.6 m tank within 5–10 min (Sims and Ingle, 1966). Rimmer and Phillips (1979) have calculated minimal mean net rates of ascent and descent for *P. cygnus* early stage phyllosoma larvae of 13.7 and 13.0 m/hr, respectively, while the rates for midstages were 16.0 and 16.6 m/hr, and for late stages 19.4 and 20.1 m/hr. P. F. Berry (personal communication) has pointed out that the final stage phyllosoma larvae of *P. homarus* has functional pleopods and uropods, i.e., a fully formed abdomen, and could be capable of substantially more movement than earlier stages.

Satisfactory methods for rearing scyllarid larvae in the laboratory are available (Robertson, 1968a), and useful investigations might be carried out on their swimming abilities and behavior of scyllarid larvae. However, the general applicability of such results would have to be considered, since the larvae of the Scyllaridae and also *Jasus* sp. lack the exopodite on the third maxilliped, usually considered to be an adaptation to a swimming existence.

The puerulus stage is capable of both forward and backward swimming movements using the abdominal appendages, and also of rapid (abdomen flex) avoidance movements similar to those of the juvenile and adult stages (Saisho, 1966b; Berry, 1974a; Serfling and Ford, 1975; Phillips and Olsen, 1975). Deshmukh (1966) recorded that the puerulus stage of *P. polyphagus* could also swim upside down. Forward swimming speeds of 6.9 cm/sec were recorded for *P. interruptus* (Serfling and Ford 1975), and 30 cm/sec maximum and 15 cm/sec mean for *P. cygnus* (Phillips and Olsen, 1975).

Differences in the morphology of the puerulus of different species, particularly the length and shape of the antennae, may be indicators of different swimming ability. Some puerulus, e.g., *P. homarus*, *P. ornatus,* and *P. polyphagus* have relatively large antennae (2.0, 2.15, and 4.0 times body length, respectively) with spatulate tips. In contrast, *P. japonicus*, *P. cygnus*, and *P. penicillatus* have relatively short antennae (just over one body length) with tapering tips (Berry, 1974a). The long antennae of *P. interruptus* are reported to "quickly break off upon settlement" (Serfling and Ford, 1975). This has not been reported in other species.

Berry (1974a) stated that the puerulus stages of *P. homarus* and *P. ornatus* in aquaria displayed strong negative phototactic responses. However, Johnson (1956) and Harada (1957) reported that in the field the puerulus of *P. interruptus* and *P. japonicus,* respectively, were attracted to a lamp at the surface. Underwater lights used to observe free swimming puerulus of *P. interruptus* and *P. cygnus,* and it was reported that the puerulus were either attracted or bemused by the light, but certainly not repelled (Serfling and Ford, 1975; Phillips, 1975a). Serfling and Ford (1975) and Phillips (1975a) found that nocturnal illumination of the collectors (discussed earlier in this chapter) resulted in increased catch of the puerulus of *P. interruptus* and of *P. cygnus,* respectively. On the basis of anatomical evidence, the eye of the puerulus stage of *P. cygnus* may be capable of extremely efficient perception of polarized light (Meyer-Rochow, 1975).

It has been shown that the rasping sound produced by juvenile and adult spiny lobsters cannot be produced by the puerulus stage of *P. cygnus,* because the physical mechanism is only partly developed in the puerulus stage (Meyer-Rochow and Penrose, 1974).

Puerulus have been found adhering to seaweed (Chittleborough, 1967) or on the ropes attached to fishing equipment. This grasping facility has been made use of by a number of workers who have developed collectors to catch the settling puerulus stage of rock lobsters (Witham *et al.*, 1968; Phillips, 1972; Anonymous, 1974; Peacock, 1974; Serfling and Ford, 1975; Little, 1977; Booth, 1979). The puerulus of different genera, i.e., *Jasus* and *Panulirus,* have been found to

1. Larval Ecology

respond to different types of collectors. Settlement data recorded from these collectors (Chittleborough and Phillips, 1975; Phillips and Hall, 1978; Serfling and Ford, 1975) have enabled seasonal patterns of settlement as well as the relative strengths of puerulus settlement from year to year to be determined.

IV. CONCLUSIONS

Although much work has been published on adult lobsters because of their considerable economic importance, information on the early stages of life history, particularly of the larval phase, is still fragmentary. Considerable progress has been made in recent years, but a great many questions, particularly on larval behavior and the effects of environmental variables on the larvae, remain unanswered.

In spite of many efforts over a prolonged period, the inability to collect larval clawed lobsters in appreciable numbers with plankton tows has hampered a detailed understanding of their distribution and abundance. Although larger numbers of the phyllosoma larvae of spiny lobsters have been caught, this is not true of the puerulus stage. There is a need for further data on this important stage in the life cycle. Modern techniques of capture and anlysis are speeding up the solutions, but adequate sampling entails enormous logistic problems, due to the large distances traveled by the larvae during their development.

Because of the paucity of data, the behavior of many species is inferred from the results of the studies of other species. However, this type of generalizing is not always correct. As noted previously, the absence of an exopodite and the different feeding behavior of phyllosoma larvae of the Scyllaridae and *Jasus* sp., by comparison to the *Panulirus* larvae, suggests other differences are possible. Indeed, there are known variations in the length of larval life and rates of development of many species. In addition, the puerulus stage of *Jasus* and *Panulirus* have been found to respond to different types of collectors. Another important area of ignorance is the almost total lack of information on stage-specific mortality rates in the wild, so that any relationships that may exist between larval abundance and recruitment to populations cannot be established.

It is of interest to speculate on the adaptive significance of a planktonic larval stage in the life cycle, as well as the characteristic phyllosoma shape, the reason for the vertical migration behavior, and why the larval life is so long in the palinurids and scyllarids. Although a wide distribution is effected by the presence of a planktonic larva in the life cycle, this would seem to be a spin-off benefit, rather than a reason for long larval life. Thus, the principal objective must be successful recruitment, rather than dispersal. The significance of planktonic larvae in the life cycle has been discussed by Crisp (1974), Vance (1973a,b), and Logan (1976). Logan has suggested that the presence of a planktonic larva among benthic organisms may be a generalized adaptation (or pre-adaptation) which has been retained to allow their exploitation of the high productivity of the plankton. In support of the hypothesis, he cites the fact that the growth of *H. americanus* increases at the maximal rate during the larval period, and is much reduced in the settled stages because of the lower productivity in the benthos. The long larval life and the behavioral adaptations associated with this stage in the palinurids and scyllarids are presumably ways of achieving successful recruitment and maximizing exploitation of the productive plankton community.

The availability of reliable techniques for culture of *Homarus* larvae affords an opportunity to study many aspects of ecology, physiology, and behavior of larval lobsters. This may provide an insight for better understanding of the ecology of lobster larvae under natural conditions. Unraveling the behavior of the larvae is a great scientific challenge. The economic significance of the animals, and the possibility in some species of larval recruitment from sources other than the region of release, are additional reasons for the continued study of this portion of the life history.

REFERENCES

Aldrich, F. (1967). Observations by the late F. Aldrich on Australian marine crayfish in captivity, ed. by R. G. Chittleborough. *West. Aust. Nat.* **10,** 162-168.

Andersen, F. S. (1962). The Norway lobster in Faeroe waters. *Medd. Dan. Fisk. Havunders.* [N.S.] 3(9), 265-326.

Anderson, J. I. W., and Conroy, D. A. (1968). The significance of disease in preliminary attempts to raise Crustacea in sea water. *Bull. Off. Int. Epizoot.* **69**(7/8), 1239-47.

Anonymous (1974). Collection of larval rock lobsters. *Catch '74* pp. 9-10.

Archey, G. (1916). Notes on the marine crayfish of New Zealand. *Trans. Proc. N.Z. Inst.* **48,** 396-406.

Austin, H. M. (1972). Notes on the distribution of phyllosoma of the spiny lobster, *Panulirus* spp., in the Gulf of Mexico. *Proc. Natl. Shellfish. Assoc.* **62,** 26-30.

Baisre, J. A. (1964). Sobre los estadios larvales de la langosta comun *Panulirus argus*. *Cent. Invest. Pesq. (Cuba), Contrib.* **19,** 1-37.

Barlow, J. (1969). Studies on molecular polymorphism in the American lobster (*Homarus americanus*). Ph.D. Thesis, University of Maine, Orono.

Batham, E. J. (1967). The first three larval stages and feeding behavior of phyllosoma of the New Zealand palinurid crayfish, *Jasus edwardsii* (Hutton, 1875). *Trans. R. Soc. N.Z., Zool.* **9,** 53-64.

Belman, B. W., and Childress, J. J. (1973). Oxygen consumption of the larvae of the lobster *Panulirus interruptus* (Randall) and crab *Cancer productus* Randall. *Comp. Biochem. Physiol. A* **44,** 821-828.

Berry, P. F. (1969). The biology of *Nephrops andamanicus* Wood-Mason (Decapoda, Reptantia). *Oceanogr. Res. Inst. (Durban), Invest. Rep.* **22,** 1-55.

Berry, P. F. (1974a). Palinurid and scyllarid lobster larvae of the Natal coast, South Africa. *Oceanogr. Res. Inst. (Durban), Invest. Rep.* **34,** 1-44.

Berry, P. F. (1974b). A revision of the *Panulirus homarus* group of spiny lobsters (Decapoda, Palinuridae). *Crustaceana* **27,** 31-42.

Booth, J. D. (1979). Settlement of the rock lobster, *Jasus edwardsii* (Decapoda: Palinuridae), at Castlepoint, New Zealand, *N.Z.J. Mar. Freshwater Res.* **13,** 395-406.

Bradford, J. R. (1978a). Incubation period for the lobster *Homarus grammarus* at various temperatures. *Mar. Biol.* **47**, 363-368.
Bradford, J. R. (1978b). The influence of day-length, temperature, and season on the hatching rhythm of *Homarus gammarus*. *J. Mar. Biol. Assoc. U.K.* **58**, 639-658.
Braine, S. J., Rimmer, D. W., and Phillips, B. F. (1979). An illustrated key to the phyllosoma stages of the western rock lobster *Panulirus cygnus* George, with notes on length frequency data. *CSIRO, Aust. Div. Fish. Oceanogr. Rep. No.* **102**, 1-13.
Caddy, J. G. (1979). The influence of variations in the seasonal temperature regime on survival of larval stages of the American lobster (*Homarus americanus*) in the southern Gulf of St. Lawrence. *Rapp. P.-V. Reun., Cons. Int. Explor. Mar.* (in press).
Capuzzo, J. M. and Lancaster, B. A. (1979). Larval development in the American lobster: changes in metabolic activity and the O:N ratio. *Can. J. Zool.* **57**, 1845-1848.
Capuzzo, J. M. and Lancaster, B. A. (1979). Some physiological and biochemical considerations of larval development in the American lobster *Homarus americanus* Milne-Edwards. *J. Exp. Mar. Biol. Ecol.* **40**, 53-62.
Capuzzo, J. M., Lawrence, S. A., and Davidson, J. A. (1976). Combined toxicity of free chlorine, chloramine, and temperature to stage I larvae of the American lobster *Homarus americanus*. *Water Res.* **10**, 1093-1099.
Chittleborough, R. G., and Phillips, B. F. (1975). Fluctuations of year-class strength and recruitment in the western rock lobster *Panulirus longipes* (Milne-Edwards). *Aust. J. Mar. Freshwater Res.* **26**, 317-28.
Chittleborough, R. G., and Thomas, L. R. (1969). Larval ecology of the western Australian marine crayfish with notes upon other panulirud larvae from the eastern Indian Ocean. *Aust. J. Mar. Freshwater Res.* **20**, 199-223.
Cobb, J. S. (1968). Delay of moult by the larvae of *Homarus americanus*. *J. Fish. Res. Board Can.* **25**, 2251-2253.
Cobb, J. S. (1970). Effect of solitude on time between fourth and fifth larval moults in the American lobster (*Homarus americanus*). *J. Fish. Res. Board Can.* **27**, 1653-1655.
Conner, P. M. (1972). Acute toxicity of heavy metals to some marine invertebrate larvae. *Mar. Pollut. Bull.* **3**, 190-192.
Crawford, D. R., and DeSmidt, W. J. J. (1922). The spiny lobster (*Panulirus argus*) of Southern Florida. Its natural history and utilization. *Bull. U.S. Bur. Fish.* **38**, 281-310.
Cresswell, G. R. (1972). Wind-driven ocean surface transport around Australia. *Rep.—CSIRO, Div. Fish. Oceanogr. (Aust.)* **52**, 1-5.
Crisp, D. J. (1974). Energy relations of marine invertebrate larvae. *Thalassia Jugosl.* **10**, 103-120.
Crosnier, A. (1972). Naupliosoma, phyllosomes et pseudibacus de *Scyllarides herklotsi* (Herklots) (Crustacea, Decapoda, Scyllaridae) récolectes par l'OMBANGO dans le sud du Golfe de Guinee. *Cah. ORSTOM, Ser. Oceanogr.* **10**, 139-149.
Dannevig, A. (1936). Hummer og Hummerkutter. *Fiskeridir. Skr., Ser. Havunders.* **4**(12), 1-60.
Davis, C. C. (1964). A study of the hatching process in aquatic invertebrates. XIII. Events of eclosion in the American lobster, *Homarus americanus* Milne-Edwards (Astacura, Homaridae). *Am Midl. Nat.* **72**, 203-210.
Deshmukh, S. (1966). The puerulus of the spiny lobster *Panulirus polyphagus* (Herbst) and its metamorphosis into the post-puerulus. *Crustaceana* **10**, 137-150.
Deshmukh, S. (1968). On the first phyllosomae of the Bombay spiny lobsters (*Panulirus*) with a note on the unidentified first Panulirus phyllosomae from India (Palinuridea). *Crustaceana, Suppl.* **2**, 47-58.
Dexter, D. M. (1972). Molting and growth in laboratory reared phyllosomes of the California spiny lobster, *Panulirus interruptus*. *Calif. Fish. Game* **58**, 107-115.
Ehrenbaum, E. (1903). Neure Untersuchungen über den Hummer. *Mitt. Dtsch. Seefischereiver.* **19**, 107-115.

Ennis, G. R. (1973a). Endogenous rhythmicity associated with larval hatching in the lobster *Homarus gammarus*. *J. Mar Biol. Assoc. U.K.* **53**, 531-538.

Ennis, G. P. (1973b). Behavioral responses to changes in hydrostatic pressure and light during larval development of the lobster, *Homarus gammarus*. *J. Fish. Res. Board Can.* **30**, 1349-1360.

Ennis, G. P. (1975a). Behavioral responses to changes in hydrostatic pressure and light during larval development of the lobster, *Homarus americanus*. *J. Fish. Res. Board Can.* **32**, 271-281.

Ennis, G. P. (1975b). Observations on hatching and larval release in the lobster *Homarus americanus*. *J. Fish. Res. Board Can.* **32**, 2210-2213.

Factor, J. R. (1977). Morphology of the mouthparts of larval lobsters *Homarus americanus*. *Am. Zool.* **17**, 971.

Factor, J. R. (1978a). Morphology of the mouthparts of larval lobsters, *Homarus americanus* (Decapoda:Nephropidae), with special emphasis on their setae. *Biol. Bull. (Woods Hole, Mass.)* **154**, 383-408.

Factor, J. R. (1978b). Preliminary Report on the development of the digestive system of larval lobsters *Homarus americanus*, *Am. Zool.* **18**, 63.

Farmer, A. S. D. (1972). The general biology of *Nephrops norvegicus* (Linnaeus, 1758) (Decapoda:Nephropidae) off the Isle of Man. Ph.D. Thesis, University of Liverpool.

Farmer, A. S. D. (1974). Reproduction in *Nephrops norvegicus* (Decapoda:Nephropidae). *J. Zool.* **174**, 161-183.

Feliciano, C. (1956). A prenaupliosoma stage in the larval development of the spiny lobster, *Panulirus argus* (Latreille) from Puerto Rico. *Bull. Mar. Sci. Gulf Caribb.* **6**, 341-345.

Figueiredo, M. J. (1971). Sobre a cultura de crustaceos decapodes em laboratorio: *Nephrops norvegicus* (Lagostim) *Penaeus kerathurus* (Camarao). *Bol. Inf. Inst. Biol. Marit.* **1**, 1-17.

Figueiredo, M. F. de, and Vilela, M. H. (1972). On the artificial culture of *Nephrops norvegicus* reared from the egg. *Aquaculture* **1**, 173-180.

Fisher, W. S., Nilson, E. H., and Shleser, R. A. (1975). Diagnostic procedures for diseases found in eggs, larvae and juvenile American lobsters. *Proc. Annu. Meet.—World Maric. Soc.* **6**, 335-346.

Fisher, W. S., Nilson, E. H., Follett, L. F., and Shleser, R. A. (1976a). Hatching and rearing lobster larvae *(Homarus americanus)* in a disease situation. *Aquaculture* **7**, 75-80.

Fisher, W. S., Rosemark, T. R., and Shleser, R. (1976b). Toxicity of malachite green to cultured American lobster larvae. *Aquaculture* **8**, 151-156.

Foxon, G. E. H. (1934). Notes on the swimming methods and habits of certain crustacean larvae. *J. Mar. Biol. Assoc. U. K.* **19**, 829-45.

Fraser, J. H. (1965). Larvae of *Nephrops norvegicus* in the Scottish area, 1935-1964. *ICES C.M.* **10**, 1-4. pp. (unpublished).

George, R. W. (1962). Description of *Panulurus cygnus* sp. nov., The commercial crayfish (or spiny lobster) of western Australia. *J. R. Soc. West. Aust.* **45**, 100-110.

George, R. W., and Cawthorn, P. (1962). Investigations on the phyllosoma larvae of the western Australian crayfish. *West. Aust. Mus. Rep. 1962*.

George, R. W., and Griffin, D. J. G. (1972). The shovel nosed lobsters of Australia. *Aust. Nat. Hist.* **17**(7), 227-230.

George, R. W., and Main, A. R. (1967). The evolution of spiny lobsters (Palinuridae): A study of evolution in the marine environment. *Evolution* **21**, 803-820.

Gilchrist, J. D. F. (1966). Larval and post-larval stages of *Jasus lalandii* (Milne Edw.), Ortmann. *J. Linn. Soc. London, Zool.* **33**, 101-125.

Gompel, M., and Legendre, R. (1927). Effects de la temperature, de la salure et du pH sur les larves des Homards. *C.R. Seances Soc. Biol. Ses Fil.* **97**, 1058-1060.

Gruffydd, L. E., Riser, R. A., and Machin, D. (1975). A comparison of growth and temperature tolerance in the larvae of the lobster *Homarus gammarus* (1) and *Homarus americanus* H. Milne-Edwards (Decapoda, Nephropidae). *Crustaceana* **28**, 23-32.

Gurney, R. (1936). Larvae of decapod crustacea. Part III. Phyllosoma. *'Discovery' Rep.* **12**, 400–440.
Gurney, R. (1942). "Larvae of Decapod Crustacea." Ray Society, London.
Hadley, P. B. (1908). The behaviour of the larval and adolescent stages of the American lobster (*Homarus americanus*). *J. Comp. Neurol. Psychol.* **18**, 199–301.
Hadley, P. B. (1909). Additional notes upon the development of the lobster. *R.I. Comm. Inland Fish., Annu. Rep.* **40**, 189.
Harada, E. (1957). Ecological observations on the Japanese spiny lobster *Panulirus japonicus* (von Siebold), in its larval and adult life. *Publ. Seto Mar. Biol. Lab.* **6**, 91–120.
Harada, E. (1958). Notes on the naupliosoma and newly hatched phyllosoma of *Ibacus ciliatus* (von Siebold). *Publ. Seto Mar. Biol. Lab.* **7**, 173–180.
Heckman, R. A., Infanger, R. C., and Mickelson, R. W. (1978). A study of the early life stages of the lobster *Homarus americanus*, using scanning electron microscopy (SEM). *Proc. World. Mariculture Society, Ninth Annual Meeting.* **9**, 481–495.
Herrick, F. H. (1894). The habits and development of the lobster and their bearing upon its artificial propagation. *Bull. U.S. Fish. Comm.* **13**, 75–86.
Herrick, F. H. (1896). The American lobster, a study of its habits and development. *Bull. U.S. Fish. Comm.* **15**, 1–252.
Herrnkind, W. F., Halusky, J., and Kanciruk, P. (1976). A further note on phyllosoma larvae associated with medusae. *Bull. Mar. Sci.* **26**, 110–112.
Heydorn, A. E. F. (1969). The rock lobster of the South African west coast *Jasus lalandii* (H. Milne-Edwards). 2. Population studies, behavior, reproduction, moulting, growth and migration. *Oceanogr. Res. Inst. (Durban), Invest. Rep.* **71**.
Hillis, J. P. (1968). Larval distribution of *Nephrops norvegicus* (L.) in the Irish Sea and North Channel. *ICES C.M.* K:**6**, 1–5.
Hillis, J. P. (1971). Studies on Dublin Bay prawns (*Nephrops norvegicus*) in the Irish Sea. *Fish. Leafl. Dep. Agric. Fish. (Dublin)* **22**, 1–11.
Hillis, J. P. (1972a). Some problems and methods in Dublin Bay prawn (*Nephrops morvegicus*) research. *Fish. Leafl., Dep. Agric. Fish. (Irel.)* **30**, 1–13.
Hillis, J. P. (1972b). Studies on the biology and ecology of the Norway lobster *Nephrops norvegicus* (l.) in Irish waters. Ph.D. Thesis, National University of Ireland.
Hughes, J. T. (1972). Lobster culture. *In* "Culture of Marine Invertebrate Animals" (W. L. Smith and M. H. Chanley, eds.), pp. 221–227. Plenum, New York.
Hughes, J. T., and Matthiessen, G. C. (1962). Observations on the biology of the American lobster, *Homarus americanus*. *Limnol. Oceanogr.* **7**, 414–421.
Hughes, J. T., Shleser, R. A., and Tchobanoglous, G. (1974). A rearing tank for lobster larvae and other aquatic species. *Prog. Fish Cult.* **36**, 129–133.
Huntsman, A. G. (1923). Natural lobster breeding. *Biol. Board Can., Bull.* **5**.
Huntsman, A. G. (1924). Limiting factors for marine animals. 2. Resistance of larval lobsters to extreme temperature. *Contrib. Can. Biol. Fish.* **2**, 91–93.
Ingle, R. M., Eldred, B., Sims, H. W., and Eldred, E. A. (1963). On the possible Caribbean origin of Florida spiny lobster populations. *Fla. Board Conserv. Mar. Res. Lab., Tech. Ser.* **40**, 1–13.
Inoue, M. (1965). On the relation of amount of food taken to the density of size of food in phyllosoma of the Japanese spiny lobster, *Panulirus japonicus* (Von Siebold). *Bull. Jpn. Soc. Sci. Fish.* **31**, 902–906.
Inoue, M. (1978). Studies on the cultured phyllosoma larvae of the Japanese spiny lobster *Panulirus japonicus*. I. Morphology of the phyllosoma. *Bull. Jpn. Soc. Sci. Fish.* **44**, 457–475.
Jackson, H. G. (1913). Decapod larvae in the Irish Sea. *Rep. Lancashire Sea-Fish. Labs.* **21**, 254–259.

Johnson, M. W. (1951). A giant phyllosoma larva of a loricate crustacean from the tropical Pacific. *Trans. Am. Microsc. Soc.* **70,** 274-278.

Johnson, M. W. (1956). The larval development of the California spiny lobster, *Panulirus interruptus* (Randall), with notes on *Panulirus gracilis* Streets. *Proc. Calif. Acad. Sci.* **29,** 1-19.

Johnson, M. W. (1960a). The offshore drift of larvae of the California spiny lobster, *Panulirus interruptus*. *Calif. Coop. Oceanic Fish., Invest. Rep.* **7,** 147-161.

Johnson, M. W. (1960b). Production and distribution of larvae of the spiny lobster, Panulirus interruptus (Randall) with records of *P. gracilis* Streets. *Bull. Scripps Inst. Oceanogr.* **7,** 413-46.

Johnson, M. W. (1968). The phyllosoma larvae of scyllarid lobsters in the Gulf of California and off Central America with special reference to *Evibacus priceps* (Palinuridea). *Crustaceana, Suppl.* **2,** 98-116.

Johnson, M. W. (1971). The palinurid and scyllarid lobster larvae of the tropical eastern Pacific and their distribution as related to the prevailing hydrography. *Bull. Scripps Inst. Oceanogr.* **19,** 1-36.

Johnson, M. W. (1974). On the dispersal of lobster larvae into the east Pacific barrier (Decapoda, Palinuridae). *Fish. Bull.* **72,** 639-647.

Johnson, M. W., and Brinton, E. (1963). Biological species, water masses, and currents. *In* "The Sea: Ideas and Observations on Progress in the Study of the Seas" (M. N. Hill, ed.), Vol. 2, pp. 381-414. Wiley (Interscience), New York.

Johnson, M. W., and Knight, M. (1966). The phyllosoma larvae of the spiny lobster *Panulirus inflatus* (Bouview). *Crustaceana* **10,** 31-47.

Johnson, P. W., Sieburth, J. McN., Sastry, A. N., Arnold, G. R., and Doty, M. S. (1971). *Leucothrix mucor* infestation of benthic crustacea, fish eggs, and tropical algae. *Limnol. Oceanogr.* **16,** 962-969.

Jorgensen, O. M. (1925). The early stages of *Nephrops norvegicus*, from the Northumberland plankton, together with a note on the post-larval development of *Homarus vulgaris*. *J. Mar. Biol. Assoc. U.K.* **13,** 870-879.

Kensler, C. B. (1970). The potential of lobster culture. *Am. Fish Farmer* **1,** 8-12.

Kurian, C. V. (1956). Larvae of decapod Crustacea from the Adriatic Sea. *Acta Adriat.* **6**(3), 1-108.

Laverack, M. S., Macmillan, D. L., and Neil, D. M. (1976). A comparison of beating parameters in larval and post-larval locomotor systems of the lobster *Homarus gammarus* (L). *Philos. Trans. R. Soc. London, Ser. B* **274,** 87-99.

Lazarus, B. I. (1967). The occurrence of phyllosomata of the cape with particular reference to *Jasus lalandii*. *S. Afr., Div. Sea Fish., Invest. Rep.* **63.**

Lebour, M. V. (1950). Notes on some larval decapods (Crustacea) from Bermuda. *Proc. Zool. Soc. London* **120,** 369-379.

Lesser, J. H. R. (1974). Identification of early larvae of New Zealand spiny and shovel-nosed lobsters (Decapoda, Palinuridae and Scyllaridae). *Crustaceana*, **27,** 259-277.

Lesser, J. H. R. (1978). Phyllosoma larvae of *Jasus edwardsii* (Hutton) (Crustacea:Decapoda:Palinundae) and their distribution off the east coast of the North Island, New Zeland. *N.Z. J. Mar. Freshwater Res.* **12,** 357-370.

Lewis, J. B. (1951). The phyllosoma larvae of the spiny lobster *Panulirus argus*. *Bull. Mar. Sci. Gulf Caribb.* **1**(2), 89-103.

Little, E. J. (1977). Observations on recruitment of post larval spiny lobsters, *Panulinus argus*, to the South Florida Coast. *Fla. Mar. Res. Publ.* **29,** 1-35.

Logan, D. T. (1976). A laboratory energy balance for the larvae and juveniles of the lobster, *Homarus americanus*. Ph.D. Thesis, University of Delaware, Newark.

Logan, D. T., and Epifanio, C. E. (1978). A laboratory energy balance of the larvae and juveniles of the American lobster, *Homarus americanus*. *Mar. Biol.* **47,** 381-389.

1. Larval Ecology

Lund, W. A., and Stewart, L. L. (1970). Abundance and distribution of larval lobsters, *Homarus americanus*, off the coast of southern New England. *Proc. Natl. Shellfish. Assoc.* **60,** 40-49.
Lyons, W. G. (1970). Scyllarid lobsters (Crustacea, Decapoda). *Mem. Hourglass Cruises* **1,** 1-74.
MacClement, W. A. (1917). Diatoms and lobster rearing. *Contrib. Can. Biol. Fish.* **2,** 11-20.
McLeese, D. W. (1974). Olfactory response and fenitrothion toxicity in American lobster (*Homarus americanus*). *J. Fish. Res. Board Can.* **31,** 1127-1131.
Macmillan, D. L., Neil, D. M., and Laverack, M. S. (1976). A quantitative analysis of exopodite beating in the larvae of the lobster *Homarus gammarus* (1). *Philos. Trans. R. Soc. London, Ser. B* **274** 69-85.
Mead, A. D., and Williams, L. W. (1903). Habits and growth of the lobster, and experiments in lobster culture. *R.I. Comm. Inland Fish., Annu. Rep.* **33,** 57-83.
Mendall, H. L. (1934). The relationship of certain sea birds to the fishing industry of the state of Maine. *Bienn. Rep. Comm. Sea Shore Fish., Maine* **8,** 37-64.
Menzies, R. A., and Kerrigan, J. M. (1978). Implications of spiny lobster recruitment patterns of the Caribbean—A biochemical genetic approach. *Proc. Gulf Caribb. Fish Inst.* **31,** 164-178.
Menzies, R. A., Kerrigan, J. M., and Kanciruk, P. (1977). A biochemical-genetic approach to the larval recruitment problem of the spiny lobster *Panulirus argus. Circ.—CSIRO, Div. Fish. Oceanogr. (Aust.)* **7,** 38 (abstr.).
Meyer-Rochow, V. B. (1975). Larval and adult eye of the western rock lobster (Panulirus longipes). *Cell Tissue Res.* **162,** 429-457.
Meyer-Rochow, V. B., and Penrose, J. D. (1974). Sound and sound emission apparatus in puerulus and postpuerulus of the western rock lobster. *J. Exp. Zool.* **189,** 283-289.
Mills, D. H. (1957). Herring gulls and common terns as possible predators of lobster larvae. *J. Fish. Res. Board Can.* **14,** 729-730.
Mitchell, J. R. (1971). Food preferences, feeding mechanisms, and related behavior in phyllosoma larvae of the California spiny lobster, *Panulirus interruptus* (Randall). M.S. Thesis, San Diego State College, San Diego, California.
Murano, R. (1957). Plankton as the indicators of water masses and ocean currents. *Oceanogr. Mag.* **9,** 53-63.
Neil, D. M., Macmillan, D. L., Robberson, R. M., and Laverack, M. S. (1976). The structure and function of thoracic exopodites in the larvae of the lobster *Homarus gammarus* (L). *Philos. Trans. R. Soc. London, Ser. B* **274,** 53-68.
Nichols, J. H. and Lawton, P. (1978). The occurrence of the larval stages of the lobster *Homarus gammarus* (Linnaeus, 1958) off the northeast coast of England in 1976. *J. Cons. Cons. Int. Explor. Mer.* **38,** 234-243.
Nilson, E. H., Fisher, W. S., and Shleser, R. A. (1975). Filamentous infestation observed in eggs and larvae of cultured crustaceans. *Proc. Annu. Meet.—World Maric. Soc.* **6,** 367-368.
Nilson, E. H., Fisher, W. A., and Shleser, R. A. (1976). A new mycosis of lobster larvae. *J. Invertebr. Pathol.* **27,** 177-183.
Ong, K. S. (1967). A preliminary study of the early larval development of the spiny lobster *Panulirus polyphagus* (Herbst). *Malays. Agric. J.* **46,** 183-190.
Pandian, T. J. (1970a). Ecophysiological studies on the developing eggs and embryos of the European lobster *Homarus gammarus. Mar. Biol.* **5,** 154-167.
Pandian, T. J. (1970b). Yolk utilization and hatching time in the Canadian lobster *Homarus americanus. Mar. Biol.* **7,** 249-254.
Peacock, N. A. (1974). A study of the spiny lobster fishery of Antigua and Barbuda. *Gulf Caribb. Fish. Inst., Univ. Miami, Proc.* **26,** 117-130.
Phillips, B. F. (1972). A semi-quantitative collector of the puerulus larvae of the western rock lobster *Panulirus longipes cygnus* George (Decapoda, Palinuridae). *Crustaceana* **22,** 147-154.
Phillips, B. F. (1975a). The effect of water currents and the intensity of moonlight on catches of the

puerulus larval stage of the western rock lobster. *Rep.—CSIRO, Div. Fish. Oceanogr. (Aust.)* **63**, 1–9

Phillips, B. F. (1975b). Effect of nocturnal illumination on catches of the puerulus larvae of the western rock lobster by collectors composed of artificial seaweed. *Aust. J. Mar. Freshwater Res.* **26**, 411–14.

Phillips, B. F., and Hall, N. G. (1978). Catches of puerulus larvae on collectors as a measure of natural settlement of the western rock lobster *Panulirus cygnus* George. *Rep.—CSIRO, Div. Fish. Oceanogra. (Aust.)* **98**, 1–18.

Phillips, B. F., and Olsen, L. (1975). Swimming behavior of the puerulus larvae of the western rock lobster. *Aust. J. Mar. Freshwater Res.* **26**, 415–17.

Phillips, B. F., Rimmer, D. W., and Reid, D. D. (1978). Ecological investigations of the late stage phyllosoma and puerulus larvae of the western rock lobster *Panulirus longipes cygnus*. *Mar. Biol.* **45**, 347–357.

Phillips, B. F., Brown, P. A., Rimmer, D. W., and Reid, D. D. (1979). Distribution and dispersal of the phyllosoma larvae of the western rock lobster *Panulirus cygnus* in the southeastern Indian ocean. *Aust. J. Mar. Freshwater Res.* **30**, 773–783.

Poulsen, E. M. (1946). Investigations on the Danish fishery for and the biology of the Norway lobster and the deep-sea prawn. *Rep. Dan. Biol. Stn.* **48**, 27–49.

Prasad, R. R., and Tampi, P. R. S. (1959). A note on the first phyllosoma of *Panulirus burgeri* (De Haan). *Proc. Indian Acad. Sci., Sect. B* **49**, 397–401.

Prasad, R. R., and Tampi, P. R. S. (1965). A preliminary report on the phyllosomas of the Indian Ocean collected by the Dana Expedition 1928–1930. *J. Mar. Biol. Assoc. India* **7**, 277–283.

Provenzano, A. J. (1968). Recent experiments on the laboratory rearing of tropical lobster larvae. *Gulf Caribb. Fish. Inst., Univ. Miami, Proc.* **21**, 152–157.

Rao, K. R., Fingerman, S. W., and Fingerman, M. (1973). Effects of exogenous ecdysones on the molt cycles of fourth and fifth stage American lobsters, *Homarus americanus*. *Comp. Biochem. Physiol. A* **44**, 1105–1120.

Rice, A. L. (1967). The orientation of the pressure responses of some marine crustacea. *In* "Proceedings of Symposium on Crustacea, Ernakulam, 1965," Symp. Ser. 2, pp. 1124–1131. Mar. Biol. Assoc. India, Cochin.

Rimmer, D. W., and Phillips, B. F. (1979). Diurnal migration and vertical distribution of phyllosoma larvae of the western rock lobster, *Panulirus cygnus* George. *Mar. Biol.* **54**, 109–24.

Ritz, D. A. (1972a). Behavioural response to light intensity of the newly hatched phyllosoma larvae of *Panulirus longipes cygnus* George (Crustacea;Decapoda;Palinuridae). *J. Exp. Mar. Biol. Ecol.* **10**, 105–114.

Ritz, D. A. (1972b). Factors affecting the distribution of rock lobster larvae (*Panulirus longipes cygnus*), with reference to variability of plankton-net catches. *Mar. Biol.* **13**, 309–317.

Ritz, D. A., and Thomas, L. R. (1973). The larval and postlarval stages of *Ibacus peronii* Leach (Decapoda, Reptantia, Scyllaridae). *Crustaceana* **24**, 5–16.

Robertson, P. B. (1968a). The complete larval development of the sand lobster, *Scyllarus americanus* (Smith) (Decapoda, Scyllaridae) in the laboratory, with notes on larvae from the plankton. *Bull. Mar. Sci.* **18**, 294–342.

Robertson, P. B. (1968b). The larval development of some western Atlantic lobsters of the family Scyllaridae. Ph.D. Thesis, University of Miami, Coral Gables.

Robertson, P. B. (1968c). A giant scyllarid phyllosoma larva from the Caribbean Sea, with notes on smaller specimens (Palinuridea). *Crustaceana, Suppl.* **2**, 83–97.

Robertson, P. B. (1969a). Biological investigations of the deep sea. No. 48. Phyllosoma larvae of the scyllarid lobster, *Arctides guineensis,* from the western Atlantic. *Mar. Biol.* **4**, 143–151.

Robertson, P. B. (1969b). Phyllosoma larvae of the palinurid lobster *Justitia longimana* (H. Milne-Edwards), from the western Atlantic. *Bull. Mar. Sci.* **19**, 922–944.

1. Larval Ecology

Robertson, P. B. (1969c). The early larval development of the scyllarid lobster *Scyllarides aequinoctialis* (Lund) in the laboratory, with a revision of the larval characters of the genus. *Deep-Sea Res.* **16**, 557-586.

Robertson, P. B. (1971). Biological Results of the University of Miami Deep-Sea Expeditions. 84. The larvae and postlarva of the scyllarid lobster *Scyllarus depressus* (Smith). *Bull. Mar. Sci.* **21**, 841-865.

Robertson, P. B. (1979). Biological Results of the University of Miami Deep Sea Expeditions. 131. Larval development of the scylland lobster *Scyllarus planorbis* Holt Luis reared in the laboratory. *Bull. Mar. Sci.* **29**, 1320-1328.

Rogers, B. A., Cobb, J. S., and Marshall, N. (1968). Size comparison of inshore and offshore larvae of the lobster, *Homarus americanus*, off southern New England. *Proc. Natl. Shellfish. Assoc.* **58**, 78-81.

Russell, F. S. (1925). The vertical distribution of marine macroplankton. An observation on diurnal changes. *J. Mar. Biol. Assoc. U.K.* **13**, 769-809.

Saisho, T. (1962). Notes on the early development of phyllosoma of *Panulirus japonicus*. *Mem. Fac. Fish. Kagoshima Univ.* **11**, 18-23.

Saisho, T. (1966a). A note on the phyllosoma stages of spiny lobster. *Inf. Bull. Planktol. Jpn.* **13**, 69-71.

Saisho, T. (1966b). Studies on the phyllosoma larvae with reference to the oceanographical conditions. *Mem. Fac. Fish. Kagoshima Univ.* **15**, 177-239.

Saisho, T., and Nakahara, K. (1960). On the early development of phyllosoma of *Ibacus ciliatus* and *Panulirus longipes*. *Mem. Fac. Fish. Kagoshima Univ.* **9**, 84-90.

Santucci, R. (1926a). Lo sviluppo e l'ecologia post-embrionali dello "scampo" (*Nephrops norvegicus* (L)) nel Tirreno e nei Mari Nordici. *Mem. R. Com. Talassogr. Ital.* **125**, 1-36.

Santucci, R. (1926b). Lo sviluppo post-embrionale dello scampo (*Nephrops norvegicus* (L)) del Quarnero. Nota preliminare. *Boll. Mus. Zool. Anat. Comp. Genova* **6**(2), 7-10.

Santucci, R. (1926c). Lo sviluppo post-embrionale dell'aragosta (*Panulirus vulgaris* Latr.) e dello "scampo" (*Nephrops norvegicus* (L)). *Mem. R. Com. Talassogr. Ital.* **128**, 1-6.

Santucci, R. (1927). Uno stadio di sviluppo non ancora descritto dello "scampo" (*Nephrops norvegicus* (L.)). *Mem. R. Com. Talassogr. Ital.*, **128**, 1-6.

Sars, G. O. (1884). Bidrag til Kundskaben om Decapodernes Forvandlinger. 1. *Nephrops, Calocaris, Gebia*. *Arch. Math. Naturvidensk.* **9**, 155-204.

Sars, G. O. (1890). Bidrag til Kundskaben om Decapodernes Forvandlinger. 2. *Lithodes, Eupagurus, Spiropagus, Galathodes, Galathea, Munida, Procellana, Nephrops*. *Arch. Math. Naturvidensk.* **13**, 133-201.

Sastry, A. N. (1977). An experimental culture-research facility for the American lobster, *Homarus americanus*. *Proc. Eur. Symp. Mar. Biol., 10th, 1975* Vol. 1, pp. 419-435.

Sastry, A. N., and Vargo, S. L. (1977). Variation in the physiological responses of crustacean larvae to temperature. *In* "Physiological Responses of Marine Biota to Pollutants" (F. J. Vernberg, A. Calabrese, F. P. Thurberg, and W. B. Vernberg, eds.), pp. 401-423. Academic Press, New York.

Sastry, A. N., and Zeitlin-Hale, L. (1977). Survival of communally reared larval and juvenile lobsters, *Homarus americanus*. *Mar. Biol.* **39**, 297-304.

Scarratt, D. J. (1964). Abundance and distribution of lobster larvae (*Homarus americanus*) in Northumberland Strait. *J. Fish. Res. Board Can.* **21**, 661-680.

Scarratt, D. J. (1968). Distribution of lobster larvae (*Homarus americanus*) off Pictou, Nova Scotia. *J. Fish. Res. Board Can.* **25**, 427-430.

Scarratt, D. J. (1969). Lobster larvae of Pictou, Nova Scotia, not affected by bleached Kraft Mill effluent. *J. Fish. Res. Board Can.* **26**, 1931-1934.

Scarratt, D. J. (1973). Abundance, survival, and vertical and diurnal distribution of lobster larvae in

Northumberland Strait, 1962-1963 and their relationships with commercial stocks. *J. Fish. Res. Board Can.* **30**, 1819-1824.

Scarratt, D. J., and Raine, G. E. (1967). Avoidance of low salinity by newly hatched lobster larvae. *J. Fish. Res. Board Can.* **24**, 1403-1406.

Scattergood, L. W. (1949). Translations of foreign literature concerning lobster culture and the early life of the lobster. *U.S. Fish. Wildl. Serv., Spec. Sci. Rep.—Fish.* No. 6.

Serfling, S. A., and Ford, R. F. (1975). Ecological studies of the puerulus larval stage of the California spiny lobster, *Panulirus interruptus. Fish. Bull.* **73**, 360-367.

Serfling, S. A., Van Olst, J. S., and Ford, R. F. (1974). A recirculating culture system for larvae of the American lobster, *Homarus americanus. Aquaculture* **3**, 303-309.

Sheard, K. (1949). The marine crayfishes (spiny lobsters), family Palinuridae, of western Australia with particular reference to the fishery on the western Australian crayfish (*Panulirus longipes*). *Aust., CSIRO, Bull.* **247**, 1-45.

Sherman, K., and Lewis, R. D. (1967). Seasonal occurrence of larval lobsters in coastal waters off central Maine. *Proc. Natl. Shellfish. Assoc.* **57**, 27-30.

Shojima, Y. (1963). Scyllarid phyllosoma's habit of accompanying jellyfish. *Bull. Jpn. Soc. Sci. Fish.* **29**, 349-353.

Silberbauer, B. I. (1971). The biology of the South African rock lobster *Jasus lalandii* (H. Milne Edwards). I. Development. *Oceanogr. Res. Inst. (Durban), Invest. Rep.* **92**.

Sims, H. W. (1965). Notes on the occurrence of prenaupliosoma larvae of spiny lobsters in the plankton. *Bull. Mar. Sci.* **15**, 223-227.

Sims, H. W. (1966). The phyllosoma larvae of the spiny lobster *Palinurellus gundlachi* Von Martens (Decapoda, Palinuridae). *Crustaceana* **11**, 205-215.

Sims, H. W., and Brown, C. L. (1968). A giant scyllarid phyllosoma larva taken north of Bermuda (Palinuridae). *Crustaceana, Suppl.* **2**, 80-82.

Sims, H. W., and Ingle, R. M. (1966). Caribbean recruitment of Florida's spiny lobster populations. *Q. J. Fla. Acad. Sci.* **29**, 207-243.

Smith, S. I. (1873). The early stages of the American lobster (*Homarus americanus* Edwards). *Trans. Conn. Acad. Arts Sci.* **2**, 351-381.

Sprague, J. B., and McLeese, D. W. (1968). Toxicity of kraft pulp mill effluent for larval and adult lobsters and juvenile salmon. *Water Res.* **2**, 753-760.

Squires, H. J. (1970). Lobster (*Homarus americanus*) fishery and ecology in Port au Port Bay, Newfoundland, 1960-65. *Proc. Natl. Shellfish. Assoc.* **60**, 22-39.

Squires, H. J., Tucker, G. E., and Ennis, G. P. (1971). Lobsters (*Homarus americanus*) in Bay of Islands, Newfoundland, 1963-1965. *Fish. Res. Board Can. Manuscript Rep. Ser. (Biol.)* p. 1151.

Stephensen, K. (1923). Decapoda Macrura. *Rept. Danish Oceanogr. Exped. 1908-1910 Medit.*, **2**(D3), 11-85.

Sweat, D. E. (1968). Growth and tagging studies on *Panulirus argus* (Latreille) in the Florida Keys. *Fla. Board Conserv. Mar. Res. Lab., Tech. Ser.* **57**, 1-30.

Takahashi, M., and Saisho, T. (1979). The complete larval development of the scyllarid lobster, *Ibacus ciliatus* (von Siebold) and *Ibacus novemdentatus* Gibbes in the laboratory. *Mem. Fac. Fish. Kagoshima Univ.* (in press).

Templeman, W. (1936). Fourth stage larvae of *Homarus americanus* intermediate in form between normal third and fourth stages. *J. Biol. Board Can.* **3**, 339-354.

Templeman, W. (1937). Egg-laying and hatching postures and habits of the American lobster (*Homarus americanus*). *J. Biol. Board Can.* **3**, 339-354.

Templeman, W., and Tibbo, S. N. (1945). Lobster investigations in Newfoundland 1938-1941. *Newfoundland, Dep. Nat. Resour., Fish Res. Bull.* **16**.

Thomas, H. J. (1954). Some observations on the distribution, biology and exploitation of the Norway lobster (*Nephrops norvegicus* L.) in Scottish waters. *Mar. Res.* **1**, 1-12.
Thomas, L. (1963). Phyllosoma larvae associated with medusae. *Nature (London)* **198**, 208.
Ting, R. Y. (1973). Culture potential of spiny lobster. *Proc. Annu. Meet.—World Maric. Soc.* **4**, 165-270.
UNESCO (1968). Zooplankton sampling. *Monogr. Oceanogr. Methodol.* **2**, 99.
Vance, R. R. (1973a). On reproductive strategies in marine invertebrates. *Am. Nat.* **107**, 339-351.
Vance, R. R. (1973b). More on reproductive strategies in marine invertebrates. *Am. Nat.* **107**, 353-361.
von Bonde, C. (1936). The reproduction, embryology, and metamorphosis of the Cape crawfish (*Jasus lalandii*) (Milne-Edwards) Ortmann. *S. Afr., Fish. Mar. Biol. Surv., Invest. Rep.* **6**, 1-25.
Walne, P. R. (1977). The potential for the culture of crustacea in salt water in the United Kingdom. *G.B., Minist. Agric., Fish. Food, Fish. Lab. Leafl.* [N.S.] **40**, 1-16.
Wells, P. G., and Sprague, J. B. (1976). Effects of crude oil on American lobster (*Homarus americanus*) larvae in the laboratory. *J. Fish. Res. Board Can.* **10**, 371-412.
Wilder, D. G. (1953). The growth rate of the American lobster (*Homarus americanus*). *J. Fish. Res. Board Can.* **10**, 371-412.
Wilder, D. G. (1954). The lobster fishery of the southern Gulf of St. Lawrence. *Fish. Res. Board Can., Biol. Stn., St. Andrews, N.B., Gen. Ser. Circ.* **24**, 1-16.
Williamson, D. I. (1956). The plankton of the Irish Sea, 1951 and 1952. *Bull. Mar. Ecol.* **4**, 87-144.
Williamson, H. C. (1905). A contribution to the life history of the lobster (*Homarus vulgaris*). *Fish. Board Scotl., Annu. Rep.* **23**, 65-107.
Witham, R., Ingle, R. M., and Joyce, A. (1968). Physiological and ecological studies of *Panulirus argus* from the St. Lucie Estuary. *Fla. Board Conserv. Mar. Res. Lab., Tech. Ser.* **53**, 1-31.

Chapter 2

Ecology of Juvenile and Adult Palinuridae (Spiny Lobsters)

P. KANCIRUK

I.	Introduction	59
II.	Palinurid Habitat	60
	A. Prepuerulus Habitat	61
	B. Puerulus Habitat	61
	C. Juvenile Habitat	62
	D. Adult Habitat	63
	E. The Den	66
III.	Palinurid Reproductive Ecology	69
	A. Sexual Dimorphism	69
	B. Sex Ratios	70
	C. Environmental Influences	71
IV.	Palinurid Fisheries and Ecology	78
V.	Behavioral Ecology	79
	A. Feeding Behavior	79
	B. Defensive Behavior	83
	C. Environmental Stimuli	84
	D. Biological Rhythms	86
VI.	Summary	91
	References	92

I. INTRODUCTION

The Palinuridae is a very successful group of benthic decapod crustacea, comprising about 49 species of spiny (or rock) lobsters distributed around the world. They are abundant and commercially important along the coasts of North and South America, Africa, the Mediterranean, India, the Orient, Australia, New Zealand, and the Pacific Islands. Although the animals are very similar morphologically, their habitat varies greatly among species. Palinurid species occupy littoral to deep (> 400 m) marine habitats in tropical, semitropical, and temperate waters. Such an extensive zoogeographic distribution reflects the great flexibility and success of the generalized morphology and behavior within the Palinuridae.

There is no adequate description of the ecology for any of the palinurid species, and in many cases the most basic information is lacking (Kanciruk and Herrnkind, 1976b). Some specifics have been acquired through fisheries research, but although such data provide the foundation of our knowledge for many of the Palinuridae, there are certain disadvantages. Research dealing with economically important species usually concentrates on specific and limited management-related goals. Adequate data are often available on population size frequency or onset of breeding season (as reflected in catch statistics and trap monitoring programs). However, little information usually is gathered on important ecological parameters such as mechanisms of larval transport and recruitment; larval, juvenile, and adult habitat selection; and social, reproductive, and migratory behavior. Often, an adequate description of the prefished, pristine population is unavailable. Even less is known about the many Palinuridae that are not economically exploited. Furthermore, monies for basic research have been sparse, some species are extremely scarce and secretive, and some exclusively inhabit deep waters (i.e., doubly expensive and difficult to study). Sometimes interesting glimpses of unique behavioral and ecological characteristics are available for these relatively unstudied species, serving to whet our appetite for additional study. Much future effort will be necessary to obtain a more adequate description of the fundamental ecology of this family, especially for the deep water and/or unexploited species. Despite these limitations, the Palinuridae possess an interesting and quite diversified ecological pattern. Being one of the most successful, and also one of the most economically exploited benthic decapod groups, they deserve detailed consideration.

II. PALINURID HABITAT

Habitat selection may be defined as the proclivity of a species for (or their adaptation to) a particular environment and the behavioral mechanisms associated with this predisposition. Habitat selection and resource use by the palinurids vary considerably. Details for species living below 200 m are presently unresolved and can only be superficially treated in this text.

Genera of Palinuridae have been grouped taxonomically by external

2. Ecology of Juvenile and Adult Palinuridae 61

morphology and can be loosely correlated with latitude, water depth, and temperature (George and Main, 1967). The eight extant genera fall into distribution patterns of high latitude (*Projasus, Palinurus,* and *Jasus*), deep-water equatorial (*Puerulus, Palinustus, Justitia,* and *Linuparus*), and shallow-water equatorial (*Panulirus*). Although useful for systematic studies, the voids that exist in our information and variations within genera weaken such a classification's effectiveness here. Instead, the ecology of the palinurids will be discussed generally by habitat, species, age, reproductive state, and behavior.

The Palinuridae occupy a circumglobal band from approximately 45°N to 45°S latitude. Species diversity is greater in tropical and subtropical waters, but densities are relatively low. Species diversity is usualy reduced in temperate waters, but densities are often higher (Morgan, 1977). In the warmer environs, as expected, molting occurs more frequently and growth rates are accelerated, although maximum adult size may be less.

A. Prepuerulus Habitat

Discussion of habitat selection usually begins with the puerulus stage (postphyllosome), commonly considered the first benthic life stage for the Palinuridae. However, reflections on the behavior and morphology of the phyllosome stage are in order.

The planktonic phyllosome may be able to postpone its transformation into the benthic puerulus according to the proximity of a suitable habitat. This has been observed in other planktonic forms, and given the dispersive capabilities of the phyllosome, postponement of transformation may be a necessary ability. There appears to be some evidence for such control. Phyllosomes of *Scyllarus* spp. (closely related and morphologically similar to palinurid phyllosomes) captured in the Bahamas attached to medusae in deep water over apparently unsuitable habitat quickly underwent synchronized metamorphosis in fingerbowls (calm water and shallow, solid substrate) when maintained in the laboratory (Herrnkind, *et al.,* 1976). It is tempting to speculate that planktonic phyllosomes are able to exert considerable control over their terminal larval molt and that this may represent their first expression of active benthic habitat selection.

B. Puerulus Habitat

Mechanisms of habitat selection for the benthic puerulus stage are unknown for any species, although we do have some information on shallow-water palinurids. The puerulus stages of deep-water species such as *Palinuris delagoae* (180–400 m), *Projasus parkeri* (600 m), *Puerulus angulatus* (300 m), and *Linuparus* spp. (300 m) are not reported in shallow inshore areas. Presumably, they settle out at the greater depths characteristic of the adults (Berry, 1971). It is not known whether the phyllosome must drift within close proximity of the substrate before settling behavior occurs, or whether suitable habitat can be recognized from higher up in the water column (exploiting information from the underlying water-column). Settlement on appropriate substrate in these forms may be entirely directed or fortuitous.

The first benthic stages of shallow-water Palinuridae (e.g., *Panulirus argus*) are found at depths of less than 3 m among mangrove roots, around pilings, in grass beds, and along grassy undercuts. Settling occurs some time after metamorphosis from the phyllosome to the puerulus stage, as the puerulus can be observed free-swimming at night when attracted to the surface by a lantern. Collection data and behavioral evidence suggest that the puerulus is able to exert considerable control over its spatial and temporal positioning in the environment to maximize its movement toward a suitable, shallow-water habitat (Phillips, 1975; Serfling and Ford, 1975).

The usual lack of late stage phyllosomes inshore has led some workers to conclude that final metamorphosis to the puerulus stage occurs offshore, followed by directed, relatively long (up to 40 km) onshore migrations or drifts (Sweat, 1968; Phillips, *et al.*, 1978). The environmental stimuli directing such movements have yet to be identified, but if such migration occurs, then it represents an impressive exhibition of habitat selection.

It is usually assumed that the preferred habitat for the settling puerulus of shallow water palinurids is characterized by very shallow depth. This may not be entirely true. There is some evidence for the settlement of the puerulus stage of shallow-living species in waters greater than 10 m in depth—e.g., *P. argus* in Jamaica (Munro, 1974). *Jasus tristani*, normally found in shallow waters around the remote mid-South Atlantic Tristan da Cunha Islands, is also found in abundance on Vema seamount, whose closest approach to the surface is 20 m. The seamount quickly slopes to great depths on all sides. This isolation by depth presumably prohibits any adult immigration to the seamount. Thus, the presence of adults can be considered an example of relatively deep water settlement of a shallow water palinurid (Heydorn, 1969c). It seems probable that for some shallow-water species, the puerulus stage is able to settle out in both shallow and slightly deeper areas.

C. Juvenile Habitat

The puerulus settles and molts rapidly, acquiring its camouflaging juvenile colors. At this stage, it is approximately 1 year old (depending on species; phyllosome drift time is still uncertain), and the carapace length (CL) measures 5–10 mm.

The habitat of juvenile shallow-water palinurids is diverse, comprising a range of small abodes characteristic of shallow bay/reef areas. Juvenile *P. argus* in the

Caribbean have been observed among sponges, on algae-fouled objects, in *Thalassia* grass beds, between mangrove roots, along undercuts, and even among the spines of groups of black sea urchins *Diadema* spp. (Davis, 1971). Prophetically, they have also been observed clinging to the fouled float lines of lobster pots in deep water (A. Craig, personal communication). Soft muddy bottoms and turbid water conditions seem to be avoided by juvenile *P. argus* as they are by adults. Juveniles older than 2 years are gregarious, but young juveniles between 1 and 2 years of age are infrequently found in groups. Phillips *et al.* (1977) have suggested that gregarious behavior is not fully developed in *Panulirus cygnus* until this spiny lobster reaches about 2 years of age. Until that time they remain secreted and recluse in weed beds or similar dense cover. Juveniles of *P. cygnus* also do not normally enter baited traps until about 2 years of age (Chittleborough, 1975). In general, the depth range for early juvenile shallow water Palinuridae is 1–4 m, although individuals are sometimes found in deeper areas, perhaps representing transients between habitats (Sweat, 1968) or older stages of deep settling pueruli.

There is some evidence that juvenile habitat is limiting for some populations of palinurids. Density dependent mortality in nursery areas, i.e., limited food supply (Chittleborough and Phillips, 1975) and limited shelter (Chittleborough, 1970) is characteristic of recruitment for *P. cygnus* in Western Australian waters, and is perhaps true for many other species. Resultant mortality apparently has a leveling effect in this species, and reduces most year classes to similar density levels. Unfortunately, carefully conducted studies on recruitment and population dynamics seem to be the patent of Australian researchers, and important information on juvenile habitat carrying capacities for other species is sorely lacking.

Shallow water palinurids usually leave juvenile habitat at or slightly before sexual maturity. In some species (i.e., *P. argus* and *P. cygnus*) there may be an impressive mass exodus to the adult habitat. For *P. cygnus,* this seems to involve synchronized molting into a "white" (unpigmented) stage and subsequent movement into offshore adult habitat (Sheard, 1962). The stimulus for the best studied mass movement in the palinurids (*P. argus*) has been shown to be the characteristically abrupt autumnal storms (Kanciruk, 1976; Kanciruk and Herrnkind, 1978b; Chapter 9). However, long-term photoperiod, temperature, and/or dietary cues may play important roles for other migratory species.

D. Adult Habitat

Deep water adult habitats differ considerably from the juvenile mangrove and shallow bay nursery areas. Habitat for adult *P. argus* has been fairly well studied and is typified by rock/coral outcrops, large undercuts, and sponge and soft coral aggregations associated with reefs. Adults have been found from 1 to >100 m, a typical adult depth range for many of the shallow water Palinuridae. An unusually comprehensive habitat and behavioral study in the Virgin Islands, conducted in part from a manned undersea laboratory (TEKTITE II, Herrnkind and Olsen, 1971; Herrnkind, et al., 1975) estimated adult *P. argus* density in mud and sand/coral rubble habitat as 0 lobsters per hectare (ha), in coral patches as 7.1/ha, and on hard reef bottoms as 19.4/ha. However, habitat selection was not routinely predictable. Lobster distribution was found to be nonuniform even between four adjacent (and seemingly similar) habitat sites, where observed densities unexplicably ranged from 0.4/ha to 20.4/ha. Davis (1977), in a careful diver survey, estimated a standing crop of 58 kg/ha (approximately 80–90 lobsters/ha) for large adult *P. argus* in a protected (nonexploited) shallow reef area. Peacock (1974) noted lagoon densities of *P. argus* of 3.9–7.0/ha. At the high end of the scale, Morgan (1974) reported that densities of *P. cygnus* at the Abrolhos Islands reached 2000/ha, a truly impressive figure but also matched by *P. argus* in selected areas under migratory conditions (Herrnkind, 1969; Herrnkind et al., 1973; Kanciruk and Herrnkind, 1978b).

Adult habitat preference varies among the Palinuridae, from shallow intertidal surf zones to great ocean depths characterized by perpetual darkness and soft mud-ooze substrates. Lobster species diversity is relatively high along the East African coast, where palinurid existence has been examined with success. Berry's (1971) extensive study surveyed the sympatrio palinurid species off the African coast and separated five species of *Panulirus* into habitat types classified by water temperature, turbidity, bottom type, depth, tidal range, and geographic area. *Panulirus homarus rubellus* was found to be the most tolerant of surge and turbulence, inhabiting the surf zone and its immediate environs (distribution correlated with the occurrence of its primary food source, the mussel *Perna perna*). The robust build and stout legs of *P. homarus* enable it to navigate with sure footing among the surge-washed rocks. This species is tolerant of turbid water and suspended sand and surprisingly, it was often found buried with only its antennae and anterior cephalathorax exposed, a behavior wholly uncharacteristic for most Palinuridae. *Panulirus penicillatus* was found in the inshore reef zone, but its habitat was less characterized by turbulence and suspended sand. It also possesses robust legs and would not retreat upon diver approach, as is usual for palinurid lobsters, but clung tightly to the substrate. *Panulirus longipes longipes* was found adapted to a variety of habitats, from clear water reef areas (>18 m) to the turbid shallow waters of Durban Harbor ($\simeq 1$ m). However, it seemed to prefer deep, inaccessible reef holes. Juvenile *Panulirus ornatus* were found in rocky, shallow (1–4 m) waters, whereas adults were found in deeper waters more characterized by soft, muddy bottoms, and turbid conditions. There is evidence that this species undergoes a mass migration between habitats, which might explain the marked contrast between juvenile and adult habitats (Pyne, 1977; Moore and MacFarlane, 1977). Adult *Panulirus versicolor* were a rare catch in Berry's study. They seemed to prefer undercut ledges close to strong tidal currents and did not shun turbid conditions.

2. Ecology of Juvenile and Adult Palinuridae

At least one deep-water species surveyed by Berry (1971) appeared to exhibit behavioral adaptation for the soft open bottom characteristic of deep water. Even upon placement in holding tanks provided with adequate shelter, *P. delagoae* aggregated in open areas and conspicuously avoided artificial dens. This behavior is discordant with that displayed by shallow water palinurids in general, which actively seek shelter under stress. Other deep water species in this survey did not appear to overlap much in habitat selection, at least by depth. *Panulirus delagoae* was captured along the edge of the continental shelf in 180–400 m on high organic content mud substrate, while *Palinurus gilchristi* was found in shallower waters (50–100 m). *Projasus parkeri* was a rare catch on soft substrates at 400 m. *Puerulus angulatus* was also captured in similar soft mud areas, but its distribution was restricted to 280–320 m depths.

The commercially important shallow water, hard-substrate species of spiny lobsters are not wholly representative of the family Palinuridae. Perhaps the lack of numerous predators and the relatively static conditions at great depths alleviate the necessity for hard substrate shelter, yet still provide an adequate habitat. However, our assumption that these deep water species exclusively inhabit areas of open, soft substrate must be examined critically. Our information comes mainly from commerical fishing trawlers, but as fishermen avoid rocky substrate in their efforts to minimize expensive equipment damage, correlations of deep water species with soft mud substrates are possibly biased.

George (1974) identified species of reef-dwelling *Panulirus* spp. in the Indo-West Pacific region according to habitat preference. He concluded that four habitats could be distinguished. *Panuluris penicillatus* inhabited the rock shelters characteristic of the outer reef zone; three subspecies of *P. cygnus* sheltered in the coral crevices in the reef flat and seaward reef face (inhabiting slightly different portions of the habitat, depending on extent of oceanic exposure); *P. versicolor* resided in the delicate coral formations within the protected lagoon waters and along the deeper reef face; and *P. ornatus* occupied the longshore area (Fig. 1). Goerge (1974), one of the few workers commenting on palinurid

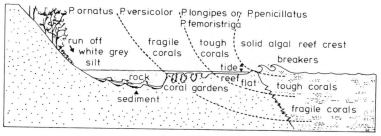

Fig. 1. Habitat utilization by the genus *Panulirus* on the Pacific barrier reefs. Available littoral habitat is partitioned by closely related *Panulirus* spp. according to environmental parameters such as depth, and turbulence. (After George, 1974, with permission.)

evolution, suggested that the coral reef zone is a historically unstable environment. As expected, the more recently evolved species (*P. ornatus* and *P. versicolor*) were observed to reside within this zone.

E. The Den

The most characteristic feature of the habitat for many of the palinurids is the residence area or den. Dens are not usually constructed by the Palinuridae, but are opportunistically chosen from the surrounding habitat. The burrowing ability of most Palinuridae is usually assumed to be minimal. Crawford and DeSmidt (1922) indicated that *P. argus* is unable to burrow and must make use of whatever shelter is available. In the field, there is little obvious evidence of burrowing or burrow maintenance for most palinurids. However, as the use of scuba enables more *in situ* observations to be made, this assumption may change. Captive lobsters (*P. argus*) will enlarge and otherwise modify their dens by pushing sand and small stones that obstruct the den entrance (P. Kanciruk, personal observation). The lobsters accomplished this by digging the anterior pereiopods into the substrate and plowing the sand forward, or else by entrapping a small rock between the basal segments of the antennae and displacing it by walking forward. Trail-flap-induced sand movement was also observed for juveniles of this species in experimental tanks (W. F. Herrnkind and J. G. Halusky, personal communication) and it has been observed in another species associated with defense (See Section V, A). There is the distinct possibility that some palinurids are capable of minor den modification in the field. Deep water, soft substrate Palinuridae may also be capable of burrowing and providing their own shelter to some extent in an otherwise featureless bottom. Evidence for this will have to be provided experimentally or through *in situ* observation.

Lobsters are always distributed nonrandomly among available den sites (Chpt. 7, Vol. I), and the chosen dens may have distinguishing characteristics. In the TEKTITE II study (noted previously) no striking physical parameters such as aperture size, height, or volume were strongly correlated with chosen den sites. *Panulirus argus* dens had only one opening, as opposed to the constructed two-opening residence typical of *Homarus*. However, dens were often typified by partially restricted openings and hard substrates. Water quality, current flow, and access to feeding grounds, among other unmeasured parameters, could also have been important in den selection. Heydorn (1969a) observed that *P. homarus rubellus* often inhabited dens that were tunnel-shaped with exists at both ends, although *P. homarus rubellus* is not involved in the construction of such dens. Multiple den occupancy is common and seems to be a typical and perhaps necessary condition for some species, including *P. argus, P. cygnus, P. interruptus, Jasus lalandii,* and *P. homarus rubellus* (Herrnkind and Olsen, 1971;

2. Ecology of Juvenile and Adult Palinuridae

Lindberg, 1955; Fielder, 1965a; Berry, 1971; Heydorn, 1969a). During the TEKTITE II study (on adult *P. argus*), 73% of observed inhabitated dens held single lobsters, while 27% of the dens were multiply occupied (this lattter class, however, accounted for 55% of all lobsters). In some reef areas, up to 81% of all lobster observed were in multiple-occupant dens. Berrill (1975) described gregarious behavior for juvenile *P. argus* in both laboratory and field situations, even withstanding moderate intraspecific aggression. Furthermore, he reported that in times of stress, aggressive encounters became rare, which allowed a higher level of shelter sharing (increase in density). This is probably true for many Palinuridae, e.g., adult *P. argus, J. lalandii,* and *P. ornatus,* during mass movements.

Den sites considered to be favored during the TEKTITE II study (i.e., those having the largest percentages of continuous occupancy), were multiply occupied. Den "attractiveness" was not lost after diver de-population. When a popular but isolated den (200 m from adjacent habitat) was depopulated during the study, a 2.7% per day repopulation rate was observed. Diver (and presumably predator) harrassment, however, does influence den habitation. The TEKTITE project recorded that 84% of newly tagged and handled (harrassed) lobsters left their dens as opposed to the usual emmigration rate of 58%, possibly representing a behavioral adjustment to an unfavorable environment. True home dens were observed, to which individual lobsters will repeatedly return after nightly foraging. Home dens were also identified in an excellent, controlled field study in the Dry Tortugas (Florida) by Davis (1977). Diver-induced dispersal was also documented for the Dry Tortugas study. A protected resident lobster population in a state park was divided into three regions; unfished control, commercially trap fished, and sport diver fished. These areas were monitored before, during, and after fishing pressure was permitted. Effects of reduced abundance and increased dispersion could therefore be accurately measured. It was determined that sport diving not only reduced abundance, but increased lobster dispersion in the diver–accessible shallow reef areas. Dispersion is not only an effect of diver harrassment, but can also be the result of trap-fishing pressure (e.g., return of berried females and undersized lobsters to the environment after boatside handling) or predator harrassment. Davis suggested that dispersion probably results in increased exposure to predation and disruption of social structure. The mechanisms involved in den selection and return after night foraging (even after blinding, Chpt. 7, Vol. I) remain unclear, but illustrate the importance of the home den, and point to the probable damaging effects of fishery-induced dispersion.

Social structure within the den has not been well defined, although there is evidence for other than a random mixture of individuals and interactions. Large *P. argus* males often inhabit dens with 4–10 female lobsters (some berried) during breeding periods, to the apparent exclusion of other, smaller males (P. Kanciruk, personal observation). Den sites are frequently defended against predator or diver intrusion. The largest male (perhaps dominant) in a den will move forward and confront the intruder by anteriorly directing the antennae. This behavior is similar to that displayed by spiny lobsters to predators or divers on the open sand plain during foraging or migration. Munro (1974), studying *P. argus* in Jamaican waters, observed a den peck order in which the largest lobster occupied the favored position in the den. Berrill (1975) noted a decrease in queuing and gregariousness with increased aggressive den encounters for juvenile *P. argus*. He documented outright eviction of subordinate lobsters in experimental situations. Fielder (1965a) described a dominance order for dens of *J. lalandii*, up to and including exclusion of subordinate males from small dens. He indicated that contact was not always necessary, but that mere approach by a dominant could elicit eviction of a subordinate individual. During the spring breeding season, Davis (1977) described large male *P. argus* apparently searching out den-resident females and forcing them into the open in what appeared to be premating behavior.

Palinuridae are gregarious intraspecifically, and often interspecifically. *Panulirus argus* is sometimes found inhabiting the same den with *P. gutattus*, although usually not in direct physical contact. *Panulirus gutattus* has often been observed clinging upside down to the den roof, or withdrawn into small holes in the roof and walls (P. Kanciruk, personal observation; Aiken, 1976), with *P. argus* found more traditionally on the den floor. *Panulirus argus* is sometimes found cohabitating with the spotted moray (*Gymnothorax moringa*) or the green moray (*Gymnothorax funebris*) in the Caribbean. After using a tethered lobster to antagonize a spotted moray, which was cohabitating with a number of *P. argus* in a large undercut den, it became obvious that even a small moray (<1 m) had the ability to make a quick meal of an average-sized *P. argus* den-mate, but apparently often chooses not to (P. Kanciruk, personal observation). Berry (1971) observed *P. homarus rubellus* coexisting with the moray *Lycodentus undulatus*. He postulated a symbiotic relationship of sorts, the moray being the natural enemy of the local octopus, *Octopus granulosus*, which in turns feeds on *P. homarus rubellus*. A dead octopus (often placed by local divers on a long stick to induce lobster den evacuation, facilitating capture) placed at the mouth of a den was immediately and ferociously attacked by the coinhabiting moray. *Lycodentus undulatus* appeared to be attracted to lobster dens by the stridulatory sounds produced by the residents. The benefit of this association for the resident lobsters is obvious, and for the moray it may be attraction of octopus as prey.

The role of the den in the social structure and population dynamics for many Palinuridae cannot be overemphasized, and adds to the overall complexity of understanding the ecology of this group.

III. PALINURID REPRODUCTIVE ECOLOGY

A. Sexual Dimorphism

In general, the males of most palinurid species attain a considerably larger size than females. This fact is usually reflected in reported population size frequencies. Reliable studies on growth are uncommon, and the cause for this sexual dimorphism as observed in the field is uncertain, but is likely due to differential growth or molting rates between sexes, longer male lifespan, or increased metabolic demands on females during egg production.

Adult male palinurids possess an elongated pair of pereiopods. Reports of palinurid mating behavior often include grasping of the female with the male's elongated second or third pereiopods and pushing her over in order to deposit the spermatophore or to achieve internal fertilization (Berry, 1970). Davis (1977) reports large *P. argus* males pulling females out of dens in apparent precopulatory behavior. Larger male size along with longer second pereiopods may therefore both be adjuncts for successful copulation.

At the Virgin Islands, Herrnkind and Olsen (1971) measured *P. argus* male mean size at 120 mm CL and female size at 103 mm CL. Munro (1974) found males to average approximately 10 mm CL larger than females in adult populations around Jamaica. Similar size discrepanices are also reported for *P. ornatus, J. lalandii, Panulirus longipes longipes,* and *Jasus edwardsii,* among others. Additional morphological differences between the sexes (and perhaps between populations, thereby a potential means of population identification) include the carapace length/abdomen ratios. It has been demonstrated for some populations of *P. argus* and *J. edwardsii* that for individuals with the same carapace lengths, the females will, on average, have a heavier and longer abdomen, and thus have a longer overall length. This morphometric difference may have a reproductive function, as the female abdomen is the egg support structure after fertilization and before hatching, but this is just speculation. Chittleborough (1976) documented growth rates as being inversely related to population density in the laboratory observing stunting of adults of *P. cygnus* in high density situations at the Abrolhos Islands, Western Australia. Juveniles (< 3 years of age) raised in solitude showed reduced growth rates, perhaps reflecting the importance of gregarious interaction for the normal development of this species (Chittleborough, 1975). Measurements of growth rates and maximum size in the field are significantly influenced by fishing pressure. Cropping causes reduction of density, interfers with social structure, increases relative food abundance, and other, more subtle effects.

Longevity and maximum size for the Palinuridae are difficult to determine from available information, but can often be estimated. The palinurids certainly do not reach the large size or the great age of the American lobster, *Homarus americanus*. Munro (1974) calculated that theoretical maximum length for *P. argus* was 190 mm CL, corresponding to an age of 20 years. This is for a species that, in the Caribbean, averages 70-90 mm CL in fished adult populations, and 100-145 mm CL in unfished adult habitats. Lewis (1977) determined the asymptotic carapace length for the spiny lobster of southern Australia (*Jasus novaehollandiae*) to be 203 mm for males, but only 143 mm CL for females.

The above relationships of differential growth rates and maximum size are, at best, rough approximations in need of refinement. They represent differences that seem to exist between the sexes and which should be taken into account when population size frequencies are developed and compared. However, this is difficult to implement, for adding to the overall confusion, workers have reported size measurements as total weight, overall length, carapace length, and tail length. Comparisons of environmental conditions reflected in growth patterns are difficult, and assessment of subtle differences between environments is often impossible.

B. Sex Ratios

The sexual composition of shallow water species is almost always reported as 1:1 for postlarvae and young juveniles. However, this ratio is often observed to differ from unity for populations of larger juveniles and adults (as high as 10:1, male:female for *P. guttatus*, Sutcliffe, 1953). Olsen *et al.* (1975) reported that for older juvenile *P. argus* in the mangrove bay areas of Virgin Islands, females showed a unimodal size frequency, while males exhibited bimodality. In populations of younger juveniles, both male and female size frequencies were unimodal. They postulated selective movements by older juvenile males to adult areas as producing the observed biomodal size frequency.

Selective male and female movements may indeed be at the root of many of the disproportionate size/sex ratios that have been reported. At Elands Bay, South Africa, a sex ratio of 30% males and 70% females of *J. lalandii* was reported for a sample size of over 2000 (Newman and Pollock, 1971). Contemporary sampling at nearby Mussel Bay indicated the ratio was reversed, the population being composed of 62% males and 38% females ($n > 3000$). It was suggested that movement by females accounted for these results. The less robust females may have found more appropriate conditions at protected Elands Bay, over the exposed Mussel Bay environment. Differences in sex ratios are not temporally constant, but vary considerably during the year. Chittleborough (1974) found males of *P. ornatus* in the Torres Strait to be considerably underrepresented, perhaps due to female reproductive movements. Sex ratios are often temporally unstable even in the same locale. Over a 4 year period, Davis (1977) measured female frequencies (*P. argus*) as ranging from 50 to 80% in a single protected area. This most likely reflected female movement to the reef edge during the spring reproductive period. This movement is postulated as functioning to enhance larval dispersion.

C. Environmental Influences

The reproductive biology of the palinurids is quite varied, and unquestionably influenced by environmental conditions. The timing of mating and egg production even for only one species (*P. argus*) has been reported variously as seasonal (once per year), multiple, or essentially continuous in different portions of its range. Many workers in the Florida Keys have reported that *P. argus* has only one major mating period in the spring. But less than 200 km away in the Bahamas, Kanciruk and Herrnkind (1976a) have observed high indices of autumnal reproduction (up to 40% berried females) for certain localized reef populations of *P. argus*. This high rate of reproduction, continuing through the early fall (traditionally assumed nonreproductive for this region), indicated two reproductive peaks, or perhaps one greatly extended reproductive period per year for *P. argus* at this location. Buesa Mas et al. (1968) reported two mating seasons for *P. argus* nearby at Cuba (April–May, August) and Brazil (March, August). Aiken (1975) reported four peaks of reproductive activity throughout the year in the Caribbean. His sample sizes were small, however, which may bias this observation. Year-round spawning in *P. argus* has been reported in the Los Roques region off Venezuela (Barany et al., 1970) and for Jamaican waters by Munro (1974). The discrepancies in reported breeding seasons of this species are probably due to differences in local environmental conditions; primarily temperature, but also diet and photoperiod. In Florida, Chitty (1973) found one breeding peak during the spring (April) for *P. gutattus,* a species sympatric to *P. argus*. Heydorn (1969b) found that large female *J. lalandii* in South Africa spawn once in June–July, but that smaller females spawn later in the year (September–October). Minor reproductive peaks were also observed for this species in January and April by Lazarus (1967). These differences are most likely due to differences in the timing of molting for large and small females, which for this species correlates well with mating.

Dissimilarities in data for adjacent populations of the same species are due, in part, to diverse methods of data collection. It has been shown (Morgan, 1974a; Kanciruk and Herrnkind, 1976b) that berried females and newly molted lobsters are reclusive and not prone to entering traps. Whenever possible, it is necessary to augment trap collection data with observation and capture of individuals *in situ*, using scuba in order to estimate the trap-sampling bias. Once this is accomplished for a given locale, trap sampling may proceed with accurate interpretation.

Chittleborough (1976) studied molting and mating frequencies in natural and aquarium-maintained populations of *P. cygnus*. He estimated that only 12% of females spawned twice a year in natural populations, but in aquarium situations, where surplus food was available and conditions ideal, 77% of the females underwent at least two spawnings per year. He was able to induce six spawnings and three molts per year for some individuals under controlled conditions. It is probable that the number of spawnings, molts per year, and overall fecundity for many palinurid species are a consequence of the interaction between many factors, such as temperature, food availability and quality, substrate suitability, population density, water chemistry, etc. Given ideal conditions, many tropical species might have the ability to spawn continuously. Chittleborough (1976) notes that reproduction in natural populations of *P. cygnus* is markedly seasonal. While a direct correlation with temperature and reproduction has not been shown, it is likely that temperature is an important factor controlling the timing of reproduction in this and other palinurid species. The factors regulating size at first breeding in *P. cygnus* (Chittleborough, 1976) are, however, obscured by the density-dependent effects on growth rate. In the field, the frequency of multiple spawnings is influenced by many factors, and might be a useful rough estimate of overall habitat suitability within the range of a species.

Sexual activity coincides with molting in some palinurids. Newman and Pollock (1971) observed such a relationship in *J. lalandii* in South Africa. There is some evidence that females in a premolt condition will evoke increased activity in male *J. lalandii* (Silberhauer, 1971; Rudd and Warren, 1976). *Panulirus argus* undergoes external fertilization and is not necessarily bound to mate while the female is in the soft condition. However, Kanciruk and Herrnkind (1976a), using external fouling as an indicator of time since last molt, determined that females do mate soon after molting (as evidenced by fairly clean carapaces), although probably after carapace hardening (see Section III, D).

Size at sexual maturity for a given species is difficult to define or measure for a population, and reports are often conflicting. Size is probably as dependent on environmental conditions as the number of spawnings per year, and just as variable throughout a given species distribution. It appears that for many species of Palinuridae in a given area, the average size of reproductive females does not vary greatly (Kanciruk and Herrnkind, 1976b; Kanciruk, 1976; Chittleborough, 1976). Differences are evident and sometimes striking between separated populations, often even when the separating distance is small (e.g., Bimini and the Florida Keys, ~150 km). In various portions of the Caribbean, the smallest size of *P. argus* at sexual maturity has variously been reported as 45 mm CL (Smith, 1948); 57 mm CL (Buesa Mas and Mota-Alves, 1971); 65 mm CL (Cobo De Barany *et al.*, 1972); 74 mm CL (Kanciruk and Herrnkind, 1976a); 74–83 mm CL (Munro, 1974; Davis, 1975); and 80–90 mm CL (Dawson, 1949; Dawson and Idyll, 1951; Peacock, 1974). These differences in observed size at female maturity may be due to differences in fishing pressure, diet, density (causing

2. Ecology of Juvenile and Adult Palinuridae

stunting of females), or temperature (Chittleborough, 1976; Bradstock, 1950; Sutcliffe, 1952; Street, 1969). Maturity of male palinurids has been an infrequent subject of investigation, perhaps due to the need for gonad examination. It appears that for most species of palinurids, the males mature at a slightly larger size than the females and exhibit a similar range of minimum mature size throughout their distribution.

Aside from environmental factors influencing maturation, methods of defining and reporting size at maturity have differed greatly. Among the variously used definitions of a mature female are the smallest female observed with ripe ovaries; the smallest carrying a spermatophore; the smallest female carrying eggs; the size class of females that is most frequently found to carry eggs; the mode of all females with eggs or spermatophore; and the size class possessing certain external female morphological characteristics. Compounding this confusion, some surveys are conducted on pristine populations by diving, while others are based on dockside measurements of slat-trap catches taken from overfished waters. Careful interpretation is necessary if (as has been the case) such data are used to draft fisheries management policies on minimum female legal size (Kanciruk and Herrnkind, 1978a).

Although average male and female size is often correlated with depth (larger lobsters are usually found in deeper areas), the size of females at first breeding and the average size of all breeding females is usually reported as constant, regardless of capture depth. This indicates that ontogenetic factors probably predominate within a locale. However, the number (percent and total) of berried females usually increases with depth (Chittleborough, 1976; Kanciruk, 1976; Kanciruk and Herrnkind, 1976a) (Fig. 2). It has been variously suggested that the shallow-water palinurids generally move to deeper regions of their habitat for spawning activities. This behavior probably functions in to maximize larval dispersal, as deep reef areas are usually more closely allied with alongshore oceanic currents. It has also been suggested by George *et al.* (1978) that the timing of spawning coincides with the occurrence of upwelling and increased food supply (*J. lalandii* on the West African coast) as well as cyclic, nearshore currents, which allow local larval deployment (*P. homarus rubellus* on the East African coast).

The reproductive potential of a lobster population per breeding season is related to female egg-carrying capacity, the propensity of females to carry eggs, and the female size frequency. Egg-carrying capacity and propensity to carry eggs are not constants, but vary with the female size and the area under consideration. Bradstock (1950) stated that *L. lalandii* matures at a smaller average size in regions of warmer temperature, which may also be the case of *P. argus* in the Caribbean and *P. cygnus* at the Abrolhos (Chittleborough, 1976). Heydorn (1965) placed size at sexual maturity for male *J. lalandii* at 60–65 mm CL, and for females at 70 mm CL. Newman and Pollock (1971) found 70% of all female

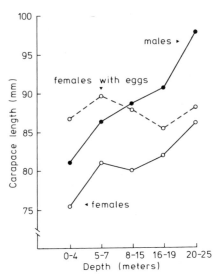

Fig. 2. Average size of males, females, and reproductive females in relation to depth. Typically both male and female average size increases with depth. However, reproductive female size is frequently independent of depth. Data for *Panulirus argus* is from the Bahamas. (After Kanciruk and Herrnkind, 1976b.)

J. lalandii were sexually mature at 70 mm CL. Street (1970) observed a gradation in size at maturity for *J. edwardsii* at New Zealand correlated with water temperature: 80 mm at Fiordland, 105-115 mm at Foreau Straight, and 110-120 mm at North Ortago. *Panulirus versicolor* females have been shown to attain maturity at ~66 mm, and males at 72 mm (George and Morgan, 1976).

Egg-carrying capacity is obviously related to female size. The larger a female, the more eggs produced and carried (attached to the abdomen). Estimates of number of eggs produced and carried range from 600 to 1200 eggs per gram body weight, with an average of perhaps 800 eggs per gram for *P. argus* (Munro, 1974; Chitty, 1973). The propensity to be berried, as observed in the field, is not uniform for all females of larger than mature size, but varies greatly with size class. This phenomenon is often reported but rarely discussed or taken into consideration when developing management regulations. Although females of *P. argus* at Bimini in the size range from 74 mm to over 120 mm CL were found to carry eggs, females in the 96-100 mm CL size class were most frequently observed as berried. Therefore, a method that considers female size frequency structure as well as other pertinent parameters should be utilized when estimating total population fecundity. A simple index of reproductive potential per spawning was recently developed and applied (Kanciruk and Herrnkind, 1976a, 1978a;

Kanciruk, 1976), which utilizes field data and takes these three factors into account.

$$\text{Index of reproductive potential} = (A \times B \times C)/D$$

where A is the proportion of females in the size class per total number of females in sample, B is the proportion of berried females in the size class, C is the egg-carrying capacity of average female in the size class, and D is a constant. Estimates of A, B, and C are made from field data. Suitable sampling techniques must be used (e.g., scuba diving capture or trap-sampling, with appropriately applied correction factors) in obtaining data to estimate A and B. Egg-carrying capacity was derived from data reported in Mota-Alves and Bezerva (1968). The constant D was chosen for the *P. argus* population at Bimini to set the 76–80 mm CL size-class index (which is the smallest "Florida legal" size class) to 100 as a workable standard ($D = 31.27$).

At Bimini, *P. argus* larger than 81 mm CL were shown to be more valuable in egg production than their proportion would indicate. Females in the 96–100 mm CL size-class (which represented only 3.6% of the population) provided 13.9% of the total egg production. Newly "mature" females (70–75 mm CL), protected under Florida's 76 mm CL legal size, made up 14.9% of the population, yet provided only an estimated 2.3% of the total egg production, making them (as a class) approximately 25 times less productive than females in the 96–100 mm CL size class. This relative importance of larger females in contributing to a population reproductive potential, and the relatively poor contribution of legally protected females, has been confirmed for a pristine population at Dry Tortugas, Florida (Davis, 1975). A similar relationship has been observed for *P. homarus rubellus* by Berry (1971a).

In order to preserve a population's overall ability to contribute to the fishery's future recruitment, we must take into account all factors that affect this ability, not just considering the small, sexually active females when determining legal size limits. Simultaneous treatment of multiple variables can allow the formulation of more realistic policies. It must be stressed, however, that data for such treatment must be collected from populations that are not heavily fished. If it is not, the skewing bias in favor of small females (due to commercial cropping of larger individuals) must be factored into the analysis. It is interesting to note that for *P. argus* at Bimini (Kanciruk and Herrnkind, 1976a), Dry Tortugas (Davis, 1974), and Jamaica (Munro, 1974), and also for *P. homarus rubellus* in Africa (Berry, 1971), very large females were proportionately less frequently observed to carry eggs than intermediate-sized females. It is not known whether this is due to a "sexual senility" that overcomes large females or to less frequent molting of older females.

The concept of relative reproductive importance should influence our cropping efforts, as well as legal size regulations, and could become useful in developing

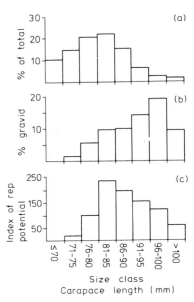

Fig. 3. Palinurid population reproductive potential by size class. Data for *Panulirus argus*, but importance of larger females to the total reproductive effort is typical of most palinurid species. (a) Female size class; (b) Frequency of gravid females by size class; (c) The index of reproductive potential by size class takes into account female size frequency, propensity to carry eggs, and average number of eggs carried per female. See text for full discussion. (After Kanciruk and Herrnkind, 1976a, with permission.)

future lobster management policies. However, some populations of palinurids may be habitat, and not recruitment, limited. In such cases this approach would be less useful. However, for most populations limiting factors are unknown, and conservative treatment should force us to assume that total egg production is an important factor. In this light, many local "protective" lobster management policies have need of reassessment.

Estimation of egg-carrying time in the field is difficult. Tag return data are often statistically inadequate, and the release times inappropriate. Captive conditions often do not accurately reflect field conditions and perhaps undermine the reliability of laboratory data. A method has been offered that examines the percent of fouled females (i.e., with external growth on the carapace—primarily worm tubes, barnacles, and algae) observed for each reproductive state and the frequency of each reproductive state within the sample population in order to estimate time spent in each state *in situ* (Kanciruk and Herrnkind, 1976a; Kanciruk, 1976).

2. Ecology of Juvenile and Adult Palinuridae

Comparison of the degree of fouling in different molt states of *P. argus* indicates that as the time since the last molt increases the degree of fouling and the percent of lobsters showing fouling also increases. These data were based on a large nonslective *in situ* sampling effort, and can be used to determine the temporal sequence of each reproductive stage. It had been suggested that females

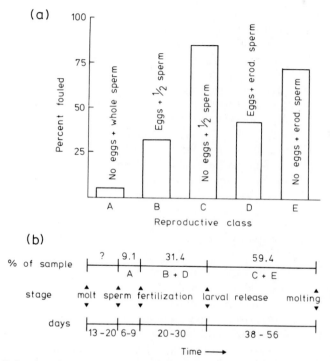

Fig. 4. Estimation of temporal sequence of reproduction in *Panulirus argus* from field data. (a) Test of hypothesis of two fertilizations per spermatophore. Carapace fouling is known to increase with time since last molt. If the reproductive sequence in *Panulirus argus* included two utilizations of the spermatophore the temporal sequence would be A, spermatophore deposit; B, fertilization (half of the spermatophore unused); C, larval release; D, second fertilization with remaining portion of spermatophore; and E, larval release. However, significant reduction of observed fouling in the D class indicates that in its correct temporal sequence it should come before C, and thus, in this population there appears to be only one utilization of the spermatophore. This justifies grouping of B and D, and C and E. (b) Estimated *in situ* duration of each reproductive stage. Sequence is molting, spermatophore deposit by male, extrusion of eggs and fertilization, larval release, and molting. The percentage of each stage in the field collection was used to estimate the time spent in each stage, based on a 20-30 day egg-carrying time for *P. argus*. First stage (after molting and before spermatophore placement), represented by a "?", can not be measured by this method. (After Kanciruk and Herrnkind, 1976a, with permission.)

might be able to use their spermatophore twice in successive fertilization, since only partially spent spermatophores are often observed on *P. argus* females. The temporal sequence (if this were true) would be molt—copulation—first egg extrusion—larval release—second egg extrusion—larval release—molt.

Reproductive females were classified as follows: females with whole spermatophore, those with partially eroded (used) spermatophore, and those with completely eroded spermatophore (with or without eggs). Based on visual observation, a subjective measure of external fouling was assigned to each female. The various states of reproductive condition (with their associated percentage of fouling), and the temporal sequence of reproduction are presented in Fig. 4. Frequencies of observed fouling for each state indicated that the postulated sequence of events, including double duty by the spermatophore, was probably incorrect for this species at Bimini under natural conditions (Fig. 4a). Frequency of occurrence of each reproductive stage in the sample was then used as a gross indicator of relative time spent in each state. The calculated figures of the time spent in each reproductive state generally agree with those obtained by more traditional tag return and tank studies of molting and reproduction. This technique (albeit approximate) allows inferences about the reproductive cycle to be drawn directly from field data.

IV. PALINURID FISHERIES AND ECOLOGY

There is a wealth of information on the fisheries biology of the Palinuridae. Full treatment of the subject is found in Chapter 7 of this volume. Here we can focus on one of the interesting, but seldomly documented indirect ecological effects of both commercial and sport fisheries, i.e., population dispersion due to displacement and harrassment. Fishing pressure on the Palinuridae can be divided into sport and, the far more important, commercial fishery. There has been little research into the effects of heavy fishing pressure on pristine lobster populations. Research usually commences after the population has shown the deleterious effects of overfishing, and often assumptions can only be made about preexploited conditions.

In a unique controlled study at Dry Tortugas, Florida (Davis, 1977), the size frequency of a previously protected lobster (*P. argus*) population exposed to new commercial pressure was significantly and quickly reduced to below legal size. A second area, similarily exposed to a new sport fishery harvest (two lobsters per diver per day), was also significantly reduced in shallow regions, but not in areas deeper than 10 m. Interestingly, the effect of the sport fishery was twofold: the overall lobster density was reduced (as it was for the commercially fished area), and the frequency of multiple occupancy was also reduced (dispersion of the lobsters). Dispersion is probably detrimental to the population, since it adversely affects the social structure (Section II, also Chap. 7, Vol. I). The entire experimental area was returned to its protected state. After 16 months, the behavioral effects on the fishery (dispersion) were not as evident (a return to 70% of the prefishery multiple den occupancy rate was observed), but the overall density remained much lower than the original level.

Such studies on protected populations are as rare as they are valuable. A reasonable understanding of the ecology of economically exploited species will be greatly augmented through the maintenance and study of strictly protected populations, uninfluenced even by sport diving. Although it may appear wasteful of an economic resource, the development and maintenance of such areas will repay the initial investment many times over by providing otherwise unavailable data.

V. BEHAVIORAL ECOLOGY

Social behavior is treated elsewhere (Chap. 7, Vol. I). Here, we will concentrate on those behavioral patterns which closely relate to the overall ecology of this group.

A. Feeding Behavior

The early literature reflects the long-held belief that spiny lobsters are scavengers, feeding opportunistically upon dead animal material (Crawford and DeSmidt, 1922). This belief was perpetuated in part by the extensive use of fish heads, rotting cowhide, etc., in baited traps (although unbaited traps also attracted lobsters). Stomach content analysis, *in situ* and laboratory observations revealed this to be generally untrue. Most material is eaten alive or freshly killed, and lobsters are selective feeders when possible (Carlberg and Ford, 1977). Because the Palinuridae are characterized by the lack of chela and by slow movement (except in tail-flap retreat), they are usually limited in their diet to sluggish, easily captured animals or organisms fixed to the substrate.

Palinurids are primarily considered carnivores. In certain habitats they may, in fact, be the major benthic carnivore (Berry, 1977; Davis, 1977). They feed upon various organisms, usually calcareous, but including Porifera, pelecypod and gastropod mollusks, echinoid and asteroid echinoderms, crustaceans, and fish. The infrequent but documented occurrence of fish scales in stomachs may represent change encounters with wounded or recently killed animals, rather than entirely true captures, per se. Although plant material has been found in palinurid stomachs, the extent to which this represents direct feeding or chance ingestion while feeding on epiphytic organisms is not always discernible. Munro (1974) reports *P. argus* in Jamaican waters feeding partially upon algae. The same species has been observed pulling *Thallassia* spp. grass blades through their mouthparts, possibly feeding on the thick epiphitic growth of hydroids on the grass, since ingestion of the blade itself was not observed (P. Kanciruk, personal observation). Barroso-Fernandes (1971) described finding calcareous algae in *P. argus* stomachs, in amounts proportionally greater than the availability of the algae in the habitat and concluded that selective foraging on vegetative material was occurring. Beurois (1971) described *Jasus paulensis* in the Indian Ocean as partially vegetarian. Heydorn (1969c) reported that at Tristan da Cunha, *Jasus tristani* was observed to be entirely vegetarian, feeding on red and brown seaweeds. This dietary fare, however, may not have been necessarily one of choice, but was likely due to a lack of prey, since these same lobsters eagerly consumed fish when it was offered as bait in traps.

Fielder (1965b) and Chittleborough (1975), among others, indicate that there is active selection between food types by *J. lalandii* and *P. cygnus* in the laboratory, and presumably food selection preferences are also expressed in the environment. The palinurids in general can be considered omnivores, with great (if not total) emphasis on animal food items.

Herrnkind *et al.* (1975) and Olsen *et al.* (1975) described nightly foraging movements for *P. argus* and reported that foraging began at dusk and ended in early morning. This has been corroborated in field and laboratory studies at Bimini (Kanciruk, 1976; Kanciruk and Herrnkind, 1978b). The majority, but usually not every member of the den, participate in nightly foraging on the reef and the surrounding grass and sand flats. The primary method of food identification in the Palinuridae is chemoreception, using the antennules and tips of the pereiopods (dactyls). Heydorn (1969c) describes an unusual "dancing gait," (i.e., antennae stretched forward and antennules twitching), associated with foraging in *J. lalandii*. The antennules confer distant chemoreception, and the dactyl tips contact chemoreception and perhaps movement detection in *P. argus* (Ache, 1977). The pointed dactyls, by being able to penetrate loose substrate to a slight degree, also allow recognition of partially buried food items. The role of vision in feeding has been poorly investigated, but it apparently plays little, if any, role in feeding for the majority of the Palinuridae.

Herrnkind *et al.* (1975) describe predation by *P. argus* on hermit crabs in the U.S. Virgin Islands in some detail. Usually two or three pairs of anterior pereiopods and the maxillipeds were observed to manipulate the hermit shell, bringing the aperture to the mouthparts. Lobsters were selective, discarding empty shells rapidly after preliminary inspection. Shell fragments were bitten off as the shell was rotated, until the hermit crab was left no retreat. It was then easily withdrawn and consumed. An 85 mm CL lobster could dispose of a crab lodged in a 5–7 cm *Fasciolaria* spp. shell in about 30 min. Ritualized feeding behavior associated with mussel opening has been described for *Panulirus inter-*

2. Ecology of Juvenile and Adult Palinuridae

ruptus (Carlberg, 1975), and specialized movements are associated with the feeding of *J. lalandii* on black mussels (Heydorn, 1969b).

Stomach analysis during the TEKTITE study indicated that *P. argus* has a varied diet. On the average, eight different food items were identified per analysis. Juvenile and adult lobsters exhibited distinctly different feeding patterns (Fig. 5a, b). It is not certain whether this can be attributed to differing preferences between juveniles and adults or food availability/suitability in the differing environments. Conspecific carapace material has also been found in *P.*

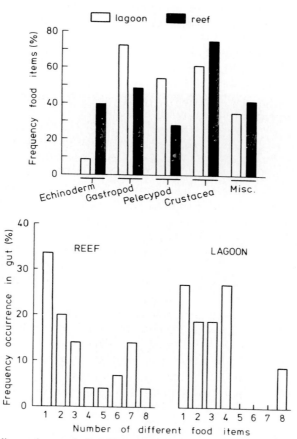

Fig. 5. Feeding preferences in juvenile and adult *P. argus* (a) By food type. Lagoon lobsters are smaller than reef animals and rely more on gastropods and pelecypods for their food supply. (b) By diversity. Reef lobsters exhibited a more varied diet. Data may represent differences in availability rather than actual preferences. (After Herrnkind *et al.*, 1975, with permission.)

argus stomachs, but this probably does not represent true cannibalism (as has been reported for laboratory maintained animals) but rather, ingestion of molt casts. Well-fed captive populations of *P. argus* do feed upon fresh molt casts, but rarely cannibalize freshly molted den mates (P. Kanciruk, personal observation). Cast injection may provide trace substances limited in the lobsters usual diet (especially for newly molted lobsters) or provide an efficient way to recover substances without energy-consuming foraging.

Some species have more restrictive diets than does *P. argus*. Berry (1971) describes the distribution of *P. homarus rubellus* in southeast Africa as being correlated with the availability of the brown mussel, *Perna perna,* on the inshore reefs. It is quite probable that other palinurid species with restricted distribution (i.e., deep sea and reef forms) have a similarly limited dietary fare reflecting their habitat specialization.

Fig. 6. Predation on *P. argus* in an open area. A lone lobster is relatively defenseless to persistant attacks by triggerfish (*Bolistes vitula*), in spite of defensive postering of antennae. The triggerfish first concentrated its attack at the eyestalks. After blinding the lobster, the fish easily attacked the soft abdomen. Daylight travel in open areas by queuing and pod formation in *P. argus* is effective against such predation. (Photo by W. F. Herrnkind.)

2. Ecology of Juvenile and Adult Palinuridae

B. Defensive Behavior

Although the palinurids are protected from small, weak, or slow predators by their spiny exoskeleton and their characteristic tail-flap retreat, and many of the larger or more powerful organisms are able to overcome these defenses. For *P. argus*, these predators (typical for many other Palinuridae) include sharks, skates, large snappers, groupers, jewfish, octopuses (Buesa Mas, 1965), dolphins, loggerhead turtles (Munro, 1974), triggerfish (W. F. Herrnkind, personal observation) (Fig. 6), a small boring whelk (*Murex pomum*) which was reported to kill lobsters in slat traps (Food and Agriculture Organization, 1968, cited in Munro, 1974) and, of course, man. It appears that natural mortality is dominated by predation and may be density-dependent (*P. argus*, Munro, 1974; *P. cygnus*, Morgan, 1974; Chittleburough, 1970).

Principle methods of palinurid defense include prodding or raking with the antennae while moving towards the intruder, retreating into the den, quick tail flapping and rearward retreat (for a maximum distance of 10–50 meters until apparently abdomenal muscle fatigue), creation of a turbid water cloud by tail flapping (*J. lalandii*, Berry, 1971; Heydorn, 1969c), bracing within the den

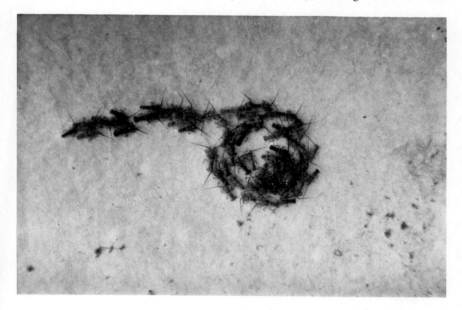

Fig. 7. Defensive pod formation. Upon harrassment by divers on migrating, queuing lobsters (*P. argus*) form a defensive pod with all members eventually facing outward with antennae pointing toward the predator. The resultant "pincushion" is an effective defense for lobsters caught away from the den. (Photo by P. Kanciruk.)

using the pereiopods and cephalothoric spines to engage the den walls and substrate, autospasy and autotomy of appendages, queuing and pod formation on open substrate (Fig. 7), and perhaps symbiotic den cohabitation (*P. homarus rubellus* with the moray *Lycodontus undulatus*, Berry, 1971; *P. argus* and morays in the Caribbean, P. Kanciruk, personal observation). Stridentes are Palinuridae with the ability to make rasping stridulatory sounds with the base of their antennae. Their stridulation is correlated with diver harrassment and tail-flex retreat, and some researchers (Meyer-Rochow and Penrose, 1974; Hindley, 1977) believe that it also functions in defense, although this has not been demonstrated. The most complete study of stridulation in the Palinuridae has been by Mulligan and Fisher (1977). They were able to distinguish and sonically characterize three stridulatory sounds: "popping," "fluttering," and "rasping." Each was composed of discrete, wide-spectrum pulses. By monitoring isolated and group situations they were able to correlate flutter, popping, and rasping sounds to low, moderate, and high levels of arousal and specific behaviors, respectively. Rasping stridulation was associated with harrassment. Responses to stridulation (either by tankmates or during electronic playback), however, were rarely observed. One may speculate that stridulatory sounds made by lobsters in dens may serve as a "sonic beacon" for lobsters returning from nocturnal foraging. Further research is needed in this most interesting aspect of social behavior.

C. Environmental Stimuli

Animal behavior is intimately associated with the environment. Of the many possible influential stimuli, temperature and light are the most obvious, and perhaps the best documented for the palinurids. Environmental temperature fluctuations often correlate with behavior and/or population distribution. Davis (1977) documented density changes in protected areas for *P. argus* that correlated closely with seasonal temperature fluctuations over a 3-year period (Fig. 8). These fluctuations were due to reproductive immigration/emigrations from deep reef areas. Chittleborough (1976) described a direct correlation of breeding frequency and an inverse relationship of egg incubation time to temperature for *P. cygnus*. Sheard (1962) postulated that species of Palinuridae which naturally experience large yearly temperature fluctuations are more inclined to exhibit ecological correlates with temperature peaks and lows. As evidence, he cites *P. argus* in Bermuda (exposed to a 12°C yearly temperature fluctuation) and *P. interruptus* off California (exposed to 7°C fluctuations). These palinurids breed in the spring when the temperature rises, and their eggs hatch in the summer when the temperature peaks. In contrast, the reproductive cycle of *P. cygnus* in Western Australia (4°C fluctuations) does not correlate as closely with temperature. In this case, mating occurs before the minimum winter temperature is reached and the eggs hatch well before the maximum summer temperature is

2. Ecology of Juvenile and Adult Palinuridae

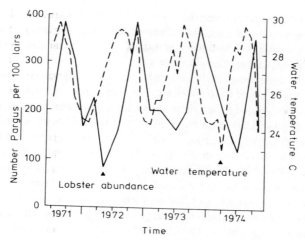

Fig. 8. Lobster abundance vs. temperature. *Panulirus argus* abundance correlates positively with temperature. Changes in density reflect movements in and out of the shallow water habitat. Temperature is probably the most important environmental parameter influencing the ecology of the Palinuridae, affecting not only movements and migrations, but also reproduction and overall distribution. (After Davis, 1977, with permission.)

attained. *Panulirus argus* in equatorial regions experience mild seasonal temperature fluctions and show no strong temperature peak reproduction correlations. Instead, they tend to reproduce throughout the year.

Temperature affects molting and growth in palinurids, as it does for other crustacea. Ford (1977) noted that *P. interruptus* raised in power plant thermal effluent molted more frequently. However, the size increase per molt was less than for controls, and overall growth was essentially equivalent between both groups. Bradbury (1977) demonstrated a cline of slow to fast growth rates correlated with temperature for *J. lalandii*. Bradstock (1950) found that *J. lalandii* matured at a smaller size in warmer regions.

It is reasonable to assume that temperature plays a role in the timing of breeding, molting, population movements, etc., in those locales where its fluctuations are substantial and somewhat regular (seasonal). The total influence of temperature on the biology of Palinurids has yet to be defined, although a relationship with a particularly spectacular behavior pattern (mass migration in *P. argus*) has been documented in laboratory and field experiments (Herrnkind, 1969; Herrnkind *et al.*, 1973; Kanciruk, 1976; Kanciruk and Herrnkind, 1978b).

Thermal changes associated with autumnal frontal storms are highly correlated with the migratory behavior of *P. argus* and are dealt with in detail in Chpt. 7, Vol. I. Interestingly, temperature decreases affect both the duration of activity (i.e., the activity period lengthens into the usually inactive daytime), and the quality of behavior expressed (i.e., the induction of queuing behavior). Most probably, temperature plays a correspondingly direct role on the activity patterns and behavior of other Palinuridae, especially those migratory species naturally exposed to large temperature fluctuations.

Other environmental stimuli influencing the ecology of the Palinuridae are varied but less understood. Newman and Pollock (1971) document inshore movements of *J. lalandii* as a direct response to the upwelling of water low in dissolved oxygen (2 ml/liter) from offshore areas. Distributions of *Puerulus sewelli* in deep water (200–350 m) off the Indian and Burmese coasts have been correlated with low oxygen levels to which they are reportedly tolerant (Chekunova, 1972). He also associates large, commercially exploited populations of palinurids on the western continental shelf of Africa with relatively closed circulations and uniform hydrological conditions. George *et al.*, (1978) was able to correlate distribution and reproductive behavior to temporal oceanographic characteristics along the east and west coasts of South Africa for *J. lalandii* and *P. homarus rubellus*.

Data substantiating such generalizations are scarce, however. The studies by Berry (1971) and by George *et al.* (1978) in which species of palinurids were divided by ecological parameters such as temperature, depth, turbidity, habitat, and wave energy are exceptional and can serve as models for research on other palinurid species.

Few studies on salinity tolerance have been conducted for the Palinuridae. However, indications are that this family is generally stenohaline, and thus primarily restricted to oceanic or near oceanic waters. Stead (1973) found that *J. edwardsii* could acclimate to salinities lowered by 3% from their norm (33%), but that a 7% salinity differential would produce high mortality. However, some environs that fluctuate somewhat in salinity (e.g., estuaries) do not exclude inhabitation by certain palinurids (*P. argus*) (Witham *et al.*, 1968). Chittleborough and Thomas (1969) speculated that low salinity masses of water or "wedges," may influence pueruli settlement along certain areas of the Western Australian coast.

Undoubtedly the interaction of many environmental parameters other than oxygen levels, turbidity, salinity, and temperature play an important role in the determination of palinurid distribution and behavior. Unfortunately, the influences of many of these factors are subtle and defy cursory investigation.

D. Biological Rhythms

The functioning of life in a rhythmic fashion is a typical attribute of all organisms. The detailed study of rhythmic activity has been only a recent phenomenon. In general, rhythmic patterns in the Palinuridae have been poorly documented. Studies on physiological or biochemical rhythms are rare, and only

2. Ecology of Juvenile and Adult Palinuridae

a few researchers have integrated laboratory with field studies to present a clear picture of the ecological significance of rhythmic behavior.

Daily and seasonal movements, as well as activity and feeding rhythms, have been reported primarily from field observations, trap-catch data, and laboratory investigation (Sutcliffe, 1952, 1956; Fielder, 1965b; Newman and Pollock, 1971; Okada and Kato, 1946; Kanciruk and Herrnkind, 1973, 1978b; Kanciruk, 1976). Field observations and trap-catch success for virtually every member of the Palinuridae studied indicate that spiny lobsters are nocturnally active foragers with few exceptions (e.g., *P. argus* will become active and feed during the day throughout autumnal mass migration).

Activity rhythms have rarely been the subject of extensive laboratory or field study. Studies have been typically of short duration and thus unable to elucidate long-term changes in activity patterns. The best studied species is *P. argus*. Extensive coordinated and simultaneous field and laboratory studies have been conducted on usual and migratory locomotor activity patterns for this species. Information is available on the daily, seasonal, and migratory activity patterns, as well as the probable associated environmental stimuli (Herrnkind *et al.*, 1975; Kanciruk, 1976; Kanciruk and Herrnkind, 1978b). The best *in situ* study of the daily activity patterns in a spiny lobster are those of Cooper and Herrnkind (1971) and Herrnkind *et al.* (1975). Lobster dens were monitored by diver observation over long periods of time, from shore and from the TEKTITE II manned underwater habitat. Den leaving and returning times were noted. It was documented that *P. argus* is indeed nocturnal, leaving the den at and just after sunset, to return in the early morning before sunrise (Fig. 9). Laboratory studies have confirmed this basic activity pattern. Sutcliffe (1956) visually monitored *P. argus* in an outdoor tank and noted the percentage within shelter. Although Sutcliffe monitored only between noon and midnight for 3 days and used a flashlight to observe in the darkness (possibly disturbing the lobsters' activity), he was able to determine that groups of *P. argus* were nocturnal in captivity, and additionally that their activity was somewhat suppressed by strong moonlight. The effect of moonlight on lobster activity is an observation long held by fishermen for many species of palinurids. Statistical correlations of moonlight with reduced catches have been reported for *P. cygnus* (Morgan, 1974) and *Panulirus japonicus* (Kubo and Ishiwata, 1964; Yoza *et al.*, 1977).

Fielder (1965b) studied the laboratory locomotor activity and feeding patterns of *J. lalandii* and concluded that this species was nocturnal in both feeding and activity rhythms. However, his study, while the most extensive up to that time, was conducted under conditions that were not innocuous to the lobsters (they were tethered about the abdomen and attached, through a set of pulleys, to a recording device). Fielder's study was not conducted long enough to detect any seasonal components in the rhythms, but it did show for the first time that palinurid rhythms would persist under constant conditions. This enabled the term

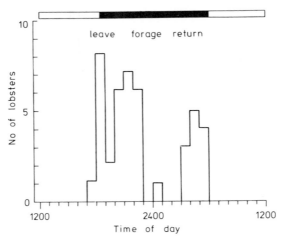

Fig. 9. Nocturnal activity patterns of *P. argus* as observed in the field. Solid bar indicates darkness. In general, Palinurids are nocturnal foragers feeding in the reef area and on adjacent sand flats. (After Herrnkind *et al.*, 1975, with permission.)

circadian to be applied to the rhythms, inferring the existence of an internal timing mechanism.

Daily, seasonal, and migratory activity patterns of *P. argus* have been studied under laboratory and field conditions (Kanciruk and Herrnkind, 1973, 1978b; Kanciruk, 1976). *Panulirus argus* was nocturnal and exhibited two distinct biorhythms—a winter and a summer activity pattern. The winter pattern consisted of locomotor activity commencing at sunset, continuing for a few hours, but not maintained throughout the night. Sunrise had no observable effect on the activity expressed. Over a period of 4 weeks in the spring, this pattern slowly transformed into the summer pattern which consists of activity commencing at sunset, being maintained through the night, and ceasing abruptly at sunrise. This summer pattern remained stable through the next autumn (Fig. 10). The major difference between these two activity rhythms is the number of active hours per night and response to sunrise. The summer pattern expressed in the laboratory is extensively corroborated by observations in the field (Herrnkind *et al.*, 1975; P. Kanciruk, personal observation). These preliminary results were somewhat surprising, since the two patterns differ greatly. This seasonal component to the activity pattern is unique for the Palinuridae, and for benthic crustacea in general. The function of these two patterns is unclear, but reduced metabolic needs during the winter (due to cooler temperatures, reduced reproductive activity, lack of migratory movements, etc.) may alleviate the necessity for the extended periods of nightly foraging that expose the lobsters to predatory pressure.

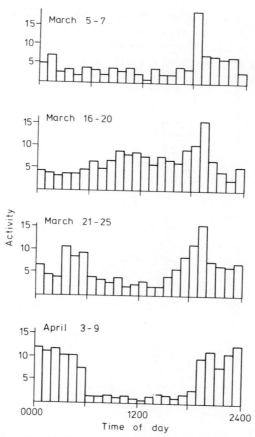

Fig. 10. Transition of the laboratory activity pattern for *P. argus* from typical winter pattern (top graph) to summer pattern (bottom graph). Temperature in the laboratory was maintained constant, but photoperiod was allowed to varry naturally. (After Kanciruk. 1976.)

The expression of activity in *P. argus* in the laboratory has a definite gregarious component. Isolated individuals varied from being arhythmic to expressing the identical parent population activity pattern. Again, these data underscore the importance of social interactions in this highly gregarious species.

The nocturnal activity pattern in *P. argus* is abruptly and dramatically abandoned during the autumnal mass migration. The activity period not only is lengthened (extending well into daylight), but the expression of activity is also modified. Queuing behavior, which is sometimes expressed during usual nocturnal foraging (P. Kanciruk, personal observation), becomes much more prevalent, and movement by single lobsters becomes rare. It is postulated that the changing autumn photoperiod, and perhaps long-term temperature changes, act to prime the lobster populations for mass migration. The sharp temperature declines associated with the passage of autumnal frontal storms act as the trigger for the migration. Decreasing temperature has been shown to modify the usual nocturnal locomotor activity pattern and stimulate migratory activity and queuing behavior

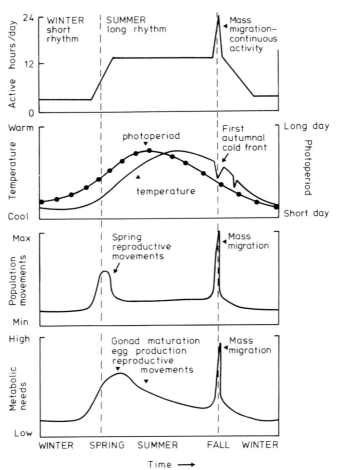

Fig. 11. Annual activity pattern of Panulirus argus correlated with environmental factors. Winter activity pattern changes over to the summer pattern in the spring, correlated with increasing photoperiod, warming temperatures, spring reproduction, population movements, and increased metabolic needs. Fall migratory activity pattern (continuous activity over a 2–6 day period) correlates with passage of first cold autumnal storm, shortening photoperiod, and partial population movement to deeper, adult habitats. (After Kanciruk, 1976.)

2. Ecology of Juvenile and Adult Palinuridae

(Kanciruk, 1976; Kanciruk and Herrnkind, 1978b). A summary of the seasonal and migratory activity pattern is given in Fig. 11, and a complete discussion of the migratory phenomenon is in Chap. 7, Vol. I.

There is overwhelming evidence that shallow water palinurids are nocturnally active, both in general movement patterns and in foraging behavior. Daylight and bright moonlight inhibit activity under most nonmigratory conditions. The intensity necessary for inhibition is unknown, but has been estimated for *P. argus* from observations in the field to be approximately 1.5 microeinsteins/m^2/sec (Herrnkind, 1977). Light levels at which activity commences in laboratory-maintained populations of *P. argus* were much greater than corresponding light levels in the field, perhaps reflecting lack of harassment from predators and/or acclimation to experimental situations.

Studies in constant light and constant dark (Fielder, 1965a) support the existance of an internal circadian clock. Hunger as a motivation for nocturnal foraging can be assumed, although starvation for short (5 days, Kanciruk, 1976) or long (20 days, Fielder, 1965b) periods did not increase the level of nocturnal foraging activity. There is no present evidence for innate lunar activity rhythms in the Palinuridae since statistically convincing evidence is difficult to obtain. However, catch statistics often correlate with moon phase, and those species that typically inhabit inshore areas affected by tidal water movements have not been extensively studied in the laboratory under the careful conditions that could identify lunar activity rhythms. It seems probable that, as for other inshore benthic crustaceans, tidal periodicities might be expressed in some palinurids.

VI. SUMMARY

The Palinuridae are of cosmopolitan distribution and range in habitat selection from turbid and turbulent beach zones to the perpetual darkness of abyssal depths. In spite of such habitat diversity, the morphology of this group is surprisingly similar within a species and generalized for a benthic foraging existence. The wide distribution of the palinurids, along with evidence of their recent (evolutionary) invasion of available habitats, underscores the success of their generalized theme.

Our knowledge of the ecology of the Palinuridae ranges from barely adequate for the shallow water economically important species (e.g., *P. argus, P. cygnus, J. lalandii*) to virtually nonexistent for rare, deep water forms (*Puerulus carinatus, P. angulatus*). The life cycle of the palinurids includes a long-lived planktonic form, which greatly complicates our attempts at population recruitment models (Sims and Engle, 1966). Our lack of knowledge and our inability to define juvenile and adult habitat limitations hinders efforts at population modeling and fisheries management.

Juvenile and adult habitats almost always include singular and multiple den residency. Except for short periods of migratory movement, palinurids are nocturnally active foragers feeding primarily on slow moving and sessile animals, with varying reliance on vegative food items. Behaviorally, their most distinguishing and unique characteristic is their gregariousness, which is expressed not only in den residency patterns, but also in the foraging and migratory movements of some species. The palinurids represent the pinnacle of social behavior for benthic decapods.

Interactions between environment (temperature, light, photoperiod, salinity, etc.) and palinurid behavior are complex, and evidence indicates that such environmental parameters strongly affect activity patterns, migratory movements, reproduction, growth, and regional and local distribution. Based on current research, the most important of these factors seems to be light (controlling activity patterns) and temperature (influencing or triggering migratory behavior and breeding activity).

Additional *in situ* observation and experimentation, as well as controlled laboratory investigations, are necessary in order to complete our understanding of this family. The Palinuridae are an economically important and ecologically interesting group. Efforts to study them will be greatly rewarded.

REFERENCES

Ache, B. W. (1977). The sensory physiology of spiny and clawed lobsters. *Circ.—CSIRO, Div. Fish. Oceanogr. (Aust.)* **7**, 103–122.

Aiken, K. A. (1975). "Marine Fisheries Research Report Spiny Lobster Investigations Annual Report, December, 1975" (mimeo.). Jamaican Ministry of Agriculture.

Aiken, K. A. (1976) "Marine Fisheries Report Spiny Lobster Studies (Port Royal Cays), June-November, 1976" (mimeo.). Jamaican Ministry of Agriculture.

Barroso-Fernandes, L. M. (1971). Sobre a alimentacao da lagosta *Panulirus argus* (Latr.) 1804 (Crustacea Reptantia). *Comision Asesora Regional da Pesa Para et Atlantica Sudoccidental (CARPAS) 5th Ses, Mar del Plata Doc.* CARPAS/5/D Tec. 31.

Berrill, M. (1975). Gregarious behavior of juveniles of the spiny lobster, *Panulirus argus* (Crustacea: Decapoda). *Bull. Mar. Sci.* **25**(4), 515–522.

Berry, P. F. (1970). Mating behavior, oviposition, and fertilization in the spiny lobster *Panulirus homarus* (Linnaeus). *S. Afr., Assoc. Mar. Biol. Res., Invest. Rep.* **24**, 1–16.

Berry, P. F. (1971). The spiny lobsters (Palinuridae) of the east coast of southern Africa. Distribution and ecological notes. *Oceanogr. Res. Inst., Invest. Rep.* **27**, 1–23.

Berry, P. F. (1977). A preliminary account of a study of biomass and energy flow in a shallow subtidal reef community on the east coast of South Africa, involving the rock lobster *Panulirus homarus. Circ.—CSIRO, Div. Fish. Oceanogr. (Aust.)* **7**, 24.

Beurois, J. (1971). Régime alimentaire de la langouste *Jasus paulensis* (Heller, 1862) des îles Sant-Paul et Amsterdam (Ocean Indien). Résultats préliminaires. *Tethys* **3**(4), 943–948.

Bradbury, J. H. (1977). Growth rates of Tasmanian rock lobster. *Circ.—CSIRO, Div. Fish. Oceanogr. (Aust.)* **7**, 32.

Bradstock, C. A. (1950). A study of the marine spiny crayfish *Jasus lalandii* (Milne-Edwards) including accounts of autospasy. *Victoria Univ., Coll. Zool. Publ.* **7**, 1–38.

2. Ecology of Juvenile and Adult Palinuridae 93

Buesa Mas, R. J. (1965). Biology and fishing of spiny lobster *Panulirus argus* (Latreille). *In* "Soviet-Cuban Fishery Research" (A. S. Bogdanov, ed.) [transl. from Russian by Israel Program for Scientific Translations, Jerusalem, 1969 (TT69-59016), 62–77].

Buesa Mas, R. J., and Mota-Alves, M. I. (1971). Escala de colores para el estudio del ciclo reproductor de la langosta *Panulirus argus* (Latreille) en el area del Mar Caribe. *FAO Fish. Rep.* **71**, 9–12.

Buesa Mas, R. J., Paiva, M. P., and Costa, R. S. (1968). Comportamiento biologica de la langosts *Panulirus argus* (Latreille) en el Brasil y en Cuba. *Rev. Bras. Biol.* **28**(1), 61–70.

Carlberg, J. M. (1975). Food preferences, feeding activity patterns, and potential competion of the American lobster, *Homarus americanus*, and ecologically similar crustaceans native to California. M..S. Thesis, San Diego State University, San Diego, California.

Carlberg, J. M., and Ford, R. F. (1977). Food preferences, feeding activity patterns, and potential competion between *Homarus americanus*, *Panulirus interruptus*, and *Cancer antennarius*. *Circ.—CSIRO, Div. Fish. Oceanogr. (Aust.)* **7**, 23.

Chekunova, V. I. (1972). Geographical distribution of spiny lobsters and ecological factors determining their commercial concentrations. *All Union Res. Inst. Mar. Fish: Oceanogr. (VNIRO) Proc.* No. 77, p. 2.

Chittleborough, R. G. (1970). Studies on recruitment in the Western Australian Rock Lobster *Panulirus longipes cygnus*. Density and natural mortality of juveniles *Aust. J. Mar. Freshwater Res.* **21**(2), 131–148.

Chittleborough, R. G. (1974). Western rock lobsters reared to maturity. *Aust. J. Mar. Freshwater Res.* **25**, 221–225.

Chittleborough, R. G. (1975). Environmental factors affecting growth and survival of juvenile western rock lobsters *Panulirus longipes* (Milne-Edwards). *Aust. J. Mar. Freshwater Res.* **26**, 177–196.

Chittleborough, R. G. (1976). Breeding of *Panulirus longipes cygnus* George under natural and controlled conditions. *Aust. J. Mar. Freshwater Res.* **27**(3), 499–516.

Chittleborough, R. G., and Phillips, B. F. (1975). Fluctuations of year-class strength and recruitment in the western rock lobster *Panulirus longipes* (Milne-Edwards). *Aust. J. Mar. Freshwater Res.* **26**, 317–28.

Chittleborough, R. G., and Thomas, L. R. (1969). Larval ecology of western Australian marine crayfish with notes upon other palinurid larvae from the eastern Indian Ocean. *Aust. J. Mar. Freshwater Res.* **20**, 199–223.

Chitty, N. (1973). Aspects of the reproductive biology of the spiny lobster *Panulirus guttatus* Latreille. M.S. Thesis, University of Miami, Miami, Florida.

Cobo De Barany, T., Ewald, J., and Cadima, E. (1972). La pescade la langosta en el archiplelago de Los Rogues, Venezuela. *Inf. Tech. Proc. Invest. Desarollo Pesquero, Venezuela* **43**, 1–34.

Crawford, D. R., and DeSmidt, W. J. J. (1922). The spiny lobster, *Panulirus argus* of southern Florida. Its natural history and utilization. *Bull. U.S. Bur. Fish.* **38**, 281–310.

Davis, G. E. (1971). Aggregations of spiny sea urchins, *Diadema antillarus*, as shelter for young spiny lobsters, *Panulirus argus*. *Trans. Am. Fish. Soc.* **100**(3), 586–587.

Davis, G. E. (1974). Notes on the status of spiny lobsters, *Panulirus argus*, at Dry Tortugas, Florida, 1974. *In* "Conference Proceedings: Research and Information Needs of the Florida Spiny Fishery" (W. Seaman and D. Y. Aska, eds.), Fla. Sea Grant Program, SVSF-SG-74-201, pp. 22–32.

Davis, G. E. (1975). Minimum size of mature spiny lobsters, *Panulirus argus*, at Dry Tortugas, Florida. *Trans. Am. Fish. Soc.* **104**(4), 675–676.

Davis, G. E. (1977). Effects of recreational harvest on a spiny lobster, *Panulirus argus*, population. *Bull. Mar. Sci.* **27**(2), 223–236.

Dawson, C. E. (1949). Florida crawfish research. *Gulf Caribb. Fish. Inst.*, **1**, 21–28.

Dawson, C. E., and Idyll, C. P. (1951). Investigations on the Florida spiny lobster, *Panulirus argus* (Latrielle). *Fla. Board Conserv., Tech. Ser.* **2**.

Fielder, D. R. (1965a). A dominance order for shelter in the spiny lobster *Jasus lalandii* (H. Milne-Edwards). *Behaviour* **24**(3-4), 236-245.

Fielder, D. R. (1965b). The spiny lobster, *Jasus lalandii*, (H. Milne-Edwards) in south Australia. III. Food, feeding and locomotor activity. *Aust. J. Mar. Freshwater Res.* **16**, 351-367.

Food and Agriculture Organization (1968). Report to the government of British Honduras (Belize) on investigations into marine fishery management, research and development policy for spiny lobster fisheries. Based on the work of W. H. L. Allsopp. *FAO/TA Mar. Fish. Biol. Rep. FAO/UNDP (TA)* **2481**.

Ford, R. F. (1977). Effects of thermal effluent on survival, growth, moulting and reproductive condition of the California spiny lobster, *Panulirus interruptus*. *Circ.—CSIRO, Div. Fish. Oceanogr. (Aust.)* **7**, 83-102.

George, R. W. (1974). Coral reefs and rock lobster ecology in the Indo-West Pacific region. *Proc. Int. Coral Reef Symp.*, **1**, 321-325.

George, R. W., and Main, A. R. (1967). The evolution of spiny lobsters (Palinuridae): A study of evolution in the marine environment. *Evolution* **21**(4), 803-820.

George, R. W., and Morgan, G. R. (1976). Linear growth stages in the rock lobster, *Panulirus versicolor* as a method for determining size at first physical maturity. *Rapp. P.-V. Reun., Cons. Int. Explor. Mer.*

George, R. W., Heydorn, A. E. F., and Berry, P. F. (1970). Climatic events, coastal zonation and the biology of marine animals in South Africa. S. Afr. Nat. Cent. Ocean. Res. (G3), 1-17.

Herrnkind, W. F. (1969). Queuing behavior of spiny lobsters. *Science*, **164**, 1425-1427.

Herrnkind, W. F. (1977). Mass migration of *Panulirus argus:* Casual Stimuli, Behavioral Specializations, and Evolutionary Significance. *Circ.—CSIRO, Div. Fish. Oceanogr. (Aust.)* **7**, 31.

Herrnkind, W. F., and Olsen, D. (1971). Ecological study for the development of lobster management techniques. *Final Rep. Ser. Grant* **GH-86**.

Herrnkind, W. F., Kanciruk, P., Halusky, J. G., and McLean, R. (1973). Descriptive characterization of mass autumnal migrations of spiny lobster, *Panulirus argus*. *Gulf Caribb. Fish. Inst., Univ. Miami, Proc.* **25**, 79-98.

Herrnkind, W. F., Vanderwalker, J., and Barr, L. (1975). Population dynamics, ecology, and behavior of spiny lobster, *Panulirus argus*, of St. John, U.S. Virgin Islands: Habitation and pattern of movements. Results of the Tektite Program, Vol. 2. *Sci. Bull., Nat. Hist. Mus., Los Angeles Cty.* **20**, 31-45.

Herrnkind, W. F., Halusky, J. G., and Kanciruk, P. (1976). A further note on phyllosoma larvae associated with medusae. *Bull. Mar. Sci.* **26**(1), 110-112.

Heydorn, A. E. F. (1965). The rock lobster of the South African west coast *Jasus lalandii* (H. Milne-Edwards). I. Notes on the reproductive biology and the determination of minimum size limits for commercial catches. *S. Afr., Div. Sea Fish., Invest. Rep.* **53**, 1-32.

Heydorn, A. E. F. (1969a). Notes on the biology of *Panulirus homarus* and on length weight relationships of *Jasus lalandii*. *S. Afr., Div. Sea Fish., Invest. Rep.* **69**.

Heydorn, A. E. F. (1969b). The rock lobster of the South African west coast *Jasus lalandii* (H. Milne-Edwards). 2. Population studies, behavior, reproduction, moulting, growth and migration. *S. Afr., Div. Sea Fish., Invest. Rep.* **71**.

Heydorn, A. E. F. (1969c). The South African rock lobster *Jasus tristani* at Vema Sea Mount, Gough Island, and Tristan da Cunha. *S. Afr., Div. Sea Fish., Invest. Rep.* **73**.

Hindley, J. P. R. (1977). A review of some aspects of the behavior of juvenile and adult palinurids. *Circ.—CSIRO, Div. Fish. Oceanogr. (Aust.)* **7**, 133-142.

2. Ecology of Juvenile and Adult Palinuridae

Kanciruk, P. (1976). Daily, seasonal, and migratory locomotor activity patterns, migratory Zeitgebers, and ecology of the spiny lobster, *Panulirus argus*. Ph.D. Dissertation, Florida State University, Tallahassee, Florida.

Kanciruk, P., and Herrnkind, W. F. (1973). Preliminary investigation of the daily and seasonal locomotor activity rhythms of the spiny lobster *Panulirus argus*. *Mar. Behav. Physiol.* **1**(4), 351–359.

Kanciruk, P., and Herrnkind, W. F. (1976a). Autumnal reproduction in *Panulirus argus* at Bimini, Bahamas. *Bull. Mar. Sci.* **26**(4), 417–432.

Kanciruk, P., and Herrnkind, W. F., eds. (1976b). "An Indexed Bibliography of the Spiny Lobsters, Family Palinuridae," Fla. Sea Grant Program, Univ. Florida, Gainesville, No. 8, 1–101.

Kanciruk, P., and Herrnkind, W. F. (1978a). Reproduction potential as a function of female size in *Panulirus argus*. *In* "Proceedings of the Sea Grant Key West Lobster Conference" (W. Seaman and D. Y. Aska, eds.), Fla. Sea Grant Program, Univ. Florida, Gainesville, Tech. Paper No. 4, 44–47.

Kanciruk, P., and Herrnkind, W. F. (1978b). A field and laboratory study of the mass migration of spiny lobster, *Panulirus argus* (Crustacea: Palinuridae): Behavior and environmental correlates. *Bull. Mar. Sci.* **28**(4), 601–623.

Kubo, I., and Ishiwata, N. (1964). On the relationship between activity of Japanese spiny lobster and underwater light intensity. *Bull Jpn. Soc. Sci. Fish.* **30**, 884–888.

Lazarus, B. I. (1967). The occurrence of phyllosomata off the cape with particular reference to *Jasus lalandii*. *S. Afr., Div. Sea Fish., Invest. Rep.* **63**, 1–38.

Lewis, R. K. (1977). Studies on juvenile and adult stages of western populations of southern rock lobster (*Jasus novaehollandiae*) in the southeast region of South Australia. *Circ.—CSIRO, Div. Fish. Oceanogr. (Aust.)* **7**, 36.

Lindberg, P. G. (1955). Growth, population dynamics, and field behavior in the spiny lobster, *Panulirus interruptus*. *Univ. Calif., Berkeley, Publ. Zool* **59**, 157–248.

Meyer-Rochow, V. B., and Penrose, J. D. (1974). Sound and sound emission apparatus in puerulus and postpuerulus of the western rock lobster. *J. Exp. Zool.* **189**(2), 283–289.

Morgan, G. I. (1974). Aspects of the population dynamics of the western rock lobster, *Panulirus cygnus* George. I. Estimation of population density. *Asut. J. Mar. Freshwater Res.* **25**, 235–248.

Morgan, G. R. (1977). Adult ecology and population dynamics of the Palinuridae—a review. *Circ.—CSIRO, Div. Fish. Oceanogr. (Aust.)* **7**, 245–258.

Mota-Alves, M. I., and Bezerra, R. C. F. (1968). Sobre o numero do ovos da lagosta *Panulirus argus* (Latr.). *Argent. Est. Biol. Mar. Univ. Fed. Ceara* **8**(1), 33–35.

Moulton, J. M. (1957). Sound production in the spiny lobster *Panulirus argus* (Latreille). *Biol. Bull. (Woods Hole, Mass.)* **113**(2), 286–295.

Mulligan, B. E., and Fisher, R. B. (1977). Sounds and behavior of the spiny lobster *Panulirus argus* Decapoda: Palinuridae. *Crustaceana* **32**(2), 185–199.

Munro, J. L. (1974). The biology, ecology exploitation, and mangement of Caribbean reef fishes. *Univ. West Indies Zoo. Dep. Res. Rep.* **3**, 1–57.

Newman, G. G., and Pollock, D. E. (1971). Biology and migration of rock lobster *Jasus lalandii* and their effect on availability at Elands Bay, South Africa. *S. Afr., Div. Sea Fish., Invest. Rep.* **94**.

Okada, Y., and Kato, K. (1946). Studies on the Japanese spiny lobster, *Panulirus japonicus* (V. Siebold). I. On the migration of *P. japonicus*. *Short Rep. Res. Inst. Nat. Res. (1)*, 1–10.

Olsen, D., Herrnkind, W. F., and Cooper, R. (1975). Population dynamics, ecology, and behavior of spiny lobster, *Panulirus argus*, of St. John, U.S. Virgin Islands: Introduction. *Sci. Bull., Nat. Hist. Mus., Los Angeles Cty.* **20**, 11–16.

Peacock, N. A. (1974). A study of the spiny lobster fishery of Antigua and Barbuda. *Gulf Caribb. Fish. Inst., Univ. Miami, Proc.* **26,** 117-130.

Phillips, B. F. (1975). The effect of water currents and the intensity of moonlight on catches of the puerulus larval stage of the western rock lobster. *Rep.—CSIRO, Div. Fish. Oceanogr. (Aust.)* **63,** 1-9.

Phillips, B. F., Campbell, N. A., and Rea, W. A. (1977). Laboratory growth of early juveniles of the western rock lobster, *Panulirus longipes cygnus. Mar. Biol.* **39,** 31-39.

Phillips, B. F., Rimmer, D. W., and Reid, D. D. (1978). Ecological investigations of the late-stage phyllosoma and puerulus larvae of the western rock lobster *Panulirus longipes cygnus. Mar. Biol.* **45**(4), 347-358.

Rudd, S., and Warren, F. L. (1976). Evidence for a pheromone in the South African rock lobster *Jasus lalandii. Trans. R. Soc. S. Afr.* **42**(1), 103-105.

Serfling, S. A., and Ford, R. F. (1975). Ecological studies on the puerulus larval stage of the California spiny lobster *Panulirus interruptus. U.S. Natl. Mar. Fish. Serv., Bull.* **73**(2), 360-377.

Sheard, K. (1962). "The Western Australian Crayfishery, 1944-1961." Peterson Brokensha Pty. Ltd., Perth, Australia, 1-107.

Silberbauer, B. I. (1971). The biology of the South African rock lobster *Jasus lalandii* (H. Milne-Edwards). 1. Development. *S. Afr., Div. Sea. Fish., Invest. Rep.* **93.**

Sims, H. W., and Ingle, R. M. (1966). Caribbean recruitment of Florida's spiny lobster populations. *Q. J. Fla. Acad. Sci.* **29**(3), 207-242.

Smith, F. G. W. (1948). The spiny lobster industry of the Caribbean and Florida. *Caribb. Res. Counc., Fish. Ser.* **3,** 1-49.

Stead, D. H. (1973). Rock lobster salinity tolerance. *N.Z. Minist. Agric. Fish., Fish. Tech. Rep.* **122,** 1-16.

Street, R. J. (1969). The New Zealand crayfish *Jasus edwardsii* (Hutton). *N.Z. Mar. Dep. Fish., Tech. Rep.* **30,** 1-53.

Street, R. J. (1970). New Zealand rock lobster *Jasus edwardsii* (Hutton) South Island fisheries. *N.Z. Mar. Dep. Fish., Tech. Rep.* **54,** 1-62.

Sutcliffe, W. H. (1952). Some observations of the breeding and migration of the Bermuda spiny lobster, *Panulirus argus. Gulf Caribb. Fish. Inst., Univ. Miami, Proc.* **4,** 64-69.

Sutcliffe, W. H. (1953). Notes on the biology of the spiny lobster, *Panulirus guttatus*, in Bermuda. *Ecology* **34**(4).

Sutcliffe, W. H. (1956). Effect of light intensity on the activity of the Bermuda spiny lobster, *Panulirus argus. Ecology* **37**(1), 200-201.

Sweat, D. E. (1968). Growth and tagging studies on *Panulirus argus* (Latreille) in the Florida Keys. *Fla. Board Conserv., Tech. Ser.* **57,** 1-30.

Witham, R., Ingle, R. M., and Joyce, A. (1968). Physiological and ecological studies of *Panulirus argus* from the St. Lucie Estuary. *Fla. Board Conserv. Mar. Lab., Tech. Ser.* **53,** 1-31.

Yoza, K., Nomura, K., and Miyamoto, H. (1977). Observations upon the behavior of the lobster and the top shell caught by the bottom-set gill net. *Bull. Jpn. Soc. Sci. Fish.* **43**(11), 1269-1272.

Chapter 3

Ecology of Juvenile and Adult *Homarus*

R. A. COOPER AND J. R. UZMANN

I.	Introduction	97
II.	Distribution	98
	A. Geographic Distribution	98
	B. Habitat	99
	C. Effects of Environmental Factors on Distribution	113
III.	Population Structure	115
	A. Size Composition and Sex Ratio	115
	B. Growth—American Lobster	118
IV.	Behavior and Activity Rhythms	122
	A. Migrations and Home Range	122
	B. Nocturnal Behavior and Predation	124
	C. Feeding Behavior and Diet	127
	D. Dominance and Aggressive Behavior of the American Lobster	130
	E. Interspecific Relations of the American Lobster	131
	Appendix	133
	References	139

I. INTRODUCTION

The American lobster, *Homarus americanus* Milne Edwards, has been the subject of research since the late 1800s. It occurs in the northwest Atlantic, from intertidal depths to 700 m (McRae, 1960). The lobster fishery is the most valuable fishery on the east coast of North America. Record landings of 15,400 metric tons worth $44 million and 23,000 metric tons worth $60 million were reported for the United States (1970) and for Canada (1956, 1960), respectively.

The European lobster, *Homarus gammarus* (Linnaeus, 1758), supports a relatively small fishery with an average annual catch of 2000 tons for the period 1970-1974, down from a high of 4200 tons in 1959. In Europe, as in America, the catches from most well-established traditional fisheries have decreased markedly. Since the mid 1950's, the declines in total catches from "inshore" fishing grounds have been masked by the development of "offshore" fishing grounds.

Fundamental knowledge of the ecology of the lobster is necessary to conserve and wisely utilize this valuable resource. This chapter reviews the ecology of juvenile and adult American and European lobsters. It is structured around inshore (shallow water) and offshore (deep water) populations of *H. americanus:* those *H. americanus* occurring within a 50 km wide band of coastal water are termed inshore lobsters, and those outside this range are offshore lobsters. The review of *H. gammarus* ecology deals primarily with the inshore lobster, and the planktonic phase (stages I-IV) of the life cycle will not be discussed.

II. DISTRIBUTION

A. Geographic Distribution

3. American Lobster

The American lobster is found naturally along the east coast of North America, from North Carolina to Labrador (Fig. 1A.), being most abundant from Nova Scotia to New York (Herrick, 1895). The major population centers, and therefore inshore fisheries, are located within the Gulf of Maine and in the New Brunswick and Nova Scotian coastal waters, where over 90% of the inshore landings are made.

Schroeder (1959) defined the offshore lobster population as

> a population of lobsters, large enough to support commercial fishing off the east coast of the United States along the outer edge of the continental shelf and upper slope between the eastern part of Georges Bank and the offing of Delaware Bay. This area at depths of roughly [110-450 m] is about [650 km] long and [8-16 km] wide. Lobsters are more plentiful along the eastern half of this stretch than to the west and south.

Since Schroeder's work in the 1950's, relatively small concentrations of lobsters have been found along the outer edge of the Nova Scotian shelf. During the spring and summer (May-September), an estimated 30-50% of the offshore population from the eastern Georges Bank to the offing of Delaware Bay moves into shoal water. At this time, some of these individuals mix with the inshore population (Cooper and Uzmann, 1971; Uzmann *et al.*, 1977a). Shoalward movements of the offshore Canadian stocks have not been demonstrated. By late fall these onshore migrants have returned to the outer shelf and upper slope. The major offshore United States fishery during winter and early spring concentrates in or near the heads of submarine canyons, suggesting that here is where the offshore lobsters tend to congregate during the cool season.

3. Ecology of Juvenile and Adult *Homarus*

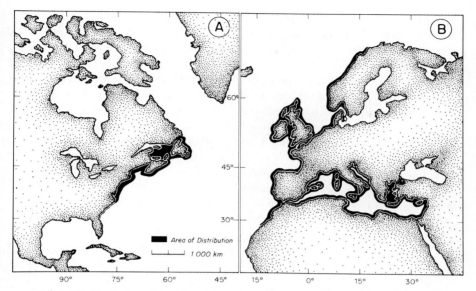

Fig. 1. (A) Distribution of American lobster, *Homarus americanus*, in western North Atlantic region. (B) Distribution of European lobster, *Homarus gammarus*, in the eastern North Atlantic and Mediterranean regions.

2. European Lobster

The European lobster, *Homarus gammarus*, has a much broader distribution than *H. americanus* (Fig. 1B). It occurs from northern Norway (Lofoten Islands) to southeastern Sweden and Denmark, where it is apparently blocked from inhabiting the Baltic Sea by lowered salinity and temperature extremes. Its distribution extends southward along the mainland European coast and around the United Kingdom and the Azores, to a southern limit of about 30° north latitude on the Atlantic coast of Morocco. This species also occurs, though less abundantly, throughout the coastal and island areas of the Mediterranean Sea and its subseas, and has been reported from the westernmost end of the Black Sea in the region of the Straits of Bosporus (Konsuloff, 1930; Havinga, 1938; Dybern, 1973; Holthuis, 1974).

B. Habitat

1. Habitat Type and Substrate of the American Lobster

Inshore lobsters, 4–120 mm carapace length (CL), are highly adaptive in their selection of habitat. Juveniles and adults occupy habitats that are characterized by mud/silt, mud/rock, sand/rock, and bedrock/rock substrates (Thomas, 1968;

Cooper, 1970; Cobb, 1971; Cooper *et al.*, 1975). The most common habitat is the sand substrate with overlying rocks and boulders (Fig. 2). Bottom gradients vary from 0 to 70°, and the ocean floor is usually characterized by rugged topography.

Lobsters of the offshore population also occupy habitats characterized by the above substrates. A large percentage of the outer continental shelf and upper slope off New England and the Mid-Atlantic Bight has a surface substrate of Pleistocene clay that is relatively free of overlying soft sediments, especially in the vicinity of submarine canyons. The clay environments range in profile from 2 to 90° of relief.

Direct observations by National Marine Fisheries Service (NMFS), scientists using scuba and research submersibles have enabled the following habitat–substrate classification to be proposed for inshore and offshore populations of American lobsters. This classification attempts to categorize the general habitat–substrate characteristics and to define the relatively abundant macrofauna and flora immediately associated with the lobster in its habitat. The Appendix presents this classification and the associated fauna and flora by habitat type. (In the table, species are designated by common and scientific name; in the text only the common name is used.)

The *sand base with rock* is the most common inshore habitat for the American lobster. In the offshore areas, the sand/rock habitat (Fig. 3) is less common, occurring primarily near the heads of certain submarine canyons (e.g., Corsair

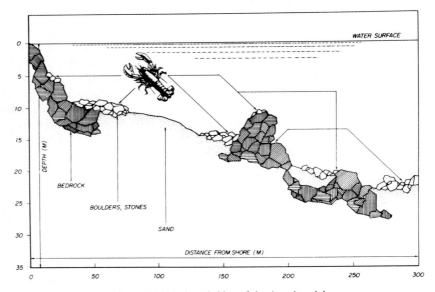

Fig. 2. Typical inshore habitat of the American lobster.

3. Ecology of Juvenile and Adult *Homarus*

Fig. 3. Offshore lobster (1 kg) at entrance to shelter excavated into sand at base of boulder, Corsair Canyon, 145 m.

Canyon off the southeast edge of Georges Bank) at depths of 150–250 m. Lobsters excavate shelters in the sand beneath the typically flattened rocks and boulders. Their shelters are generally tunnel-shaped, with a slightly greater internal width than height. An isolated rock pile on a sand base may have small, medium, and large lobsters in residence, each with its own shelter. The larger lobsters (50–100 mm CL) occupy shelters that do not extend more than 30–40 cm beneath the sediment–water interface; smaller lobsters (less than 30 mm CL) frequently tunnel 40–70 cm into the sand–rock substrate and have several shafts in their tunnel system.

The macrofauna and flora associated with this habitat inshore are listed in the Appendix. Because of the greater depth distribution of the sand–rock substrate (>40 m inshore), the biomass of the associated benthic algae is generally less than that of the algae found in the bedrock base with rock habitat, which is described below. Also, the abundance of large anemones is comparatively less, but ascidians are equally abundant in either habitat. Sponges are abundant, attached to rocks and boulders, and the brachiopod, *Terebratulina*, is common below depths of 20 m. The mollusks, *Mytilus* and *Modiolus*, are scarce, the green sea urchin and starfish are less abundant than on the bedrock substrate, and Jonah and rock crabs are very abundant.

In the offshore sand–rock habitat there are no macrobenthic algae. Very dense concentrations of anemones occur on the exposed surfaces of rocks and boulders, but sponges are scarce. A tube-dwelling polychaete and a shell-less hermit crab are very abundant. The most common fish inhabiting the sand-rock habitat are the Gulf Stream flounder, ocean pout, conger eels, greeneyes, blackbellied rosefish, tilefish, and cusk. The *bedrock base with rock* and boulder overlay is a relatively common habitat type (Fig. 4) for lobsters inshore, but not offshore. The bedrock substrate generally extends from the low tide level to a depth of 15–45 m. With few exceptions, it does not occur in the offshore lobster grounds. Encrusting and attached organisms cover nearly 100% of the exposed bedrock surface. Lobsters obtain their shelter within crevices or under rock overhangs. When sufficient rock/boulder accumulations are present, lobsters may find shelter within the rock pile. Burrowing is generally not possible on this substrate.

Green, red, and brown macrobenthic algae (Appendix) provide varying degrees of cover to depths of 42 m (Sears and Cooper, 1978). The calcified encrusting coralline algae cover up to 90% of the exposed bedrock. Large anemones, ascidians, sponges, brachiopods, and molluscs represent the majority of encrusting or attached invertebrates. The green sea urchin and starfish are the most abundant echinoderms. Cunner, tautog, sculpin, sea raven, and redfish are the most abundant fish occupying the bedrock-rock habitat.

The *mud base with burrows* is another distinctive lobster habitat. It occurs in harbors and estuaries along the coastal waters of Nova Scotia and the Gulf of Maine, in the deep basins of the Gulf of Maine, and along southern Massachusetts, Rhode Island, and within Long Island Sound (Fig. 5). Lobsters

Fig. 4. Inshore lobster taking shelter in bedrock crevice, 12 m, Boothbay Region, Maine.

3. Ecology of Juvenile and Adult *Homarus*

Fig. 5. Lobster type burrow in mud bottom of Wilkinson Basin, 250 m, Gulf of Maine. Entrance width approximately 40 cm.

excavate burrows in a soft mud bottom or occupy burrows made by the squirrel hake and the stomatopod (Cooper *et al.*, 1975). These burrows may extend below the sediment-water interface 60–80 cm, at an angle of 40–90° to the horizontal. The mud base habitat occurs at depths of 10–250 m in the Gulf of Maine basins and is rarely observed on the outer shelf and upper slope.

The macrofauna most commonly associated with the lobster in the mudburrow habitat are mud anemones, shrimp, rock and Jonah crabs, and several species of fish; i.e., squirrel and white hakes, redfish, witch flounder, four-bearded rockling, ocean pout, and goosefish. Attached or encrusting organisms and macrobenthic algae are generally not found in this habitat type, due to the absence of exposed hard surfaces.

The *clay base with burrows and depressions* is common to the outer shelf and upper slope (Fig. 6) and is similar to the inshore mud-burrow habitat. A large percentage of the outer shelf and upper slope of New England and the Mid-Atlantic Bight has a surface substrate of Pleistocene clay that is relatively free of overlying soft sediments, especially in the vicinity of submarine canyons. These clay habitats range in profile from 0 to 5° of relief. Bowl-shaped depressions are excavated primarily by lobsters in the surface of the exposed clay. These depressions are generally symmetrical in shape, extending into the clay substrate up to 1.5 m and frequently having a solid object at the center of the crater, such as a boulder, rock, coil of steel towing cable, fish net, or a fuel oil drum. Lobsters

Fig. 6. Offshore lobster (1 kg) in bowl shaped depression in clay-mud bottom, 150 m.

have been observed excavating these depressions. However, sediment removal from the vicinity of a solid object is facilitated by current flow and scouring of the bottom. Similar depressions are excavated in sand substrates inshore by onshore migrants. The offshore depressions range in diameter from 1.0 to 5.0 m and shelter from one to several lobsters at a time. Inshore, only one lobster per crater has been observed.

Mud anemones, pandalid shrimp, Jonah crabs, galatheid crabs, and blackbellied rosefish are the primary macrofauna associated with this clay habitat. Occasional squirrel hakes and Gulf Stream flounder are observed in or very close to lobster occupied depressions. At the greater depths (300–500 m) of the lobster's distribution red crabs, long-finned hakes, four-spot flounder, witch flounder, and goosefish are frequently observed in the clay-depression habitat.

The relatively featureless *mud-clay base with anemones* represents a relatively common habitat for lobsters on the outer shelf and upper slope (Fig. 7). The mud-clay substrate generally has a thin overlay of silt up to 3 cm thick. A similar habitat is rarely seen in the inshore lobster grounds. In some offshore areas "forests" of the mud anemone, *Cerianthus*, may reach densities of 3 or 4 per m^2. Generally, these individuals are 20–35 cm tall with a diameter of 5–10 cm at the base of their sheltering tube. Through excavation or current erosion, a relatively shallow depression is frequently formed at the base of the tube. These depressions serve as shelter for relatively small lobsters, generally within the

3. Ecology of Juvenile and Adult *Homarus*

50–80 mm CL size range. This is much smaller in dimension than the depressions described for the clay base with depression habitat.

The macrofauna generally associated with the clay-anemone habitat is composed primarily of Jonah crabs, bathynectes crabs, galatheid crabs, red crabs, pandalid shrimp, and also squirrel hake, blackbellied rosefish, and redfish. Unidentified invertebrates are commonly seen attached to the tube of the anemone from the substrate level up to 12–14 cm above the mud-clay base.

The most striking offshore habitat-substrate type occupied by the lobster is the *submarine canyon clay wall with burrows*, composed primarily of a Pleistocene clay (Fig. 8). The heads of many of Georges Bank's submarine canyons contain large areas of a sloping (5–70°) clay substrate that has been extensively excavated by lobsters and other macrobenthic fauna. Bioerosion is evident throughout this gradient range, being most extensive between 20 and 50° in areas we have termed "Pueblo Village" communities (Fig. 8 and 9). We define a "Pueblo Village" as a relatively localized area of a submarine canyon wall where megabenthic crustaceans and finfish have dug extensive depressions and burrows into the substrate, occupying these places as "borers" and "nestlers" (Cooper and Uzmann, 1977; Warme et al., 1978).

Figure 9 portrays a 2.5 by 2.0 m section of a "Pueblo Village" that was initially sketched by Cooper from the submersible *Nekton Gamma*, at a depth of

Fig. 7. Forest of mud anemones, *Cerianthus borealis*, frequently used as shelter by offshore lobsters. Oceanographer Canyon, 250 m.

Fig. 8. Offshore lobster (0.5 kg) in clay burrow shelter of "Pueblo Village" community, Veatch Canyon, 175 m.

Fig. 9. "Pueblo Village" community (2.5 m wide, 2 m tall), Veatch Canyon, 185 m. Distribution of species is (1) tilefish in center excavation with cleaner shrimp on roof of cave, (2) Jonah crabs in upper right and left corners, (3) galatheid crabs (long slender chelipeds) to the left and right of tilefish, (4) blackbellied rosefish in foreground, and (5) lobster in left portion of schematic. (Original drawing by R. A. Cooper.)

3. Ecology of Juvenile and Adult *Homarus*

185 m in the northeast corner of the head of Veatch Canyon. The topographic gradient was approximately 45°.

Based on size, shape, and degree of penetration into substrate, borings can be classified as one of the following three types: (1) Relatively shallow depressions that measure up to approximately 10 cm across, penetrate the substrate up to 6 cm, and vary from circular to half-moon shaped; this type is most commonly occupied by galatheid crabs and juvenile Jonah crabs. (2) Half-moon shaped depressions that project horizontally into the substrate up to 15 cm, and up to 20 cm in width; this type of depression is generally occupied by juvenile and adult Jonah crabs. (3) Borings that measure up to 1.5 m in width, 1 m in height, and 2 m or more in depth; these borings or caves are commonly occupied by juvenile and adult lobsters, tilefish, cusk, and cleaner shrimp. The blackbellied rosefish is commonly positioned near the entrance to the type (2) and (3) borings. Between 5 and 10% of the type (3) excavations have multiple openings and several caves were observed to have three.

The smallest excavations are occupied by juvenile galatheid and Jonah crabs, which we believe are the primary borers or excavators. Adult crabs and juvenile and adult lobsters occupy the next larger array of excavations. Jonah crabs and lobsters have been observed enlarging these excavations and pushing sediment to the foreground or entrance of the excavation.

2. Habitat Type and Substrate of the European Lobster

The European lobster generally selects or excavates shelter on rocky or stony bottoms where the substrate is sand or gravel. Juveniles and adults dig out hollows or tunnels under the boulders or stones with one or more openings, using the hollows as hiding places (Dybern, 1973; Berrill and Stewart, 1973; Berrill, 1974; Howard, 1977). Lobsters in Swedish waters occasionally dig saucerlike depressions in muddy substrates, far from boulders, rocks, and stones. In other areas along European coasts, lobsters construct different forms of burrows in mud substrates, without using rocks or stones as part of the shelter (Dybern, 1973). Crevices in bedrock containing shellsand are relatively common as shelters. Lobsters often inhabit cavities in rock piles where the chamber can be improved by moving small stones and smoothing out the soft substrate (Dybern, 1973). In general, the type of habitat and substrate chosen by the European lobster is similar to that for the American lobster. Shelter characteristics are also similar.

Dybern (1973) presents a classification of habitat or dwelling place types with relative frequency of lobster sightings by scuba divers along the west coast of Sweden. This classification, presented in Table I, has many similarities to the habitat-substrate classification presented for the American lobster (Appendix), with the exception of the mud-clay base with anemones and the submarine canyon clay wall with burrow, both of which are relatively deep water offshore habitats.

TABLE I

Relative Frequency of Dwelling Places of the Lobster, *Homarus gammarus*, off the Swedish West Coast[a]

Habitat or dwelling place	Relative frequency[b]
A. Rocky bottom	
1. Unsheltered on the bottom	0–1
2. In hollows or crevices	1–2
3. Among and under boulders and stones on bedrock	2–3
4. In excavated hollows or crevices in rocks containing soft bottom material	1–2
B. Mixed bottom	
1. Unsheltered on the bottom	1
2. In natural hollows under boulders and stones, etc.	4
3. In excavated burrows under boulders and stones, etc.	4–5
C. Soft bottom	
1. Unsheltered on the bottom	0–1
2. In excavated burrows or tunnels	0–1

[a] From Dybern (1973), with permission.
[b] Data based on observation of 260 lobsters and their associated habitat and shelter. 0, Very low; 5, very high frequency.

Information on the habitat, substrate, and shelter for offshore populations of the European lobster is not available.

Habitat types A-4, B-3, and C-2 (Table I) are the three types of dwelling places where lobsters have actively improved on natural conditions through burrowing (Dybern, 1973). The type B-3 dwelling, which the lobster most frequently occupies, is generally found at depths where the algal vegetation is less dense than in shoal waters. Lobsters seem to avoid shelter where algae excessively conceals the opening. Many shelters face downward on a sloping bottom. Dybern *et al.* (1967) and Dybern (1973) suggest that burrows are used not only as hiding places but also as lookouts, thus the general habit of the lobster to lie halfway out of its burrow. Burrows may have one, two, or three openings. The ratio of single to multiple entrance burrows is probably 4 or 5:1, similar to that of American lobster burrows.

3. *Burrowing American and European Lobsters*

Burrowing behavior of *Homarus* has been described by Cobb (1971, 1977), Berrill and Stewart (1973), Berrill (1974), Dybern *et al.* (1967), and Dybern (1973). In general, this behavior is similar for the American and European lobsters.

3. Ecology of Juvenile and Adult *Homarus*

Newly settled stage IV lobsters are very proficient at burrowing rapidly into soft substrates under objects such as stones or shells. Aquarium studies have demonstrated that most or all of the newly settled stage IV larvae burrowed and took shelter in the burrows before molting into stage V. These newly constructed burrows can be U-shaped, or have two openings 180° apart at opposite edges of a stone. Tunnels or shelters are excavated by one or a combination of several methods: "bulldozing" or plowing, i.e., making a scoop of its third maxillipeds and second, third, and occasionally fourth pereiopods; "backward digging," or scooping material back through the last pair of legs followed by "pleopod fanning" of the swimmerets to force material from the shelter; and "tail carry," where material is scooped up with the tail fan and carried to the burrow entrance while bulldozing (Cobb, 1971; Berrill, 1974).

Adult lobsters demonstrate the same methods of burrowing. These methods are used in differing degrees depending on the nature of the substrate, i.e., mud, sand, gravel, and shell fragments. During the winter, lobsters may take shelter in a mud burrow and close off the entrance with a partition of sediment and debris, remaining enclosed in this shelter for weeks or months at a time (Stewart, 1972; Thomas, 1968).

Figure 10A (Dybern, 1973) shows a longitudinal section of a typical European lobster burrow in soft substrate under a boulder with one opening. A wall of excavated bottom material is generally in front of the entrance, being most conspicuous in front of recently excavated burrows. Movements into and out of

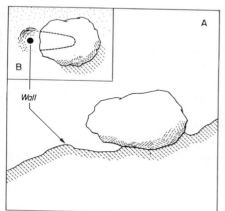
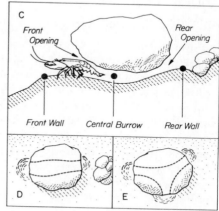

Fig. 10. Schematic of burrow types for the European lobster. (A) longitudinal section of typical burrow with one opening, (B) semicircular shaped walls of burrow, (C) longitudinal section of a burrow with two openings, (D) extension of burrow under rock, and (E) burrow with three openings. (From Dybern, 1973, with permission.)

the shelter tend to level off the mound. Where bottom configuration is not limiting, the walls of the burrow are generally semicircular in shape (Fig. 10B). The lobster typically rests against the inner wall with its antennae directed toward the entrance. The height of the burrow entrance is usually less than the width, and the inner portion is enlarged into a chamber (Cobb, 1971; Dybern, 1973).

Figure 10C (Dybern, 1973) shows a longitudinal section of a burrow with two openings, and Fig. 10D shows the extension of the burrow under a boulder. Generally, there is a larger front opening and a smaller rear opening, where the lobster can escape when disturbed. Typically, a small wall or other type of construction is found in front of the rear opening (Cobb, 1971; Dybern, 1973). Relatively few burrows have a third opening (Fig. 10E).

Burrowing of American lobsters in estuarine mud has been described by Thomas (1968) off Prince Edward Island and Cooper *et al.* (1975) in the Gulf of Maine. Thomas reported lobster burrows at depths of 1.3–4 m, sloping into the bottom at about 15° to the horizontal. These were roughly circular in cross section, having a diameter about twice that of the lobster. Burrows were spaced 2 m apart and extended into the mud substrate more than 30 cm. Most lobsters were facing the opening, but a few lacking claws were oriented head in and tail out. Many were covered with silt indicating negligible activity. Cooper reported lobster burrows at depths of 30–60 m on a mud substrate in the Sheepscot Estuary, Gulf of Maine. These burrows were circular and extended into the substrate 60–80 cm at a relatively steep angle. One adult lobster was observed "bulldozing" mud from its burrow. In contrast to burrows developed in coarse-grained substrates, little or no mounding persists at the burrow entrance.

4. Shelter Occupancy by American and European Lobsters

Inshore American lobsters generally choose or excavate a shelter that is dark and of sufficient size to enable the lobster to maintain physical contact with the walls and roof of the shelter. Darkness appears to be the major criterion of shelter selection. Using lobsters held under laboratory conditions, Cobb (1971) demonstrated that "negative phototaxis was more important than positive thigmotaxis in a lobster's choice of shelter." Cooper and Uzmann (1977) found that during the warm summer and fall months (July–September) this criterion of shelter selection partially breaks down. Lobsters will occasionally take up shelter during daylight hours in open craters on sand or mud substrates and under the overhanging fronds of macroscopic algae. This may be a result of a greater proportion of the day being spent foraging for food and less time spent in their primary shelters.

Lobsters of the inshore population are basically solitary. Multiple occupancy of a given shelter is rarely observed and occurs only during the coldest period of the winter (January–March) when the bottom water temperatures are $-1.8°$–

1.0°C. At this time of the year, at depths of 10–30 m, 2–4% of the shelters examined along the coast of Maine on a sand/rock substrate had two and occasionally three lobsters of 40–90 mm CL occupying the same shelter (Cooper et al., 1975). Cohabitants always varied considerably in size.

Sheehy (1976) described rates of occupancy of single- and triple-chambered shelters constructed of pumice concrete and placed in 6–7 m water off southern Rhode Island in areas devoid of natural shelter. These shelters were occupied throughout a 20-month period by juvenile and adult lobsters. The mean number of lobsters per shelter for the first 12 months of the study was 1.25 for single and 2.12 for triple chamber shelters. Several lobsters would make separate burrows under different parts of the shelter. Such multiple occupancy reached a maximum in late December and early January, with a second peak in July and August. Multiple occupancy exceeded 50% in one of the single-chamber surveys and consistently exceeded 65% for the triple-chamber surveys. These rates of occupancy are probably atypical for natural lobster habitat. "The frequency of multiple occupancy in the single chamber shelters suggests that the critical nearest-neighbor distance for juvenile lobsters was not reached and probably lies within the dimensions of the shelter itself" (Sheehy, 1976). The relatively high densities of juvenile lobsters ($1-20/m^2$) on the coast of Maine, reported by Cooper and Uzmann (1977), also suggest that the nearest-neighbor distance is considerably less than the 32.5 cm entrance width of Sheehy's single-chamber shelters.

Offshore American lobsters tend to choose or excavate shelters that are relatively dark and of sufficient size to enable the individual to maintain physical contact with the boundaries of the shelter (Cooper and Uzmann, 1977). However, with the relative scarcity of suitable substrate for excavation and solid objects associated with them, the shelters chosen have a wider range of physical characteristics, i.e., the bowl-shaped depressions and anemone forests described previously. Many of these shelters do not permit seclusion of the lobster.

Lobsters of the offshore population are more social in behavior than their shallow-water counterparts, perhaps due to the relative scarcity of shelters. Frequently, several lobsters are located at the bottom of depressions, in close contact with one another. In areas of suitable shelter, such as the excavated walls of submarine canyons and dense boulder fields, multiple occupancy of shelters is less frequently observed.

Man-made objects, such as 120- and 200-liter (30- and 50-gallon) oil drums and paint cans have become a part of the lobster's deep water environment. Some of the largest lobsters observed occupied drums missing one or both ends. These drums were usually occupied by a single large lobster with an occasional smaller lobster occupying a burrow beneath the drum. Ghost pots (lost pots) or discarded lobster traps are frequently seen in areas of lobster fishing and effectively serve to increase the number of shelters available to the lobster. Most of these pots have

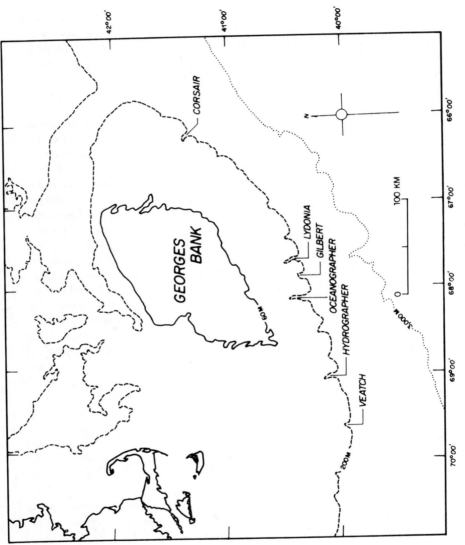

Fig. 11. Georges Bank submarine canyons.

been biodegraded to the point that they do not result in entrapment. Discarded sections of fishing nets and lengths of steel towing cable are occasionally seen in depressions and usually have one or more lobsters in occupancy.

The European lobster selects or excavates a shelter with basically the same dimensions and characteristics as the American lobster. Multiple occupancy of shelters has not been described, suggesting that the European lobster is a more solitary animal. Similarly, the availability or effectiveness of man-made objects as shelter for the lobster has not been described.

5. Population Density of the American Lobster

In the Gulf of Maine, inshore sandy substrates with overlying flattened rocks support the greatest concentrations and biomass of lobsters, juveniles and adults combined (3.25 lobsters/m^2, average CL=40 mm, total weight = 178 g/m^2). Average density and biomass for the sand/rock habitat, which is prime lobster fishing ground throughout the year, is 1.2 lobsters/m^2 or 63 g/m^2 (Cooper et al., 1975; Cooper and Uzmann, 1977). Scarratt (1968, 1973) reported a mean of 12.6 g/m^2 on "good natural lobster grounds" (i.e., sand/rock substrate) in Northumberland Strait, Gulf of St. Lawrence. Scarratt (1972), sampling Irish moss beds from Prince Edward Island on the Canadian Atlantic coast, found lobster densities of 0.007 to 0.11/m^2, with the lowest densities on smooth, flat bedrock ledges and the highest densities on boulder-rock substrates. Cobb (1971) reported densities of 0.1 lobsters/m^2 on bedrock and 0.15/m^2 on a mud-shell/rock substrate off Rhode Island. In Long Island Sound, monthly biomass estimates ranged from 5.6 to 24.3 g/m^2, with a 2-year average of 15.6 gms/m^2 for a mud/rock substrate (Lund et al., 1971, 1973; Stewart, 1972). Mud substrates mixed with solid objects (rocks, stones, man-made objects, etc.) at depths of 5-15 m in harbors and estuaries on the Maine coast support densities of small juveniles (stages IV-X) of up to 20/m^2 (Cooper and Uzmann, 1977).

Estimates of offshore lobster density have been made during the NMFS submersible dives from Corsair to Veatch Canyons (Fig. 11), over a depth range of 100-700 m. Average density was 0.001 lobsters/m^2 (or 0.085 g/m^2) for the open shelf and upper slope environments and 0.005 lobsters/m^2 (or 4.25 g/m^2) for the submarine canyon environments. These estimates of lobster abundance were made during June, July, and August when an estimated 30-50% of the offshore population had migrated into shoal (10-175 m) water (Cooper and Uzmann, 1971; Uzmann et al., 1977b). All lobster habitats were assessed from submersibles and probably represent minimal estimates of abundance. Comparable estimates of abundance for the European lobster are not available.

C. Effects of Environmental Factors on Distribution

The geographic distributions of *H. americanus* and *H. gammarus* (Fig. 1), like all marine aquatic organisms, are broadly determined by their physiological tolerance to the major environmental variables of temperature, salinity, and dissolved oxygen. Currents, light, and hydrostatic pressure may affect the behavior and successful metamorphosis of the planktonic larvae. An acceptable regime must also include suitable habitat for benthic stages, as well as food sources for both larval and adult forms on a seasonal and perennial basis. Local distributions and their permanence are ultimately determined by interspecific competition for the available food and niche space, and the scale of predator-prey interactions.

Modern man has affected the historical balance of the evolutionary determinants of distribution through excessive predation (fishing) and gross alteration of large areas of the coastal marine environment. Heavy metals, pesticides, pulp mill effluents, petroleum hydrocarbons, and other pollutants have been shown to have adverse effects on lobster physiology and behavior. Discussion of these modern environmental constraints is beyond the scope of this chapter.

Extensive experiments by McLeese (1956) with pre-adult American lobsters demonstrated that the effects of temperature, salinity, and oxygen are broadly interdependent. In addition, tolerance to extremes of any of these factors singly or in combination was significantly enhanced by prior acclimation or physiological conditioning at sublethal levels of the factor being tested. Given realistic acclimation such as might be expected within natural fluctuations of the temperature cycle, pre-adult lobsters can tolerate temperatures from 1.8° to 30.5°C at optimal oxygen concentration and salinity. Tolerance to both high and low temperature was positively correlated with acclimation temperature. Temperature, salinity, and oxygen acclimation all had highly significant effects on the upper lethal temperature range, which was raised as acclimation temperature was raised, but lowered as salinity and oxygen levels were lowered.

Lobsters are surprisingly tolerant to low oxygen concentrations, without regard to the level of oxygen acclimation. McLeese (1956) found that prior oxygen concentration did not have a significant effect on lethal oxygen concentration, but that temperature and salinity acclimation were highly significant variables. He showed that the lethal oxygen level was raised by an increase in temperature acclimation and also by a decrease in the level of salinity acclimation. The range of tolerance to low dissolved oxygen was from 1.72 mg/liter in lobsters acclimated at 25°C and 20 °/oo salinity (high temperature, low salinity) to 0.20 mg/liter in lobsters acclimated at 5°C and 30 °/oo salinity (low temperature, high salinity).

Tolerance to low salinity varied with acclimation temperature, salinity, and dissolved oxygen concentration. Lobsters acclimated at 25°C, 30 °/oo salinity, and 6.4 mg/liter oxygen (high temperature, high salinity, and high dissolved oxygen) tolerated 16.4 °/oo, while those acclimated at 5°C, 30 °/oo salinity, and 6.4 mg/liter oxygen (low temperature, high salinity, high oxygen) tolerated as low as 6.0 °/oo.

III. POPULATION STRUCTURE

A. Size Composition and Sex Ratio

1. American Lobster

Extensive sampling of the inshore catches of lobsters by American and Canadian scientists have been conducted over the past 20 years. These catch statistics indicate striking regional differences in the sizes of lobsters caught in the inshore waters of Canada and in those of the United States. Lobsters in the southern Gulf of St. Lawrence are smaller than those caught elsewhere; even within the Gulf there are marked differences between fishing grounds separated by only 20 miles (Chouinard et al., 1975). Similar regional differences occur between the Gulf of Maine, Cape Cod Bay, Rhode Island Sound, Long Island Sound, and Long Island's south shore (Wilder, 1954; Cooper, 1970; Lund et al., 1971, 1973; Krouse, 1973; Thomas, 1973; Cooper et al., 1975; Krouse and Thomas, 1975; Pecci et al., 1978). These size differentials represent a range in mean carapace length of 65–85 mm.

The differences in mean size of inshore commercial catches or regional fisheries are probably a function of (1) varying degrees of trap selectivity, (2) different levels of fishing pressure, (3) seasonal influx of significantly larger lobsters migrating onshore from the offshore population (Cooper, 1970; Uzmann et al., 1977a), and (4) shelter sizes and physical habitat characteristics (Howard, 1977), rather than actual differences in growth. *In situ* collections by divers from the central coast of Maine yielded over 2000 lobsters ranging from 5 to 90 mm CL (Fig. 12), with an average CL of 42 mm (Cooper et al., 1975). Mean size was similar, season by season, over a depth range of 6–24 m. Similar collections from the Gulf of St. Lawrence (Scarratt, 1968) yielded mean carapace lengths of 40–42 mm. *In situ* collections from Long Island Sound (Lund et al., 1971, 1973; Stewart, 1972) gave similar results, while our diver and commercial trap collections (unpublished data) from the south shore of Long Island averaged 100–105 mm CL, representing a high percentage influx of large onshore migrants. From a depth of 30–60 m, in estuaries on the central coast of Maine, the size range was 85–100 mm (Cooper et al., 1975), suggesting that small juveniles of the inshore population do not frequent the deeper depths on mud substrates. The diver-collected samples are considered representative of the natural population.

Sex ratios of the diver-collected inshore samples demonstrate equal numbers of males and females up to at least 80 mm CL (Scarratt, 1968; Cooper, 1970; Stewart, 1972; Cooper et al., 1975). Commercial catches and experimental trap catches substantiate the 1:1 ratio for lobsters of 90 mm and smaller CL (Krouse, 1973; Pecci et al., 1978). For lobsters larger than 90 mm CL, there are insufficient *in situ* collections to generalize on sex ratio.

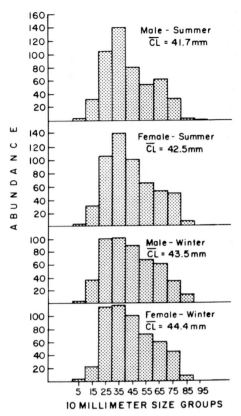

Fig. 12. Carapace length frequency distributions and mean carapace lengths of male and female lobsters sampled by divers during September 1967 and February 1968, coast of Maine. (From Cooper et al., 1975.)

Skud and Perkins (1969) and Skud (1969) present data on the size composition and sex ratio of offshore lobsters (research catches) from five submarine canyon fishing grounds. They concluded that the canyon fishing grounds closest to shore are the most heavily fished commercially. There, the catch per unit effort (CPUE) was relatively low, lobsters were small, and total mortality rates were highest. The canyons farthest from shore supported the largest lobsters and the highest CPUE. Mean carapace length at Veatch Canyon (200 km from shore) was 120 mm in 1956 and less than 90 mm in 1967. In three consecutive years of fishing, the contribution of small lobsters had increased to where 90% of the legal-sized lobsters in 1967 were between 80 and 100 mm CL. The mean size of lobster from Lydonia and Oceanographer Canyons (250–300 km from shore) was

125–130 mm, while lobsters from Corsair Canyon (400 km from shore) averaged 150 mm. Since 1967 these mean sizes have decreased (Uzmann et al., 1977a). For lobsters smaller than 80 mm CL, the ratio of females to males was about 1:1. This ratio increased gradually through a CL range of 80 to 130 mm and then declined (Skud and Perkins, 1969). Since females molt less frequently than males, large females are less frequently observed.

In situ observations of offshore lobsters in these canyons by Cooper and Uzmann (1977) suggest that both juveniles and adults are distributed throughout the 100–450 m depth range. Males and females are unevenly distributed during periods of migration, and lobster size distributions by canyon are generally similar, with the possible exception of Corsair Canyon (easternmost United States submarine canyon). We feel that surface-operated catch gear (otter trawl) may not provide representative samples of the lobster population (Uzmann et al., 1977a). Thus, the above statistics on sex ratio and mean size might be suspect.

2. European Lobster

In situ collections of the European lobster by divers and a definition of their size composition and sex ratio have not been reported, with the exception of small samples collected by Dybern et al. (1967) off the Swedish west coast in July and August 1966. Sex ratio of 59 lobsters, collected during daytime and nighttime dives, was 1:1. Individual lobster sizes and a sample mean size was not presented. Therefore, a comparison of size composition for American and European lobster populations is not possible.

In the absence of diver-collected lobster samples, the nature of size and sex compositions are best acquired through analysis of trap or creel samples, realizing that traps are probably selective by lobster size and by sex versus time of year. Mean lobster sizes and sex ratios from trap catches along the Irish and Scottish coasts have been reported by Gibson (1967, 1971), Thomas (1954, 1957, 1962, 1969), and Watson (1974). The size composition of the commercial catch depends, in part, on the fishing grounds. In good weather fishermen generally fish the more exposed areas where the lobsters are larger (Thomas, 1958). It is difficult to distinguish this seasonal characteristic of the fishery from the effect of the winter decrease in catch of the larger lobsters that are carrying eggs in an advanced stage of development. The latter results in the mean size of females of the catch being greater in the summer than in the winter. Mean carapace length of males and females from the coast of Scotland from 1966 through 1969 was highest (104–109 mm) in May, June, and July, and lowest (97–100 mm) in September, October, and November. Sex ratio varied according to size. In the 80–89 mm CL group, females, which were mostly immature, comprised 52% of the catch as a yearly average. At sexual maturity (90–109 mm) females comprised 60% of the catch, then decreased progressively to 39% in the 140–149 mm CL group. From samples collected throughout the year for 1934–1938, females predominated during April through June, comprising 60-64% of the catch and were least abundant during September and October, comprising 46-50% of the catch (Thomas, 1954).

Along the northern coast of Ireland the retainable commercial catches from traditional shallow water grounds were compared with those from the less traditional deep water grounds (Watson, 1974). Males and females averaged 93-94 and 94-97mm, respectively, for deep water grounds and 85 and 85-88 mm for shallow water grounds from catches taken during February through October during 1972 and 1973. Females comprised 51-61% of the catches.

B. Growth—American Lobster

The subject of the physiology of molting and growth of the lobster is covered in detail in Chapter 2 of Volume I. In addition to this information, we shall (1) present newly acquired results of growth studies on the offshore lobster (Cooper and Uzmann, 1977), (2) compare their growth with the growth of inshore lobsters, and (3) discuss the significance of their growth patterns with the very distinctive ecological characteristics of inshore and offshore populations.

An extensive mark and recapture study, using a long-term tag (Cooper, 1970), provided growth data for lobsters from 55 to 180 mm CL over a time interval of up to 4 years and 6 months from release to recapture. Comparable growth data for smaller lobsters do not exist. The recovery of several hundred lobsters with precise data on time of recapture and carapace length has permitted an estimation of annual probability of molting, regardless of number of molts, as a function of size at time of release and the average annual increase in size at time of release. These relationships were defined separately for males and females of the offshore population. The probability of molting was determined as follows:

$$\text{Males} \quad M = 125.975 - (0.489)(CL)$$
$$\text{Females} \quad M = 128.524 - (0.544)(CL)$$

where M is the probability of molting per year and CL is the carapace length (mm) at release. The average growth increment per year was

$$\text{Males} \quad G = 25.937 - (0.091)(CL)$$
$$\text{Females} \quad G = 23.347 - (0.094)(CL)$$

Where G is the average growth increment (mm) per year for lobsters demonstrating growth, and CL is the carapace length at release. L_∞ values (size of zero growth) of 270 mm CL (males) and 240 mm CL (females), were estimated from an examination of length data for 60,000 lobsters and used in the computation of the above linear regressions.

The probability of molting for any given year multiplied by the expected annual increase in length, regardless of the number of molts, equals the average

3. Ecology of Juvenile and Adult *Homarus*

increase in carapace length for all males or females of a given year. To accurately scale the time axis so that length at age values can be computed, the age of lobsters at 60 mm CL (smallest size of recaptured lobsters) was set at 3 years for males and 4 years for females (Cooper and Uzmann, 1977). Thus length at specific ages can be computed. The age of lobsters at 60 mm CL (smallest size of recaptured lobsters) was set at 3 years for males and 4 years for females (Cooper and Uzmann, 1977). Growth curves for male and female offshore lobsters were computed (Fig. 13) and the corresponding von Bertalanffy growth parameters that best described these specific curves were then calculated, using the von Bertalanffy growth equation:

$$l_t = L_\infty [1-e^{-K(t-t_0)}]$$

where, t is the age (years), l_t is the carapace length (mm) at age t, L_∞ is the maximum carapace length (mm) attained, K is the rate at which l_t approaches the asymptote (L_∞), and t_0 is the theoretical age (years) at which l_t equals zero.

The following equations characterize growth of the offshore American lobster reasonably well:

Males $\quad l_t = 270 [1-e^{-0.096(t-0.50)}]$
Females $\quad l_t = 240 [1-e^{-0.074(t-0.30)}]$

Thomas (1973), using size frequency modes, estimated von Bertalanffy growth parameters for the inshore Gulf of Maine lobster. The pertinent parameter estimates are $L_\infty = 266.770$, $K = 0.048$, and $t_0 = -0.772$. Cooper (1970) presented growth increment data for inshore lobsters taken from their natural environment in the Gulf of Maine. Increase in carapace length for 90 mm lobsters ranged from 12 to 13%. Growth for males and females was similar. In contrast, offshore lobsters 90 mm in carapace length increased 19 and 17% for males and females, respectively (Cooper and Uzmann, 1971). The frequency of shedding for inshore lobsters of a given size is less than that for offshore lobsters. Thus, the rate of growth for inshore males and females is significantly less than for their deep water counterparts.

These significant growth differences are probably directly related to the differences in ecology of inshore versus offshore lobsters. Cooper and Uzmann (1970) and Uzmann *et al.* (1977a) hypothesize that at least 20% and probably 30–50% of the offshore population migrates seasonally to maintain a temperature regime of 8°–14°C (Fig. 14). In contrast, the inshore population is subjected to temperatures considerably colder than 8°C for at least half of the year. Direct observations of the nocturnal activities of lobsters and their foraging for food (Cooper *et al.*, 1975; Pecci *et al.*, 1978) showed reduced rates of nocturnal activity and feeding at temperatures below 4°–5°C. Lund *et al.* (1971, 1973) and Stewart (1972) demonstrated a similar change in nocturnal activity and feeding through analyses of lobster stomachs. Thus, it seems logical that the inshore

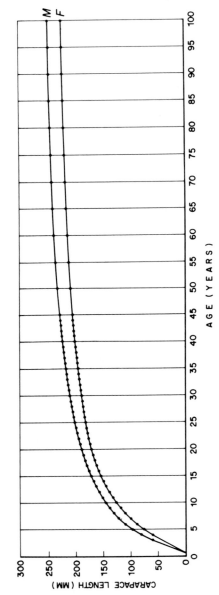

Fig. 13. Calculated CL (mm) at age (years) for offshore American lobsters, males and females. Growth curves are scaled to age 3 for males and age 4 for females at CL of 60 mm.

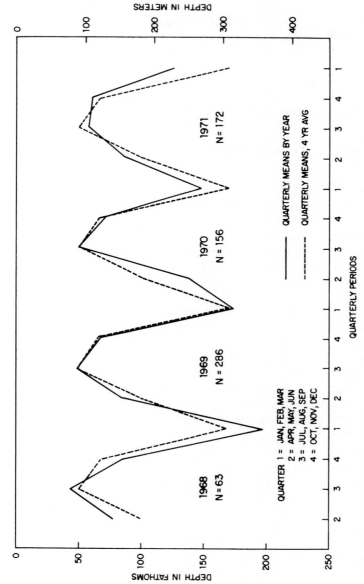

Fig. 14. Mean depths of recapture for tagged American lobsters by quarterly periods, from April 1968 through March 1972.

lobster shows less annual growth due to a period of relative inactivity during the colder months of the year. In comparison, the offshore lobster feeds and grows during most of the year. Some of the largest offshore trap catches are composed of newly shed lobsters during the winter months (personal communication with commercial fishermen). Shedding inshore during the winter is virtually nonexistent.

IV. BEHAVIOR AND ACTIVITY RHYTHMS

A. Migrations and Home Range

1. American Lobster

American lobsters of the inshore population(s) are generally endemic (Templeman, 1935, 1940; Wilder and Murray, 1956; Wilder, 1963; Zeitlin-Hale, 1963; Cooper, 1970; Cooper et al., 1975). Once a lobster has left the planktonic stage to begin the benthic stage of its life, horizontal movements are probably restricted to several meters or less. Juvenile lobster, ≥ 45 mm CL, leave shelter sites at night, traversing distances less than 2 km and usually less than 300 m.

Inshore lobsters typically move from shoal water (5–20 m) to deep water (30–60 m) when strong winds generate heavy seas and considerable vertical turbulence at the substrate–water interface (Cooper et al., 1975). These movements are irregular in timing, but associated with storms, and usually represent horizontal distances of 100 m or less.

Large lobsters, 100 mm CL and larger, of the inshore population, may represent an exception to the endemic rule. A small percentage of the large lobsters tagged and released along the Maine coast have demonstrated extended migrations along the coast in a southwesterly direction for distances in excess of 270 km (Dow, 1974).

Tagging studies by Cooper and Uzmann (1971) and Uzmann et al. (1977a) at various inshore and offshore locations, by Lund et al. (1971), Lund and Rathbun (1973), and Stewart (1972) in Long Island Sound, and by Morrissey (1971) off Cape Cod have demonstrated a mixture of local, endemic stocks with onshore migrants from the offshore population(s). Mixing takes place from May to September in coastal areas ranging from southern New York to Cape Cod Bay. Recaptures from these studies clearly characterize endemic stocks from migratory stocks.

In general, inshore lobsters exhibit a rather limited home range. The extent of daily or weekly movements is established during the night when foraging for food occurs. Dispersal from the home shelter generally does not exceed 300 m, with a very small percentage ranging up to 2 km (Lund and Rathbun, 1973;

3. Ecology of Juvenile and Adult *Homarus*

Stewart, 1972; Cooper and Uzmann, 1977). These nighttime excursions are more frequent and extensive during the summer and fall than in the winter and spring. Lobsters exhibiting these nightly excursions tend to return to the same or nearby shelter on consecutive nights (Cooper, 1970; Cooper *et al.*, 1975; Stewart, 1972). Comparable information on the home range of the offshore lobster does not exist.

The migratory behavior and dispersions of tagged American lobsters from the offshore population of the New England Continental Shelf has been defined by Cooper and Uzmann (1971) and Uzmann *et al.* (1977a), based on the distribution of tag returns from a 4-year tagging and recapture study, utilizing a long-term tag (Cooper, 1970). Figure 15 demonstrates migration and dispersion patterns of tagged lobsters from Veatch Canyon, one of 29 release stations. This migratory behavior is probably elicited by temperature, since the seasonal distribution of tagged lobsters (Fig. 14) is positively correlated with bottom temperature. Through random and (or) directed movements, the offshore population maintains itself within a temperature range of 8°–14°C. We believe that the continental slope habitat lacks sufficiently high temperatures during the summer to optimize extrusion of eggs, molting, and subsequent mating. Thus, the temperature deficiency is compensated for by seasonal shoalward migration to warmer water. *In situ* observations of offshore lobsters at Corsair, Lydonia, Oceanographer, Hydrographer, and Veatch Canyons during June–July substantiate this belief. Evidence of lobster molting was observed only at depths less than 200 m, whereas lobsters were distributed to depths of at least 700 m.

The magnitude of variation in depth at recapture by month suggests that the migration toward shoal water is not a total, well-coordinated population response. Cooper and Uzmann (1971) hypothesize that some lobsters migrate

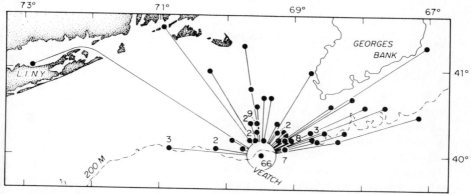

Fig. 15. Migrations and dispersions of tagged American lobsters from Veatch Canyon. (From Uzmann *et al.*, 1977a.)

early, some late, and some not at all. Superimposed upon these variations in migratory behavior is an apparent tendency of some lobsters to move laterally east or west along the outer shelf and upper slope. Hence, it is unlikely that there are discrete populations in each canyon.

For those lobsters defined as migrants (Uzmann et al., 1977a) the average speed of travel was 1.8 km per day, with a maximum speed of 11 km per day. Maximum one way distance traveled by a recaptured tagged lobster was 335 km in 71 days, or a hypothetical round trip migration of approximately 670 km. Lobsters ranging in size from 65 to 160 mm CL demonstrated extensive migrations into shoal water.

Saila and Flowers (1968) displaced berried female American lobsters from Veatch Canyon inshore to Narragansett Bay, Rhode Island. The displaced tagged lobsters remained within the bay until their eggs had hatched and then moved offshore to the general area of original capture. Lund et al. (1971) and Stewart (1972) conducted a tag and recapture program with the American lobster in Long Island Sound, off Connecticut. They demonstrated a net easterly movement out of Long Island and into Block Island Sound. The few lobsters that returned to the edge of the Continental Shelf they assumed to be previous onshore migrants. These studies support the findings of Cooper and Uzmann (1971) and Uzmann et al. (1977a).

2. European Lobster

Marking and recapture experiments off Ireland, Scotland, and Norway suggest that the European lobster does not undertake extensive migrations alongshore or inshore-offshore as has been described for the offshore American lobster (Thomas, 1954; Gibson, 1967; Gundersen, 1969; Watson, 1974). Maximum distances traveled were 8–12 km, with an average of approximately 2 km. Information on the home range of *H. gammarus* is not available.

B. Nocturnal Behavior and Predation

1. European Lobster

Weiss (1970), Cooper and Uzmann (1977), and Pecci et al. (1978), using scuba, described the nocturnal behavior of the American lobster of the inshore population from Long Island Sound and the coast of Maine. Juvenile lobsters became nocturnally active (left their shelters) at approximately 45 mm CL. Individuals became "restless" 1–2 hr before sunset, moving back and forth in the entrance to their shelter. Within the first hour after sunset, 10% or more of the lobsters had left their shelters. Underwater light intensity when the lobsters first emerged was less than 2×10^{-2} μW/cm^2 in June through November and 0.02×10^{-2} μW/cm^2 in January and February (Weiss, 1970). By the fourth to fifth hour of darkness 80-90% were out of their shelters in July and 40-50% in February. The time associated with the return of the lobsters to their shelters was highly variable, but occurred at about the same light intensity. These findings are supported by Lund et al. (1971), Lund and Rathbun (1973), and Stewart (1972). The cold water of the winter ($-2°-1°C$) appears to inhibit nocturnal activity. Feeding is minimal in the winter compared to an active foraging behavior for food in the summer.

Stewart (1972) noted considerable activity of the American lobster in murky water at 20-30 m depths in Long Island Sound. Our observations in the Gulf of Maine indicate that lobsters first emerge from their shelters 1-3 hr before sunset when the sky is overcast and/or when the moon phase is new. Chapman et al. (1975) constructed a hypothetical series of curves to relate expected activity of *Nephrops norvegicus* with water depth and light intensity, suggesting a single daylight peak in activity at the greatest depths. This same relationship may exist with the American lobster and include the environmental variables of water clarity and amount of incident light reaching the sea-air interface. Laboratory measured locomotor activity, as reported by Cobb (1969), Krekorian et al. (1974), and Zeitlin-Hale (1975) for the American lobster, tends to be highest at sunset and sunrise with the peak activity at sunset.

Juveniles less than 40 mm CL appear to spend their entire time within their tunnel system or shelter (Cooper and Uzmann, 1977). The walls and roofs of these tunnels are generally heavily encrusted with soft bodied invertebrates which probably comprise the major source of food for these juveniles. When individuals reach the 35-40 mm CL size they become partially active at night, moving back and forth within the entrance to the shelter and occasionally just outside the entrance. Most of the observed predation on lobsters occurred at this time, on individuals within this size range. Lobsters 50 mm CL and larger appear to be difficult prey for most of the predators observed on the lobster grounds. The Atlantic cod, the wolffish, the goosefish, and several species of shark can and do consume lobsters up to about 100 mm CL, but the magnitude of predation by these relatively large animals is considered minimal. The major predators on the juvenile lobster are sculpin, cunner, tautog, black sea bass, and sea raven, (Cooper and Uzmann, 1977). The nature of predation on lobsters less than 35-40 mm CL is not known.

> The defensive posture assumed by lobsters is common in the clawed crustacea; the claws are raised in the meral spread posture and the lobster stands high on its legs. If a lobster outside its shelter is pursued by a diver it will bring the large claws together in front of the body forming a streamline shape and flex the abdomen forcefully several times, moving rapidly backwards through the water (Cobb, 1977).

Our observations indicate that the lobster will not swim a distance greater than 10-15 m, at which point it will stand its ground and defend itself against the diver with its large claws.

The acts of predation by a longhorn sculpin, on a surface-released 50 mm CL lobster and a 30–35 mm CL lobster at the entrance to its shelter were observed during July and September, respectively, at depths of 15 and 20 m on the coast of Maine (Cooper and Uzmann). The first observation dealt with a juvenile male lobster released at the surface by a commercial trap fisherman. The lobster swam to the bedrock bottom (15 m depth) and immediately began walking toward a group of large boulders. Time of day was 11 AM (EST) and the horizontal visibility was approximately 10 m. Within 10 sec after the lobster reached the bedrock substrate a longhorn sculpin, approximately 30 cm total length, rapidly swam to within 1 m of the lobster and stopped. It then spread or extended its pectoral fins and opercles in an outward direction while facing the lobster. The lobster turned to face the sculpin with its antennae directed toward the predator and its large claws in a defensive position. The sculpin made two quick approaches toward the lobster, attempting to come in to one side of the large claws. On both occasions the lobster grabbed at the sculpin with its pincher claw and the sculpin retreated. After the second attempt the sculpin swam from the immediate area. The second observation was made at sunset on a sand-rock-boulder substrate 20 m deep. A juvenile lobster was demonstrating a "restless" mode of nocturnal behavior, moving in and out of its shelter that projected into a sand substrate beneath a large rock. At the entrance to the shelter was a large mound of sand and shell fragments. A longhorn sculpin approximately 35 cm total length approached the shelter to within 1.5 m of the entrance mound. The sculpin remained motionless for several minutes watching the juvenile lobster move back and forth over the mound. When the lobster retreated into the shelter, the sculpin, screened by the mound of sediment, moved forward 10–15 cm using its pectoral fins until the lobster reappeared on the mound. This cycle repeated itself until the sculpin was within 8–10 cm of the front edge of the mound. As the lobster emerged onto the mound the sculpin lunged forward and caught the lobster head on so that only its abdomen projected from the sculpin's mouth.

Information on the nocturnal behavior of the offshore lobster is minimal. *In situ* observations from submersibles were made during daylight hours during June through August at depths of 100–500 m (Cooper and Uzmann, 1977). Throughout this depth range, most of the lobsters seen were within shelters. With all submersible lights off, observers can see general shapes of objects to distances of 10–30 m at depths between 100–300 m. At greater depths, available light drops off rapidly and horizontal visibility is greatly reduced. In July and August, the relative light levels offshore (100–300 m), corresponding to the period of maximum nocturnal activity of the inshore lobster and its associated light levels occur in the early morning (1–3 hr after sunrise, 7–9 AM, EST) and in the late afternoon (1–3 hr before sunset, 5–7 PM). It is not known if offshore lobsters are active during the darkest hours of the day. Commercial trap fishermen have reported good catches over an 8- to 10-hour period for both daytime and nighttime sets, suggesting that some portion of the population is out of their shelters and foraging for food at all times of the day. The relatively small percent of out-of-shelter daysightings by submersible divers and the daytime catches by trap fishermen could, however, be directed at migrants, which might confuse the diurnal periodicity issue.

Social conditions may effect lobster activity. Juvenile American lobsters held communally showed lower levels of activity than the same animals held individually. When these lobsters were provided with shelters in communal conditions their activity decreased further (Zeitlin-Hale, 1975). Krekorian *et al.* (1974) observed nearly seven times as much roaming (a measure of activity) by single adults without shelter as by those with shelter. With the relative scarcity of available shelters for lobsters on the offshore grounds compared to inshore grounds (Cooper and Uzmann, 1977), offshore lobsters may be more active and forage for food a greater portion of the day than their inshore counterparts.

Offshore lobsters are subject to predation by the same predator species that prey on the inshore lobsters, with the exception of black sea bass, cunner, tautog, and sea raven. Several species of sharks, the tilefish, and the conger eel are known predators of the deep water lobster.

2. European Lobster

Observations by Hallback and Waren (1972), Dybern (1973), and Berrill (1974) indicate that the European lobster hides during the day within its shelter and forages for food at night. There are no published data available on the timing of this foraging. Like the American lobster, a larger percentage (60–70%) of the population leaves their shelters during the summer and fall than during the winter (Hallback and Waren, 1972). Lobsters normally do not feed in the winter, but remain in their shelters when the water temperature falls below 5°C. When water visibility is poor at depths of 10–15 m, lobsters will leave their shelters during the daytime to forage for food. Lobsters leaving their shelters off the Swedish coast generally move into somewhat shallower water to feed, returning to deeper water during the daytime. Information on predators of the European lobster are not available, but it seems likely that the same predators or related species that occur in the northwest Atlantic also prey upon the European lobster in the northeast Atlantic and off the coasts of Norway, Sweden, Denmark, and Scotland.

C. Feeding Behavior and Diet

1. American Lobster

A description of the diet of the American lobster in Long Island Sound, New York (Weiss, 1970), and Bonavista Bay, Newfoundland (Ennis, 1973) are presented. Changes in food and feeding activity in relation to season, carapace length, sex, molting condition, and state of maturity of females are discussed. Composition of the diet is compared to the lipid content of the hepatopancreas (Weiss, 1970), and the blood serum protein concentration is used as an index of physiological condition (Ennis, 1973).

The American lobster is an omnivorous feeder and a predator. The majority of its natural food comprises a variety of bottom invertebrates, mainly crabs, polychaetes, mussels, periwinkles, sea urchins, and starfish (Herrick, 1911; Squires, 1970; Weiss, 1970; Miller *et al.*, 1971; Ennis, 1973). Fish, plants, hydroids, ascidians, and ectoprocts comprise a relatively minor part of the diet (Weiss, 1970). Weiss suggests that much of the ingested plant material is taken accidentally as lobsters sift through detritus to obtain small crabs and other invertebrates. Mather (1894) concluded that lobsters were incapable of digesting plant tissue.

The relative contribution of the various prey species to the diet of the American lobster varies considerably from one area to another. Lobster stomach contents generally reflect the relative abundance of prey species in the habitat (Weiss, 1970; Miller *et al.*, 1971; Ennis, 1973). The seasonal variation in the frequency of ingestion of many of the prey species appear to be related to their availability rather than to a shift in the diet preferences of the lobster (Weiss, 1970). However, Ennis (1973) reported:

> Although crabs made up the bulk of the food throughout the year, they made up a much smaller proportion of the total amount of food during July, August, and September than at other times of the year. During this period sea stars and sea urchins made up a much greater proportion than at other times. This change to what is probably a more calcium rich diet coincided with the molting season and it did not appear to reflect a relative change in abundance of the prey species. It may have resulted from active food selection by the lobsters.

Stomach fullness was considered by Weiss (1970) and Ennis (1973) as an adequate index of seasonal changes in feeding activity. Generally, the proportion of lobsters with empty stomachs was highest during January, February, and March, and lowest during July through November. The amount of contents in the lobster stomachs decreased throughout the intermolt cycle and most lobsters in the proecdysial stages had empty stomachs (Weiss, 1970). There were no significant differences in the diets of male and female lobsters. Ennis (1973) found no difference in diet between size groups, and Weiss (1970) found only minor differences between 3 and 10 cm CL groups.

Changes in the blood serum protein concentration were related to the molting cycle of the American lobster by Ennis (1973). From December through June (intermolt period) serum protein averaged 35–40 mg/ml and in August (premolt period) it averaged 60–65 mg/ml. During September and October (postmolt period) serum protein averaged 20–25 mg/ml. "Physiological recovery from molting was fairly rapid and the intermolt condition was reached by mid-

November about 2½ months after the peak molting period" (Ennis, 1973). To accomplish this recovery, lobsters maintain a high level of feeding activity throughout the fall. Females feed more than males into early winter, perhaps because of the greater physiological demands on the female that are due to gonad development.

Breen and Mann (1976) and Mann (1977) hypothesize that the American lobster exerts control through predation over the green sea urchin, which indirectly preserves the sheltering kelp beds. They note that when lobster populations are reduced below a given level, the sea urchins concentrate in the vicinity of kelp beds and destroy them through feeding. The lobsters, thus deprived of kelp cover, are exposed to increased predation. Conversely, Himmelman and Steele (1971) state that in Newfoundland coastal waters, predation by lobsters on the green sea urchin is not effective in controlling the abundance of the urchin. A 5:1 dietary preference of crabs over sea urchins by the lobster was demonstrated by Evans and Mann (1977). However, this may have been a result of the holding conditions in the tanks, which limited the escaping ability of the crabs. Hirtle and Mann (1978) demonstrated that lobsters are sensitive to the odor of live crabs and sea urchins and also to the sight of a moving crab or urchin. They conclude that a lobster presented with the choice of crab versus urchin will invariably choose the crab. Evans and Mann (1977) state that when urchins outnumber crabs by 10:1 or more, the lobster will eat two urchins for every crab, since the urchins are easier to catch. On the Atlantic coast of Nova Scotia the ratio of sea urchins to crabs is from 20:1 to 30:1 (Hirtle and Mann, 1978). Accordingly they hypothesize that sea urchins outnumber crabs in the diet of lobsters in spite of the lobsters' preference for crabs.

Information on the feeding of the offshore American lobster is not available. However, our direct observations of the offshore bottom fauna of the New England and Mid-Atlantic Bight habitats indicate an ample supply of crabs, polychaetes, starfish, hydroids, and fish, which could serve as food prey.

Feeding behavior of the American lobster has been described by Herrick (1911), Weiss (1970), and Smith (1976). Lobsters can dig clams out of the bottom and crack the shell using the crusher claw. They can obtain small crabs by sifting through clumps of algae where the crabs congregate. Lobsters can attack larger crabs, such as *Cancer irroratus*, by grasping a walking leg, removing many of the remaining legs, and crushing the cephalothorax (Weiss, 1970). Smith (1976) described a behavior of burying uneaten food near the burrow at the end of a feeding session. The greater the time since the last feeding, the greater the tendency to bury the food and at a closer distance to the shelter mound. Smith suggests that the burial behavior was a hoarding technique, preventing competitors from eating the food. The frequency of food burial varied with season, water temperature, and food type.

2. European Lobster

Hallback and Waren (1972) conducted diver investigations of the feeding behavior and diet of the European lobster off the west coast of Sweden. The major food items consumed were crabs, gastropods, and polychaetes, with mussels and starfish comprising a minor portion of the diet. During the molting season lobsters ate a lot of calcareous material. Berried females had the same feeding behavior and diet as other lobsters. Small lobsters preferred polychaetes, small crabs, and gastropods. The diet of males and females was similar as was the choice of food along the coast.

D. Dominance and Aggressive Behavior of the American Lobster

The American lobster demonstrates strong aggressive behavior when held under controlled laboratory conditions. Generally, with two or more individuals in the same aquarium, one lobster displays dominant behavior and the rest are subordinate. Molt state affects the probability of an individual being dominant (Tamm and Cobb, 1978). Juvenile lobsters in molt states D_1 and D_2 (hard-shelled) were dominant or equal in their agonistic encounter classification and became subordinate in molt states A, B, and D_3 (within 96 hr after molting and just before molting). Eight of the 10 acts of agonistic encounters varied as a result of molt state. Lobsters in postecdysis states A and B displayed evasive actions when encountering their state C (hard-shelled) opponents. "Fights were rarely initiated, never won, and, for state A animals, usually resulted in appendage loss and death" (Tamm and Cobb, 1978). Lobsters in states D_1 and D_2 were the most aggressive. These individuals maintained an exaggerated meral spread, a greatly elevated stance with the chelae held very high and widely separated. Tamm and Cobb hypothesize that the submissiveness of A, B, and D_3 lobsters is attributable to limited shell hardness. In the absence of shelter, aggressive behavior is accentuated. If the diet is insufficient, cannibalism occurs (Krekorian *et al.*, 1974; Van Olst *et al.*, 1975).

Interpretation of behavioral studies with the American lobster under controlled laboratory conditions is difficult. We have concluded from studies of the lobster in its natural environment that (1) inshore and offshore lobsters are rarely cannibalistic or aggressive toward one another, (2) in general, inshore and offshore lobsters enjoy an ample food supply, but abundant shelter is available only to the inshore species, and (3) displays of dominance do occur in the natural environment when lobsters are nocturnally active, but to a lesser degree than observed under seminatural conditions. Information on the dominance and aggressive behavior of the European lobster is not available.

E. Interspecific Relations of the American Lobster

The American lobster of the inshore population(s) lives during much or all of the year in close proximity to several species of macrobenthic fauna and certain bottom-oriented finfish. *In situ* observations of the interspecific relations between the lobster and these associated species have been made by diver scientists along the Canadian and New England coasts. The Jonah crab is probably the primary competitor for space with the lobster (Lund *et al.*, 1971; Lund and Rathbun, 1973; Stewart, 1972; Cooper and Uzmann, 1977). Jonah crab and lobster burrows are similar in shape and a given burrow is frequently occupied by a lobster one night and a Jonah crab the following night. The crab may displace or be displaced by the lobster, depending on their relative sizes (Fogarty, 1975). It may also exclude the lobster by passive displacement rather than active defense (Stewart, 1972).

Two species of bottom-oriented fish, tautog and cunner, frequent the rocky areas of inshore ocean bottom where lobsters occur. During the spring, summer, and fall these fish actively feed in the immediate vicinity of lobster shelters. In the winter when bottom water temperatures reach $-2°$ to $1°C$, both species of fish go into semihibernation, deeply penetrating unoccupied crevices, burrows or tunnels into the ocean floor at depths of 20 m or more (Cooper, 1966). These hiding places are similar to the shelters occupied by lobsters. In the winter, a cunner and tautog may occupy the same shelter as a lobster or may displace a lobster from its shelter (Lund *et al.*, 1971; Stewart, 1972).

Several other species of bottom-oriented finfish live in close proximity to inshore lobsters during major portions of the year. They frequently come within physical contact, or nearly so, at nighttime when both lobster and finfish are active. These species are the sculpin, sea raven, goosefish, eel pout, redfish, codfish, pollock, winter flounder, and the conger eel. The eel pout, redfish, cod, and conger eel are all burrow or crevice dwellers and compete with the American lobster for shelter space. Cohabitation of a given shelter space generally does not occur, although openings to shelter spaces may be less than 0.3 m apart.

Lobsters of the offshore population live in close proximity to at least 19 species of megabenthic crustaceans, bottom oriented fishes, and cephalopods (Cooper and Uzmann, 1977). Of these, there are eight species that compete with the lobster for shelter space, primarily in areas of rugged bottom topography. These are the Jonah crab, galatheid crab, octopus, eel pout, tilefish, squirrel hake, blackbellied rosefish, cusk, and the conger eel. The smaller species (Jonah crab, galatheid crab, and octopus) compete with juvenile lobsters and small adults for the relatively small shelters and the larger animals (eel pout, tilefish, cusk, squirrel hake, and conger eel) for the larger shelters. Lobsters, blackbellied rosefish, and cleaner shrimp commonly occupy the same shelter (Fig. 3), and a commensal relationship appears to exist between the lobster and the rosefish. Rosefish commonly dart in among the chelipeds or walking legs of a sheltered lobster when approached by a submersible. Due to the relative scarcity of shelter spaces in the offshore environment, multiple occupancy of shelters has occasionally been observed between lobsters, Jonah crabs, galatheid crabs, tilefish, rosefish, eel pout, and cusk. Conger eels and lobsters have never been observed within the same shelter.

In the more open, less rugged habitats there appears to be a less intense competition for space between the lobster and the bathynectes crab, the redfish, the squirrel fish, and the longfinned and spotted hakes. These species commonly occupy the bowl-shaped depressions generally associated with solid objects, or the dense "forests" of mud anemones.

There is a fairly narrow depth zone of 350–400 m where the lobster and red crab populations overlap. Cooper and Uzmann (1977) hypothesize a strong competition for shelter space between these two species, since both occupy the same types of shelters; i.e., excavations in clay substrates or discarded objects such as fishing nets and lobster traps. Which species is dominant is not known. Multiple occupancy of similar shelters or even nearby shelters, excluding "ghost traps," is rarely observed. Information on the interspecific relations of the European lobster and its associated fauna is not available.

APPENDIX

Macroflora and Fauna Generally Associated with Lobsters by Habitat–Substrate Category

Species (common name)[a]	Sand base with rock	Bedrock base with rock	Mud base with burrows	Clay base with burrows and depressions	Mud–clay base with anemones	Submarine canyon clay wall with burrows
Algae						
Ulva lactuca * (sea lettuce)	X	X				
Monostroma sp.*	X	X				
Porphyra sp.* (Nori)	X	X				
Dumontia incrassata *	X	X				
Chondrus crispus * (Irish moss)	X	X				
Gigartina stellata *	X	X				
Rhodymenia palmata * (dulse)	X	X				
Ptilota serrata *	X	X				
Phycodrys rubens *	X	X				
Fucus vesiculosus * (sea wrack)	X	X				
Ascophyllum nodosum * (knotted wrack)	X	X				
Agarum cribrosum * (sea colander)	X	X				
Alaria esculenta * (winged kelp)	X	X				
Laminaria digitata * (fingered kelp)	X	X				
Laminaria saccharina * (kelp)	X	X				
Laminaria agardhi * (kelp)	X	X				
Lithothamnium glaciale * (encrusting coralline)	X	X				
Lithothamnium lemoineae * (encrusting coralline)	X	X				
Pseudolithophyllum sp.* (encrusting coralline)	X	X				

(continued)

APPENDIX—Continued

Species (common name)[a]	Sand base with rock	Bedrock base with rock	Mud base with burrows	Clay base with burrows and depressions	Mud-clay base with anemones	Submarine canyon clay wall with burrows
*Clathromorphym compactum** (encrusting coralline)	X	X				
*Clathromorphym circumscriptum** (encrusting coralline)	X	X				
Sponges						
*Halichondria panicea** (bread sponge)	X	X	X			
Haliclona oculata (eyed finger sponge)	X	X				
Cladocroce ventilabrum	X	X				
*Cliona celata** (sulfur sponge)	X	X				
Polymastia sp.	X	X	X			
Suberitechinus hispidus	X	X				
Anemones						
Cerianthus borealis (mud anemone)	X		X	X	X	X
Ceriantheopsis americanus (mud anemone)	X		X	X	X	
Edwardsia sp.	X		X	X	X	
Metridium senile (plumose anemone)	X	X				
Tealia felina	X	X				
*Corymorphya pendula** (nodding nosegay)	X	X				
Tubularia crocea (pink hearted hydroid)	X	X				
*Diadumene leucolena**	X	X				
Polychaetes						
Onuphis sp. (tube dwelling polychaete)	X		X	X	X	
Crabs						
Hyas sp. (toad crab)	X		X			X

	1	2	3	4	5	6	7
Pagurus sp. (hermit crab)	X						
Catapagurus sharreri‡ (shell-less hermit crab)	X			X			
Cancer irroratus (rock crab)	X		X		X	X	
Cancer borealis (jonah crab)	X		X		X	X	
Lithodes maia (lithodes crab)	X		X		X		
Geryon quinquedens‡ (red crab)	X		X		X		
Bathynectes superba (Bathynectes crab)			X				
Munida iris (galatheid crab)	X		X		X	X	
Munida valida (galatheid crab)	X		X		X	X	
*Carcinus maenas** (green crab)	X	X					
Shrimps							
Pandalus borealis (deep sea shrimp)	X		X		X		
Pandalus montagui (Montagu's shrimp)	X	X					
Lysmata sp.‡ (cleaner shrimp)	X			X	X	X	
Squilla empusa (mantis shrimp)	X		X				
Bivalve and univalve molluscs							
*Mytilus edulis** (blue mussel)	X		X				
Modiolus modiolus (horse mussel)	X		X				
Pecten magellanicus (deep sea scallop)	X		X		X		
*Buccinum undatum** (waved whelk)	X		X				
*Thais lapillus** (dog whelk)	X						
Ischnochiton ruber (chiton)	X						
*Acmaea testudinalis** (tortoise shell limpet)	X						
*Lacuna vincta** (periwinkle)	X						
*Littorina littorea** (common periwinkle)	X						
*Littorina obtusa** (little green periwinkle)	X					X	
*Polinices heros** (northern moon snail)	X		X				
Brachiopods							
Terebratulina septentrionalis	X	X					

(*continued*)

APPENDIX—*Continued*

Species (common name)[a]	Sand base with rock	Bedrock base with rock	Mud base with burrows	Clay base with burrows and depressions	Mud-clay base with anemones	Submarine canyon clay wall with burrows
Echinoderms						
Echinarachnius parma* (sand dollar)	X					
Strongylocentrotus droebachiensis (green sea urchin)	X	X				
Arbacia punctulata (purple sea urchin)	X	X	X			
Dorocidaris sp.‡ (pencil urchin)	X					X
Cucumaria frondosa (northern sea cucumber)	X	X				
Leptosynapta tenuis* (burrowing sea cucumber)	X		X			
Comactina echinoptera‡ (sea pen)	X			X		X
Psolus fabricii (scarlet psolus)	X	X				
Asterias vulgaris (purple starfish)	X	X				
Asterias forbesi* (common starfish)	X	X				
Henricia sanguinolenta (blood sea star)	X	X				
Gorgonocephalus agassizi (basket star)	X	X	X	X	X	X
Hippasteria phrygiana (red gold bordered sea star)			X	X	X	
Ctenodiscus crispatus (mud star)			X	X	X	X
Ophiopholis aculeata (daisy brittle star)	X	X				
Amphipholis squamata (long-armed snake star)	X	X	X	X	X	X
Leptasterias polaris (polar starfish)	X	X	X	X	X	X
Crossaster papposus (common sun star)	X	X	X	X	X	X
Solaster endeca (purple sun star)	X	X	X	X	X	X

Ascidians			
Boltenia ovifera (sea potato)	X		
Amaroucium sp.* (sea pork)	X		
Ciona intestinalis (sea vase)	X		
Molgula sp. (sea grape)	X		
Halocynthia pyriformis (sea peach)	X		
Fishes			
Citharichthys arctifrons‡ (gulfstream flounder)	X	X	X
Pseudopleuronectes americanus (winter flounder)	X	X	X
Glyptocephalus cynoglossus (witch flounder)		X	X
Paralichtys dentatus (four spot flounder)		X	X
Urophycis chuss (squirrel hake)	X	X	X
Urophycis chesteri‡ (longfinned hake)		X	X
Urophycis tenuis (white hake)		X	X
Urophycis regius (spotted hake)		X	X
Merluccius bilinearis (silver hake)		X	X
Enchelyposus cimbrius (4-bearded rockling)		X	X
Brosme brosme (cusk)	X	X	X
Myxine glutinosa (hagfish)		X	X
Omochelys cruentifer (snake eel)	X	X	X
Conger oceanicus (conger eel)	X	X	X
Macrozoarces americanus (ocean pout)	X	X	
Pholis gunnellus (rock eel)	X		
Ulvaria subbifurcata (radiated shanny)	X		
Myoxocephalus scorpius (shorthorn sculpin)	X	X	

(continued)

APPENDIX—Continued

Species (common name)[a]	Sand base with rock	Bedrock base with rock	Mud base with burrows	Clay base with burrows and depressions	Mud-clay base with anemones	Submarine canyon clay wall with burrow
Myoxocephalus octodecimspinosus (longhorn sculpin)	X					
Sebastes marinus (redfish)	X	X	X	X	X	
*Helicolenus dactylopterus** (blackbellied rosefish)			X	X	X	X
*Tautogolabrus adspersus** (cunner)	X	X				
*Tautoga onitis** (tautog)	X	X				
Lophius americanus (goosefish)	X		X	X	X	X
Anarhichas lupus (wolffish)	X	X				
Gadus morhua (cod)	X	X	X	X	X	X
Chlorophthalmus agassizii‡ (green eye)	X		X	X	X	X
Lopholatilus chamaeleonticeps‡ (tilefish)	X			X		X
Anthias nicholsi‡ (squirrelfish)	X	X				X
Sharks–Skates						
Squalus acanthias (spiny dogfish)	X	X	X	X	X	X
Scyliorhinus retifer‡ (chain dogfish)	X	X		X	X	X
Raja erinacea (little skate)	X		X	X	X	
Raja ocellata (big skate)	X		X	X	X	

[a] (*) species generally occurring only inshore; (‡) species occurring only offshore. Common names are given in parentheses.

REFERENCES

Berrill, M. (1974). The burrowing behavior of newly settled *Homarus vulgaris*. *J. Mar. Biol. Assoc. U.K.* **54**, 797-801.

Berrill, M., and Stewart, R. (1973). Tunnel-digging in mud by newly settled American lobsters, *Homarus americanus*. *J. Fish. Res. Board Can* **30**, 285-287.

Breen, P. A., and Mann, K. H. (1976). Changing lobster abundance and the destruction of kelp beds by sea urchins. *Mar. Biol.* **34**, 137-142.

Chapman, C. J., Johnstone, A. D. F., and Rice, A. L. (1975). The behavior and ecology of the Norway lobster, *Nephrops norvegicus*. *Proc. Eur. Mar. Biol. Symp.*, *9th 19DD* pp. 59-74.

Chouinard, G. *et al.* (1975). "Lobster Fishery Task Force,"—Final report. Fish. Res. Brd., Canada.

Cobb, J. S. (1969). Activity, growth and shelter selection of the American lobster. PhD. Thesis, University of Rhode Island, Kingston.

Cobb, J. S. (1971). The shelter-related behavior of the American lobster, *Homarus americanus*. *Ecology* **52**, 108-115.

Cobb, J. S. (1977). A review of the habitat-related behavior of clawed lobsters (*Homarus* and *Nephrops*) In Proc. U.S.-Australia Lobster Workshop eds. B. F. Phillips and J. S. Cobb, *Circ.—CSIRO, Div. Fish. Oceanogr. (Aust.)* **7**, 143-158.

Cooper, J. S. (1971). Migration and population estimation of the tautog, *Tautoga onitis* (Linnaeus), from Rhode Island. *Trans. Am. Fish. Soc.* **95**, 239-247.

Cooper, R. A. (1970). Retention of marks and their effects on growth, behavior, and migrations of the American lobster, *Homarus americanus*. *Trans. Am. Fish. Soc.* **99**, 409-417.

Cooper, R. A., and Uzmann, J. R. (1971). Migrations and growth of deep-sea lobsters, *Homarus americanus*. *Science* **171**, 288-290.

Cooper, R. A., and Uzmann, J. R. (1977). Ecology of juvenile and adult clawed lobsters, *Homarus americanus*, *Homarus gammarus*, and *Nephrops norvegicus*. *Circ.—CSIRO, Div. Fish. Oceanogr. (Aust.)* **7**, 187-208.

Cooper, R. A., Clifford, R. A., and Newell, C. D. (1975). Seasonal abundance of the American lobster, *Homarus americanus*, in the Boothbay Region of Maine. *Trans. Am. Fish. Soc.* **104**, 669-674.

Dow, R. L. (1974). American lobsters tagged by Maine commercial fishermen, 1957-1959. *Fish. Bull.* **72**, 622-623.

Dybern, B. I., Jacobsen, L., and Hallback, H. (1967). On the habitat behavior of lobster, *Homarus vulgaris*, in Swedish water. *ICES, C.M.* **K:3**.

Dybern, B. I. (1973). Lobster burrows in Swedish waters. *Helgol. Wiss. Meeresunters.* **24**, 401-414.

Ennis, G. P. (1973). Food, feeding, and condition of lobsters, *Homarus americanus*, throughout the seasonal cycle in Bonavista Bay, Newfoundland. *J. Fish. Res. Board Can.* **30**, 1905-1909.

Evans, P. D., and Mann, K. H. (1977). Selection of prey by American lobster, *Homarus americanus*, when offered a choice between sea urchins and crabs. *J. Fish. Res. Board Can.* **34**, 2203-2207.

Fogarty, M. (1975). Competition and resource partitioning in two species of *Cancer*. M.S. Thesis, University of Rhode Island, Kingston.

Gibson, F. A. (1967). Irish investigations on the lobster, *Homarus vulgaris*. *Ir. Fish. Invest. Ser. B* **1**, 13-45.

Gibson, F. A. (1971). Catch and effort in the Irish lobster/crawfish fisheries, 1951-1969. *ICES, C.M.* **K:4**.

Gundersen, K. R. (1969). Preliminary results of field tagging experiments on lobster, *Homarus vulgaris*, in Norwegian waters. *ICES, C.M.* **K:38**.

Hallback, H., and Waren, A. (1972). Food ecology of lobster, *Homarus vulgaris*, in Swedish waters. Some preliminary results. *ICES, C.M.* **K:29**.

Havinga, B. (1938). Krebse und Weichtiere. *Handb. Seefisch. Nordeurop.* **3**(2), 1–147.

Herrick, F. H. (1895). The American lobster: A study of its habits and development. *Bull. U.S. Fish. Comm.* **15**, 1–252.

Herrick F. H. (1911). Natural history of the American lobster. *Bull. U.S. Bur. Fish.* **29**, 149–408.

Himmelman, J. H., and Steele, D. H. (1971). Foods and predators of the green sea urchin, *Strongylocentrotus droebachiensis*, in Newfoundland waters. *Mar. Biol.* **9**, 315–322.

Hirtle, R. W., and Mann, K. H. (1978). Distance chemoreception and vision in the selection of prey by American lobster, *Homarus americanus*. *J. Fish. Res. Board Can.* **35**, 1006–1008.

Holthuis, L. B. (1974). The lobsters of the superfamily Nephropidea of the Atlantic Ocean (Crustacea: Decapoda). *Bull. Mar. Sci.* **24**(4), 723–884.

Howard (1977). The influence of topography and current on size composition of lobster populations. *ICES, C.M.* **K:31**.

Konsuloff, S. (1930). Die Hummer, *Homarus vulgaris* M. Edw., in schwartzen Meere. *Zool. Anz.* **87** (11/12), 318–320.

Krekorian, C. O., Sommerville, D. C., and Ford, R. F. (1974). Laboratory study of behavior interactions between the American lobster, *Homarus americanus*, and the California spiny lobster, *Panulirus interruptus*, with comparative observations on the rock crab, *Cancer antennarius*. *Fish. Bull.* **72**, 1146–1159.

Krouse, J. S. (1973). Maturity, sex ratio, and size composition of the natural population of American lobster, *Homarus americanus*, along the coast of Maine. *Fish. Bull.* **71**, 165–173.

Krouse, J. S., and Thomas, J. C. (1975). Effects of trap selectivity and some population parameters on size composition of the American lobster, *Homarus americanus*, catch along the Maine Coast. *Fish. Bull.* **73**, 862–871.

Lund, W. A., and Rathbun, C. J. (1973). Investigation on the lobster, *Homarus americanus*. *Comm. Fish. Res. Dev. Act.—Final Rep., Proj.* No. 3-130-R, pp. 1–190.

Lund, W. A., Stewart, L. L., and Weiss, H. M. (1971). Investigation on the lobster, *Homarus americanus*. *Comm. Fish. Res. Dev. Act.—Final Rep., Proj.* No. 3-44-R, pp. 1–104.

McLeese, D. W. (1956). Effects of temperature, salinity, and oxygen on the survival of the American lobster. *J. Fish. Res. Board Can.* **13**, 247–272.

McRae, E. D. (1960). Lobster explorations on continental shelf and slope off northeast coast of the United States. *Comm. Fish. Rev.* **22**, 1–7.

Mann, K. H. (1977). Destruction of kelp beds by sea urchins: A cyclical phenomenon or irreversible degradation? *Helgol. Wiss. Meeresunters.* **30**, 455–467.

Mather, F. (1894). What we know about lobsters. *Bull. U.S. Fish. Comm.* **13**, 281–286.

Miller, R. J., Mann, K. H., and Scarratt, D. J. (1971). Production potential of a seaweed-lobster community in eastern Canada. *J. Fish. Res. Board Can.* **28**, 1733–1738.

Morrissey, T. D. (1971). Movements of tagged American lobsters, *Homarus americanus*, liberated off Cape Cod, Massachusetts. *Trans. Am. Fish. Soc.* **100**, 117–120.

Pecci, K. J., Cooper, R. A., Newell, C. D., Clifford, R. A., and Smolowitz, R. J. (1978). Ghost fishing of vented and nonvented traps. *Mar. Fish. Rev.* **40**(5), 9–43.

Saila, S. B., and Flowers, J. M. (1968). Movements and behavior of berried female lobsters displaced from offshore areas to Narragansett Bay, Rhode Island. *Jour. Cons. perm. int. Explor. Mer* **31**, 342–351.

Sastry, A. N., and Pechenik, J. A. (1977). A review of ecology, physiology, and behavior of lobster larvae, *Homarus americanus* and *H. gammarus*. *Austr. Commonwealth Sci. Indus. Res. Org., Div. Fish. Ocean. Circ.* **7**, 159–173.

Scarratt, D. J. (1968). An artificial reef for lobsters, *Homarus americanus*. *J. Fish. Res. Board Can.* **28**, 1733–1738.

Scarratt, D. J. (1973). Lobster populations on a man-made rocky reef. *ICES C.M.*, **K:47**.

Schroeder, W. C. (1959). The lobster, *Homarus americanus*, and the red crab, *Geryon quinquedens*, in the offshore waters of the western North Atlantic. *Deep Sea Res.* **5**, 266-282.

Sears, J. R., and Cooper, R. A. (1978). Descriptive ecology of offshore, deep water, benthic algae in the temperate Western North Atlantic. *Mar. Biol.* **44**, 309-314.

Sheehy, D. (1976). Utilization of artificial shelters by the American lobster, *Homarus americanus*. *J. Fish. Res. Board Can.* **33**, 1615-1622.

Skud, B. E. (1969). The effect of fishing on size composition and sex ratio of offshore lobster stocks. *Fiskeri dir. Skr., Ser. Havunders.* **15**, 259-309.

Skud, B. E., and Perkins, H. C. (1969). Size composition, sex ratio, and size at maturity of offshore northern lobsters. *U.S., Fish. Wildl. Serv., Spec. Sci. Rep.—Fish.* **598**, 1-10.

Smith, E. M. (1976). Food burial behavior of the American lobster, *Homarus americanus*. M.S. Thesis, University of Connecticut, Storrs.

Squires, H. J. (1970). Lobster, *Homarus americanus*, fishery and ecology in Port au Port Bay, Newfoundland, 1960-1965. *Proc. Natl. Shellfish. Assoc.* **60**, 22-39.

Stewart, L. L. (1972). The seasonal movements, population dynamics, and ecology of the lobster, *Homarus americanus*, off Ram Island, Conn. Ph.D. Thesis, University of Connecticut, Storrs.

Tamm, G. R., and Cobb, J. S. (1978). Behavior and the crustacean molt cycle: Changes in aggression of *Homarus americanus*. *Science* **200**, 79-81.

Templeman, W. (1935). Lobster tagging in the Gulf of St. Lawrence. *J. Biol. Board Can.* **1**, 269-278.

Templeman, W. (1940). Lobster tagging on the west coast of Newfoundland, 1938. *Newfoundland, Dep. Nat. Res., Res. Bull. (Fish.)* **8**, 1-16.

Thomas, H. J. (1954). Observations on the recaptures of tagged lobsters in Scotland. *ICES, C.M.* No. 7.

Thomas, H. J. (1957). Some seasonal variations in the catch and stock composition of the lobster. *ICES, C.M.* No. 18.

Thomas, H. J. (1958). Lobster and crab fisheries in Scotland. *Mar. Res.* **8**, 1-107.

Thomas, H. J. (1962). The lobster fishery of the south-east Scottish coast. *ICES, C.M.* No. 22.

Thomas, H. J. (1969). Observations on the seasonal variations in the catch composition of the lobster around the Orkneys. *ICES, C.M.* **K:35**.

Thomas, J. C. (1973). An analysis of the commercial lobster, *Homarus americanus*, Fishery along the coast of Maine, August, 1966 through December, 1970. *NOAA Tech. Rep., NMFS, SSRF* **667**.

Thomas, M. L. H. (1968). Overwintering of American lobsters, *Homarus americanus*, in burrows in Biddeford River, Price Edward Island. *J. Fish. Res. Board Can.* **25**, 2525-2527.

Uzmann, J. R., Cooper, R. A., and Pecci, K. J. (1977a). Migration and dispersion of tagged American lobsters, *Homarus americanus*, on the southern New England continental shelf. *NOAA Tech. Rep., NMFS, SSRF* **705**.

Uzmann, J. R., Theroux, R. B., and Wigley, R. B. (1977b). Synoptic comparison of three sampling techniques for estimating abundance and distribution of selected megafauna: Submersible vs. camera sled vs. otter trawl. *Mar. Fish. Rev.* **39**(12), 11-19.

Van Olst, J. C., Carlberg, J. M., and Ford, R. F. (1975). Effects of substrate type and other factors on the growth, survival, and canibalism of juvenile *Homarus americanus* in mass rearing systems. *Proc. Annu. Meet.—World Maric. Soc.* **6**, 261-274.

Warme, J., Cooper, R. A., and Slater, R. (1978). Bioerosion in submarine canyons. In "Submarine Canyon, Fan Trench Sedimentation" (D. J. Stanley and G. Kelling, eds.), Chapter 6, pp. 65-70. Dowden, Hutchinson & Ross, Stroudsberg, Pennsylvania.

Watson, P. S. (1974). Investigations on the lobster, *Homarus gammarus*, in Northern Ireland—a progress report, 1972-73. *ICES, C.M.* **K:20**.

Weiss, H. M. (1970). The diet and feeding behavior of the lobster, *Homarus americanus*, in Long Island Sound. Ph.D. Thesis, University of Connecticut, Storrs.

Wilder, D. G. (1954). The lobster fishery of the southern Gulf of St. Lawrence. *Fish Res. Board Can., Biol. Stn., St. Andrews, N.B., Gen. Ser. Circ.* **24**, 1-16.

Wilder, D. G. (1963). Movements, growth, and survival of marked and tagged lobsters liberated in Egmont Bay, Prince Edward Island. *J. Fish. Res. Board Can.* **20**, 305-318.

Wilder, D. G., and Murray, R. C. (1956). Movements and growth of lobster in Egmont Bay, Prince Edward Island. *Fish. Res. Board Can., Prog. Rep. Atl. Coast Stn.* **64**, 3-9.

Zeitlin-Hale, L. (1963). Movements, growth, and survival of marked and tagged lobsters liberated in Egmont Bay, Prince Edward Island. *J. Fish. Res. Board Can.* **20**, 305-318.

Zeitlin-Hale, L. (1975). Effects of environmental manipulation on the agonistic behavior and locomotor activity of juvenile American lobster, *Homarus americanus*. M.S. Thesis, University of Rhode Island, Kingston.

Chapter 4

Ecology of Juvenile and Adult *Nephrops*

C. J. CHAPMAN

I.	Introduction	143
II.	Habitat	145
	A. Nature of the Substratum	145
	B. Other Factors Influencing Distribution	146
	C. Burrow Structure	147
	D. Density of *Nephrops* and Other Burrowing Species	150
III.	Behavior and Activity Rhythms	151
	A. Emergence from the Burrow	151
	B. Factors Influencing and Controlling Emergence	153
	C. Migrations and Local Movements	161
IV.	Population Structure and Life Cycle	162
	A. Length and Sex Composition of Stocks	163
	B. Fecundity and Reproduction	166
	C. Juvenile Phase and Recruitment	170
	D. Growth	171
	E. Mortality	173
V.	Conclusions	174
	References	175

I. INTRODUCTION

The Norway lobster, *Nephrops norvegicus** (L), is widely distributed on the continental shelf of the northeast Atlantic and in the Mediterranean Sea (Fig. 1). It is found off Greenland and Iceland in the north and as far south as Morocco

*Hereafter the generic name *Nephrops* will be used.

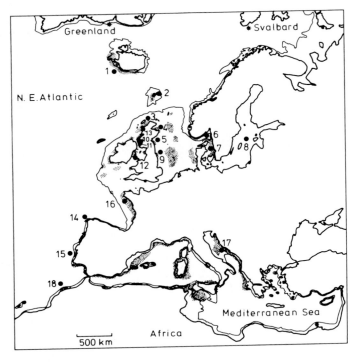

Fig. 1. Geographical distribution of *Nephrops*. The 200 m depth contour is shown, corresponding roughly with the edge of the continental shelf. The precise location of grounds marked * are not known. Some areas mentioned in the text (1) Iceland, (2) Faroe, (3) N. Minch, (4) Moray Firth, (5) Firth of Forth, (6) Skagerak, (7) Kattegat, (8) Baltic, (9) Farn Deeps, (10) Sound of Jura, (11) Clyde, (12) Irish Sea, (13) Loch Torridon, (14) N.W. Spain, (15) W. Portugal, (16) Golfe de Gascogne, (17) Adriatic, and (18) Morocco. (After Farmer, 1975, reproduced by permission of F.A.O.)

(Farmer, 1975). In the countries bordering these waters, *Nephrops* supports important trawl fisheries. Much of our knowledge of the ecology of *Nephrops* is derived from the catches in these fisheries. The main fishing grounds are in depths from 15 to 500 m, though *Nephrops* has been caught at depths of 800 m in the Mediterranean (Farmer, 1975).

Nephrops is confined to areas where the sea bed consists of fine cohesive mud suitable for burrowing. This lobster spends the greater part of its life concealed in burrows, and its periodic emergence gives rise to large diurnal and seasonal fluctuations in the magnitude and composition of the fishery catches. This makes accurate assessment of the populations by normal fishing techniques very difficult. The general biology of *Nephrops* has been reviewed by Figueiredo and Thomas (1967a). Since 1967, there has been a considerable expansion in re-

4. Ecology of Juvenile and Adult *Nephrops*

search on this species. This chapter reviews the ecology of *Nephrops* in the light of recent research. The history and management of the fisheries are dealt with in Chapter 8 of this volume. A useful synopsis and bibliography of biological data on the Norway lobster has recently been published by FAO (Farmer, 1975).

II. HABITAT

A. Nature of the Substratum

The type of bottom sediment is a major factor governing the distribution of *Nephrops*. Exploratory fishing surveys by many workers (see Farmer, 1975) have conclusively demonstrated that *Nephrops* is found only on fine cohesive muds, stable enough to support unlined burrows. However, few detailed studies have been carried out on the sediments. Figure 2 shows the results of particle size

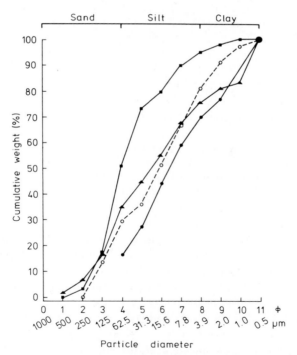

Fig. 2. Sediment analyses of mud from *Nephrops* grounds in the Irish Sea [■—■, from Farmer (1975), by permission of F.A.O.], in Loch Torridon [●—●, from Rice and Chapman (1971), by permission of Springer-Verlag] in Lough Ine [▲—▲, from Kitching *et al.* (1976), by permission of Blackwell Scientific Publications Ltd], and from *Goneplax* habitat in Fishguard harbour. [○- - -○, from Atkinson (1974a), by permission of Springer-Verlag.]

analyses on sediment samples from *Nephrops* grounds. For comparison, one sample from an area with burrows of the crab *Goneplax rhomboides* (but not *Nephrops*) is also depicted. Silt and clay fractions (particle diameter, 0.5–60 μm) account for 50–85% by weight of the sediments. Clearly, these data are too sparse to allow a "typical" *Nephrops* ground to be defined in terms of particle size distribution alone. Other characteristics of the mud, i.e., its clay mineral, organic, and water contents, are also important in determining its cohesive properties. The sediments are "thixotropic," in that they tend to become more fluid under an applied force (Rice and Chapman, 1971).

B. Other Factors Influencing Distribution

Apart from sediment, few other factors have been reliably linked to changes in the distribution of *Nephrops*. Reference to salinity, temperature, and oxygen exist in the literature, but confusion can occur because changes in these environmental factors may affect behavior and catchability of *Nephrops* rather than its distribution. Since *Nephrops* generally inhabits deep water, it is unlikely to experience sudden changes in salinity and temperature.

In the Icelandic fishery, low catches in 1968 were partly attributed to abnormally low temperature arising from the influx of Arctic drift ice during the spring (Eiriksson, 1970a). Jensen (1965) suggested that *Nephrops* could not be caught in the Skagerrak and Kattegat when the bottom temperature fell below 5°C, because they remained in their burrows. This question was recently reconsidered by Bagge and Münch-Petersen (1979), who concluded that variation in catches was more likely to be related to the amount of dissolved oxygen in the water than to temperature. These authors reported a strong negative correlation between the trawl catch of *Nephrops* in the Kattegat and the concentration of dissolved oxygen. The oxygen level in the water close to the sea bed attained a minimum level of 42% saturation during September when catches were greatest. The authors suggest that an even lower level of oxygen within their burrows forces *Nephrops* out onto the mud surface.

Following periods of exceptionally calm weather in summer, the dissolved oxygen near the sea bed may fall to very low levels in some inlets on the west coast of Scotland. These conditions occur rarely but have led to reports of dead *Nephrops* in creel catches. In Lough Ine, in the south of Ireland, *Nephrops* is found only above thermocline depth (25 m). Low oxygen levels below this destroy the entire benthic fauna during the summer (Kitching *et al.*, 1976).

There is little evidence that the distribution of *Nephrops* is influenced by pollution other than the dumping of sewage sludge. McIntyre and Johnston (1975) have described such a case in the Clyde estuary on the west coast of Scotland. About 10^6 tons/year of sewage sludge from the city of Glasgow are dumped in water of 80 m depth, on what was undoubtedly former *Nephrops* ground. The dumping started in 1904 before there was a *Nephrops* fishery. Television observations have shown a clear transition from ground containing *Nephrops* burrows to a black mud of fibrous texture which would inhibit normal burrowing. The authors estimated that about 10 km^2 of former ground had been made unsuitable for *Nephrops*.

The possible influence of light intensity in determining depth distribution is dealt with in Section III, B.

C. Burrow Structure

The method of burrow construction by *Nephrops* has been fully described from aquarium studies (Rice and Chapman, 1971; Farmer, 1975) and appears to be similar to the burrowing of *Homarus* in mud (Howard and Bennett, 1979). Initially, the endopodites of the third maxillipeds are repeatedly thrust into the mud that is to be moved. During this manipulation the mud appears to become more fluid, and when pushed forward seems to flow over the rest of the sub-

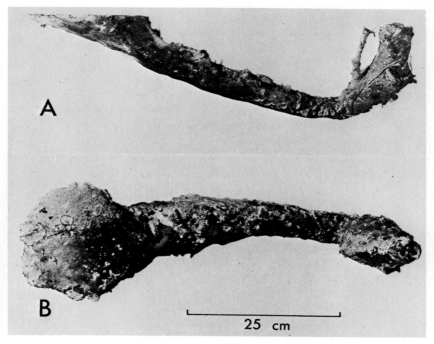

Fig. 3. (A) Side and (B) plan views of polyester resin cast of a *Nephrops* burrow in L. Torridon to show the wide flared entrance and small rear opening. (From Rice and Chapman, 1971, reproduced by permission of Springer-Verlag.)

stratum in a discrete mass. Thus, *Nephrops* appears to make use of the thixotropic property of the mud.

The first phase in burrow construction is the excavation of an elongated depression in the mud with the formation of a large mound of mud at one end. The animal then tunnels into the side of the depression opposite the mound. After sloping down at a shallow angle, the tunnel levels off and then rises almost vertically to the surface to form a rear opening. A polyester resin cast of a burrow at this stage is illustrated in Fig. 3. Further elaboration of the burrow may take place, i.e., enlargement of the rear opening to create a second entrance and the

Fig. 4. Daytime underwater photograph showing a large, two-entrance burrow occupied by two *Nephrops* that have emerged in response to the diver. (From Chapman and Rice, 1971, reproduced by permission of Springer-Verlag.)

formation of side tunnels leading to additional entrances. Fig. 4 shows a large two-entrance burrow with two similarly sized occupants. This is unusual in that most burrow systems have only one occupant. When there is more than one, the occupants are usually of different sizes, and each animal occupies its own tunnel within the burrow system (see Fig. 5). This type of burrow is discussed in greater detail in Section IV, C. The burrows generally extend 200–300 mm below the surface of the mud.

Rice and Chapman (1971) classified the burrows occupied by *Nephrops* in Loch Torridon, Wester Ross, Scotland into four types: I, single opening; II, single entrance with one rear opening (as in Fig. 3); III, two entrances (as in Fig. 4); and IV, three or more entrances. The frequency of each burrow type was 15, 15, 42, and 27% respectively. Burrows with three entrances were most common in the Irish Sea (Hillis, 1974) and in Loch Aline, Scotland (Atkinson, 1974b). In Loch Torridon there was evidence that female *Nephrops* may always inhabit burrows with at least two openings, since only males were found in tunnels with a single opening. For ovigerous females, a rear opening is probably essential in enabling the oxygen requirements of the developing embryos to be met. The backward-directed pleopod current, which keeps the eggs clean and supplied with fresh water, would be effective only in a tunnel with two (or more) openings. There is some evidence that *Nephrops* constructs burrows with fewer

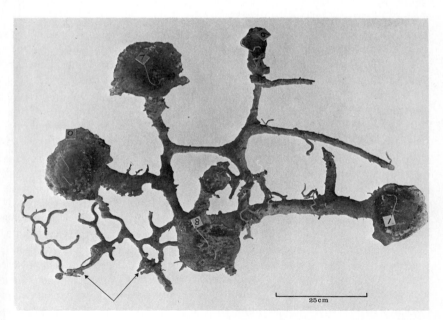

Fig. 5. Cast of burrow system from Loch Brollum consisting of three interlinked sections, each occupied by different size male *Nephrops*.

openings as it increases in size (C. J. Chapman, unpublished). According to Dybern and Höisaeter (1965), the length and depth of tunnels below the mud surface are directly related to the size of the occupants.

D. Density of *Nephrops* and Other Burrowing Species

The density and spacing of burrows has been studied by divers in shallow areas. In Swedish and Norwegian waters, Dybern and Höisaeter (1965) found that *Nephrops* burrows tended to occur in discrete groups of 5–10 tunnels, covering an area of 2–5 m in diameter, with each group well separated from its neighbors. The densities of *Nephrops* were not given, but the authors noted that many burrows were empty.

Chapman and Rice (1971) found no evidence of discrete burrow groups at 30 m depth in Loch Torridon, although the burrows were clumped to some extent. The density of burrows ranged between 0.5 and $0.7/m^2$, but only about one-third of these belonged to *Nephrops*. About 30% of the *Nephrops* burrows were empty and the densities of *Nephrops* varied from 0.13 to $0.18/m^2$. The population consisted mainly of large, mature *Nephrops*, carapace length (CL) greater than 35 mm. Many of the remaining burrows in Loch Torridon were occupied by the burrowing goby, *Lesueurigobius friesii* (Collett). The density of burrows occupied by these fish in 1969 was $0.43/m^2$, more than twice the density of *Nephrops* (Rice and Johnstone, 1972). Most of the gobies and *Nephrops* were found in separate burrows, but a few burrows contained individuals of both species. The burrowing crab, *Goneplax rhomboides* (L.) (Rice and Chapman, 1971), and the thalassinid decapod, *Calocaris macandreae* (Bell), were also found in Loch Torridon, but no density estimates were obtained.

Atkinson (1974b) studied a shallow water community of mud-burrowing species in Loch Aline, Scotland. In one area at 17–20 m depth, the density of all burrows was $0.31/m^2$. Approximately one-half were occupied by *Nephrops* and the rest by *Goneplax*. In another area at 10 m depth, the density of *Nephrops* was much lower. *Lesueurigobius* was also present in small numbers. The burrows of *Nephrops* were distributed randomly, whereas those of *Goneplax* were clumped. In Loch Aline, as in Loch Torridon, large *Nephrops* (>40 mm CL) predominated.

In the mud zone of Lough Ine the density of burrow openings was about $20/m^2$, but *Nephrops* accounted for relatively few of these ($0.1/m^2$) (Kitching *et al.*, 1976). *Calocaris* was the only other burrowing species positively identified.

Some diving observations at various depths down to 31 m in the Irish Sea were reported by Hillis (1974). The average density of burrows was $0.32/m^2$ and about 60% of the burrows were occupied by *Nephrops*. Most of the occupants were smaller than 35 mm CL. Small numbers of *Goneplax* were seen but no estimate of their density was made.

Most of the above diving observations were obtained in shallow water at the

4. Ecology of Juvenile and Adult Nephrops

periphery of the populations so that the density measurements may not be typical of each area. Dybern and Höisaeter (1965), for example, reported that the density of burrows increased with depth. Few attempts have been made to estimate the density of *Nephrops* on important fishing grounds in deep water. Peterson's tagging-recapture method was employed by Andersen (1962) at Faroe and by Gibson (1967) and A. C. Simpson (personal communication) in Irish waters. The proportion of tagged animals recaptured was low, varying from about 1 to 3%. This may have been partly due to high mortality of tagged *Nephrops* returned to the sea (Section III, C). From the tag recaptures, Simpson calculated an average density of $1/m^2$ in the Irish Sea, higher than the density estimates obtained by diving at the edge of the population (Hillis, 1974). To the south of Ireland, Gibson estimated that the minimum density was $0.4/m^2$, of which almost 75% were of marketable size. Andersen presented his results in terms of the total biomass of *Nephrops* in different areas, which cannot be directly compared with the other data.

Underwater television, photography, and a submersible have been used to study populations of *Nephrops* in Scottish waters (Chapman, 1979). Studies in the Sound of Jura revealed a very high density of burrow openings, which varied from $40/m^2$ at 65 m depth to $100/m^2$ at 130 m. Detailed examination of these openings in one area from a submersible showed two types of burrow; one occupied by *Nephrops* and the other by *Calocaris*. The latter burrows had small circular openings at the base of funnel-like depressions in the mud surface and were readily distinguishable from those of *Nephrops*. The density of these two species was very similar, i.e., $3-5/m^2$. Approximately 25% of the burrows were unoccupied. Interestingly, most of the *Nephrops* were smaller than 30 mm CL (see Fig. 9).

In one part of the Firth of Clyde, the density of *Nephrops* burrows was about $1/m^2$. In the Firth of Forth, the density of *Nephrops* and *Calocaris* burrows was roughly similar, $0.4/m^2$. These figures are likely to give overestimates of the density of *Nephrops*, since some of the burrows were probably unoccupied. *Nephrops* can be counted directly as they emerge from their burrows by using a television camera. However, this leads to an underestimate of population density because only a fraction of the population is likely to be out of their burrows at any one time (see Section III).

III. BEHAVIOR AND ACTIVITY RHYTHMS

A. Emergence from the Burrow

Detailed studies of individual *Nephrops* effected by diving and tagging (Chapman and Rice, 1971; Chapman *et al.*, 1975) show that they spend most of their time within their burrows, emerging for a comparatively short time during each 24-hr period. As indicated above, the burrows usually extend 200–300 mm below the surface of the mud, so that animals within are unlikely to be caught by trawl nets. It has been well known for some time that the commercial catches of *Nephrops* show large diurnal and seasonal fluctuations, and it has been assumed that these reflect periodic emergence from their burrows. This assumption has recently been confirmed by comparing trawl catches and television camera observations on the same ground (Chapman and Howard, 1979).

Most of the early trawling results were summarized by Farmer (1974a). More recent data may be found in papers by Aréchiga and Atkinson (1975) and Atkinson and Naylor (1976). The most striking feature of the daily patterns of trawl catches on different grounds is their variation with depth. Peak catches are generally obtained around dawn and dusk, but with increasing depth the time of these two peaks shifts toward midday. Thus, the dusk peak occurs earlier in the afternoon and the dawn peak later in the morning. The relationship between depth and the times of peak emergence, as indicated by trawling and television observation, is illustrated in Fig. 6. The diagram shows data from several areas and seasons, and the times given are relative to sunrise and sunset. This reduces seasonal variations, since for any given depth the emergence rhythm tends to maintain the same phase relationship to sunrise and sunset throughout the year. The correlation coefficients associated with the regression lines in Fig. 6 are highly significant.

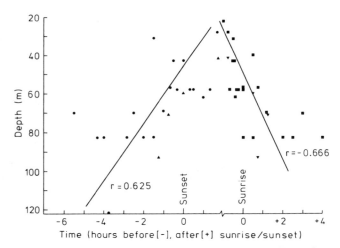

Fig. 6. Relationship between the times of peak emergence and depth. Times of P.M. peaks given relative to sunset (●, trawl catches; ▲, television observations); A.M. peaks relative to sunrise (■, trawl; ▼, television). Regression lines fitted by least squares with correlation coefficients (r) given on each graph. Note the split in the horizontal axis to separate the two graphs. Data from Irish Sea (Simpson, 1965; Hillis, 1971b; Farmer, 1974a; Aréchiga and Atkinson, 1975), Scotland (Chapman *et al.*, 1975; Chapman and Howard, 1979), and Sweden (Höglund and Dybern, 1965).

4. Ecology of Juvenile and Adult Nephrops

Extrapolation of the regression lines to depths beyond the range in Fig. 6 shows that the dawn and dusk peaks should eventually converge to single peaks, around midday in deeper water (over 150 m) and around midnight in shallower depths (less than 20 m). Direct evidence for this is provided by Atkinson and Naylor (1976). These authors studied populations off the west coast of Scotland at two extreme depths; 184 m in the Minch, where peak trawl catches were taken close to midday, and 10 m in Loch Aline, where diving revealed the highest numbers of Nephrops out of their burrows at night.

B. Factors Influencing and Controlling Emergence

1. Light

The patterns of activity shown by Nephrops at different depths (Fig. 6) lends support to earlier suggestions (Andersen, 1962; Simpson, 1965) that emergence from their burrows was related to the intensity of light reaching the sea bed. Chapman and Rice (1971) suggested that Nephrops only emerged when light intensity was optimal. By using a light meter in conjunction with television observations, Chapman et al. (1975) showed that the period of emergence at different depths corresponded to the same range of light intensity at the sea bed. This has also been confirmed from trawling experiments (Hillis, 1971b; Aréchiga and Atkinson, 1975; Atkinson and Naylor, 1976). Atkinson and Naylor (1976) showed that the light intensity on the sea bed during the period of peak emergence at 184 m and 10 m was virtually the same. Figure 7 brings together the published data on Nephrops emergence in relation to light intensity. Although the same light meter was used by all the authors, it should be mentioned that the meter was not specifically designed or calibrated for work on Nephrops, and its response curve (maximum sensitivity at 530 nm) was different from that of the Nephrops eye. Loew (1976) has shown that the visual pigment from the compound eye of Nephrops has maximum absorption at 498 nm. Figure 7 shows that the emergence of Nephrops is generally confined to the range of light intensity from 10^{-5} to 1 meter candles (mc) as measured by the meter. Clearly, more work is needed using a light meter matched to the eye of Nephrops before these intensities can be related to the vision of the animal.

It is worth noting that Nephrops do not always emerge whenever light is optimal. This is revealed in Fig. 7 by the large number of data points showing low activity within the optimum range. Thus, during periods of activity only about 10–30% of the population may emerge at the same time (Chapman et al., 1975). It is possible that the maximum depth of Nephrops in different areas is related to light penetration and may be the depth at which the light intensity around midday is optimal for emergence. Beyond this depth, light intensity on the sea bed would never reach the minimum light required for emergence.

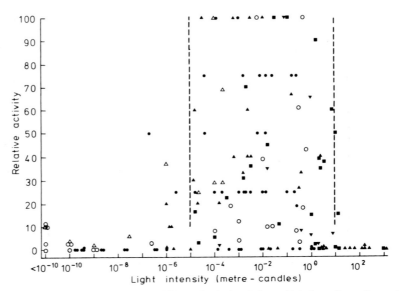

Fig. 7. Emergence of *Nephrops* at different depths in relation to light intensity on the sea bed. From variations in trawl catches; (◇) 184 m (Atkinson and Naylor, 1976), (○) 70 m (Aréchiga and Atkinson, 1975), (▼) 62–68 m, (■) 20–38 m (Hillis, 1971b). From television observations; (●) 71–93 m, (▲) 42–60 m (Chapman *et al.*, 1972). All light intensity measurements were taken with the same photomultiplier calibrated in photometric units. To facilitate comparison the original data has been converted to a standard scale from 0 to 100.

In this section, it is appropriate to mention the effects of tidal and lunar cycles on emergence, since both can influence light intensity at the sea bed. According to fishermen, catches of *Nephrops* are higher during neap tides and slack water than during spring tides or maximum tidal flow. Storrow (1912) showed some correlation along these lines from landings at North Shields, but more convincing evidence from the same fishery is provided by P. Warren (personal communication). Warren has found a significant inverse correlation between *Nephrops* landings per unit of fishing effort from the Farn Deep grounds and the tidal height in the nearby River Tyne, confirming that highest catches are taken during neap tides. The explanation for this is unknown at present, but it is possibly linked to light intensity, through the direct influence of moonlight and/or changes in turbidity caused by tidal currents. Although water currents over *Nephrops* grounds are weak, they are evidently sufficient at times to increase turbidity close to the sea bed and attenuate light (Chapman *et al.*, 1972). The claim that reduced catches follow periods of rough weather may stem from the same cause. It would be interesting to compare Warren's data with catches from a population in much deeper water where *Nephrops* emerge during the day. Here, moonlight would

4. Ecology of Juvenile and Adult Nephrops

have little influence, and any diurnal catch variations would more likely be related to the water current or turbidity.

Before closing the subject of light, the recent work on the physiology of the compound eye in *Nephrops* should be discussed. The eyes of *Nephrops* are relatively large, compared to those of other genera within the Nephropidae. They are of the "superposition" type characteristic of nocturnal crustaceans. Their structure has been described by Aréchiga and Atkinson (1975) and Loew (1976). The eyes of *Nephrops* are clearly adapted for vision in low light conditions, demonstrated by the lack of distal screening pigment migration during cycles of normal ambient light, the limited mobility of the proximal screening pigment (which never fully shields the rhabdomes), and the presence of a well developed tapetum. Measurements of the extent of proximal pigment migration suggest that the eye changes gradually from the dark-adapted condition to the fully light-adapted state between light intensities of 10^{-2} and 1 mc (Aréchiga and Atkinson, 1975; Atkinson and Naylor, 1976). This corresponds well to the upper end of the optimum light range for emergence (Fig. 7). The eye cannot cope with higher light intensities, and damage to the eyes may occur if *Nephrops* is brought to the sea surface during full daylight. A 30-min exposure to 200 mc is sufficient to impair the proximal screening pigment migration mechanism (Atkinson and Naylor, 1976). After prolonged exposure (2.5 hr) to high surface light intensity, the visual pigment becomes irreversibly bleached and the structure of the rhabdomes breaks down (Loew, 1976). These are important findings, indicating the strong possibility that *Nephrops* brought to the surface and maintained in daylight will be blinded. Although unpublished observations by Loew and Chapman showed that short exposures to full daylight of less than 10 min was not harmful, it is strongly recommended that experimental work with *Nephrops* (e.g., tagging) should be carried out under red light, which is outside the absorption band of the visual pigment in the eye. It is not known whether light-induced blindness is likely to be permanent. Further work is needed to establish this point and also to find out how the behavior of the animal is affected by blindness. This is crucial, in view of the widespread practice by fishermen of returning large numbers of undersized *Nephrops* to the sea.

2. Feeding and Predation

The restriction of emergence to an optimum range of light intensity is probably governed by two factors, the availability of suitable food and the avoidance of predators.

It is generally assumed that adult *Nephrops* emerge from their burrows to feed on prey species living on or in the mud. The diet of *Nephrops* is varied, with most invertebrate phyla represented in analyses of foregut contents. Thomas and Davidson (1962) found that most prey were fragmented, and accurate identifications of species could not be made. Their results are summarized in Table I. It is

TABLE I

Percentage Occurrence of Various Food Species in Foreguts Containing Food[a]

Species	%[b]	Species	%[b]	Species	%[b]
Polychaeta (total)	66	Crustacea (total)	67	Mollusca (total)	64	Echinodermata (total)	50
Errantia (unident.)	6	Ostracoda	15	Solenogastres	1	Ophiuroidea	16
Aphroditidae	1	*Philomedes* sp.	+	Gastropoda	8	*Amphiura* sp.	1
Aphrodite aculeata	+	Copepoda	5	Bullomorpha	2	*A. chiajei*	+
Polynoinae	+	Harpacticoida	5	*Turbonilla* sp.	+	*A. filiformis*	1
Sigalioninae	+	*Nebalia bipes*	+	Scaphandridae	+	*Ophiura* sp.	1
Nereidae	4	Cumacea	6	Philinidae	1	Echinoidea	28
Nephthyidae	2	Leuconidae	1	Scaphopoda	8	Irregular Urchins	12
Glyceridae	9	*Campylapsis costata*	+	Lamellibranchia	59	Holothurioidea	+
Goniada sp.	1	Tanaidae	1	*Nucula* sp.	20	Echinoderm (unident.)	17
G. norvegica	+	Isopoda	1	Pectenidae	+		
G. maculata	+	*Cirolana borealis*	+	*Cyprina islandica*	+	Pisces	13
Eunicidae	1	Amphipoda	7	*Mysella* sp.	+		
Lumbriconereis sp.	1	Ampeliscidae	4	*Cardium* sp.	3	Foraminifera	41
Sedentaria (unident.)	30	Phoxocephalidae	+	Veneridae	8		

Taxon	Count	Taxon	Count
Spionidae	1	Amphilochidae	+
Chlorhaemidae	1	Oedicerotidae	+
Ammotrypane aulogaster	5	Lembos longipes	+
Capitellidae	+	Caprellidae	+
Oweniidae	5	Mysidae	+
Owenia fusiformis	3	Natant decapoda	5
Myriochele sp.	2	Processa sp.	+
Pectinaria sp.	14	Crangon sp.	3
Hydroides norvegica	+	Reptant decapoda	21
Unident. polychaetes	12	Nephrops norvegicus	+
		Galantheidae	1
		Porcellana sp.	+
		Thallassinidea	+
		Paguridae	2
		Anapagurus laevis	+
		Portunus sp.	4
		Corystes cassivelaunus	+
		Unidentified crustacea	35
Abra sp.	18	Coelenterata (Hydroid)	3
Tellinidae	+	Nemertini	+
Solenidae	+	Echiuroidea	+
Corbula gibba	+	Insecta	+
Hiatella sp.	+	Phoronidea	1
Mya sp.	+	Polyzoa	3
Cephalopoda	2	Particulate matter	68
Unidentified mollusca	+		

[a] From Thomas and Davidson (1962).

[b] (+) present, but in less than 1% of foreguts examined.

possible that the periods of emergence are related to the availability of prey species, which may, themselves, have diurnal rhythms of behavior.

There is some evidence that the duration of emergence varies with the size of *Nephrops*. Andersen (1962) suggested that *Nephrops* becomes less responsive to light with increasing size. Also, it is likely that large animals need to spend relatively more time out of their burrows to obtain adequate food. This would account for the variation in size composition of trawl catches at different times of the day and night (Section IV, A). There may be times when prey densities are sufficiently great for *Nephrops* to feed near or even within their burrows. During the winter months, females carrying developing eggs are seldom caught in trawls or baited traps (Section IV, A), suggesting that they do not emerge to feed. Similarly, newly molted individuals probably remain in their burrows until the exoskeleton has hardened. It is not known whether either of these categories feeds on prey entering or present within their burrow system or whether they rely on food reserves. McIntyre (1973) showed that sediment cores taken from the mouth of *Nephrops* burrows have a higher biomass than those taken from the roof of the burrow. The presence of polychaetes (on which *Nephrops* are known to feed) in a damaged condition suggested the possibility that prey may be caught and stored at the burrow entrance and the animals may not always have to leave their burrows to feed.

The main function of the burrow presumably is to provide shelter and protection from predators, so the period of emergence is likely to be timed to minimize the risk of predation. The chief predators on *Nephrops* are fish (Thomas, 1965a; Farmer, 1975). In Scottish waters, Thomas identified *Nephrops* in the stomach contents of thirteen species of fish caught on known *Nephrops* grounds. The stomach content of the chief predators was examined, and it was found that the cod, *Gadus morhua*, contained an average of 2.7 *Nephrops* in 115 out of 143 (80%) stomachs. The lesser spotted dogfish, *Scyliorhinus canicula*, contained an average of 1.3 *Nephrops* in 63 out of 123 (50%) stomachs, and the thornback ray, *Raja clavata*, contained an average of 2.9 *Nephrops* in 16 out of 31 (50%) stomachs. Rae (1967) reported that *Nephrops* was a major component of the stomach contents of cod in areas around Scotland where the crustacean was abundant. From examination of over 4000 specimens, Rae found *Nephrops* in 3–33% of the stomachs of small cod (210–500 mm in total length) and in 30–70% of the stomachs of large cod (500–1110 mm in total length). Unpublished Scottish data show that *Nephrops* of 20–35 mm CL are most abundant in stomach contents of cod (230–870 mm in total length) (F. G. Howard, personal communication). No research has yet been conducted on the locomotor and feeding rhythms of fish preying on *Nephrops*. Such work is clearly needed in view of Symonds' (1971) suggestion that large aggregations of cod on the fishing grounds could cause *Nephrops* to remain in their burrows (see Section IV, B).

4. Ecology of Juvenile and Adult *Nephrops*

3. Endogenous Rhythms

Rhythms of locomotor activity in *Nephrops* were investigated recently under laboratory conditions, in which the movements of the animals were monitored within and without artificial burrows. The results revealed persistent endogenous activity rhythms in animals kept in constant dim light. Emergence from the

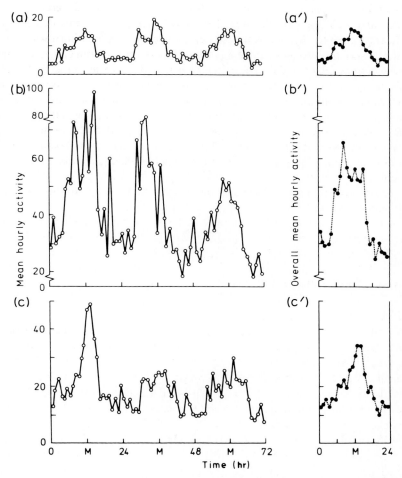

Fig. 8. Mean activity of 22 *Nephrops* freshly collected from 80 m and recorded for 3 days in continuous darkness in an actograph supplied with an artificial burrow: (a) activity outside burrow, (b) at entrance to burrow, and (c) within the burrow; (a'-c'), hourly activity averaged over the 3 days, M, midnight. (From Atkinson and Naylor, 1976, reproduced by permission of Elsevier, North-Holland Biomedical Press.)

burrow occurred during the time of "expected" night, irrespective of the depth from which the animals originally came (Atkinson and Naylor, 1973, 1976; Naylor and Atkinson, 1976). Similar rhythms were induced by artificial light cycles, but in contrast to field observations, emergence always took place during the darkest phase of the cycle irrespective of the absolute intensities (Aréchiga and Atkinson, 1975; Atkinson and Naylor, 1976; Hammond and Naylor, 1977a,b).

The relationship between emergence and light intensity in the natural environment has not yet been found in laboratory experiments. Atkinson and Naylor (1976) suggested that the endogenous nocturnal rhythm in *Nephrops* reflected burrow orientated behavior, since the rhythm was generally expressed most strongly within the burrow (Fig. 8). In nature, this activity probably includes burrow excavation and maintenance. Although the animals frequently vacated their artificial burrows (Fig. 8a), it is unlikely that this represents the emergence found in nature, which is almost certainly a feeding rhythm. Food was usually withheld from laboratory animals (R. J. A. Atkinson, personal communication), and this may partly account for the differences between laboratory and field results.

It is interesting that *Nephrops* was not responsive to light–dark cycles when light intensity was just below 10^{-5} mc (Aréchiga and Atkinson, 1975), suggest-

TABLE II

Migrations—Results of Tagging Experiments

Area	Methods			Numbers released		
	Fishing	Tagging	Dates	Male	Female	Total
Skagerrak and Kattegat	Trawl	Plastic tag on tail	April/June 1958–1962 Oct/Nov 1957–1961	Sex not given		1991
North Sea Farn Deep	Trawl	Tail clipping	April 1912	365	200	565
North Sea Farn Deep	Trawl	Colored plastic disc wired to claw	Oct/Nov 1975	109	880	989
Scotland						
Moray Firth	Trawl	As above	July/Aug 1962–1964	4469	67	4536
Clyde	Trawl	As above	July 1962, 1964	1943	57	2000
North Minch	Trawl	As above	July 1962–1964	2399	16	2415
Tiree Passage	Trawl	As above	July 1962, 1964	1118	2	1120
Sound of Jura	Trawl	As above	July 1962	280	0	280
Loch Torridon	Creel	As above	June–Oct 1970	246	163	409

4. Ecology of Juvenile and Adult Nephrops

ing that this represents the lower light threshold for the eye. This observation, together with the inability of the lobster's screening pigment migration to cope with intensities above 1 mc, provides the physiological basis for emergence in an optimum light range (Fig. 7). Laboratory experiments suggest that changing light intensity is more likely to trigger emergence than is absolute intensity. This could well be so, since *Nephrops* over most of its depth range emerges during the period of rapidly changing light intensity at dusk and dawn (but still within the optimum range). Evidently, emergence may be induced both by a decrease (at dusk) and an increase (at dawn) in light intensity. In deep water, where peak emergence occurs around noon, only an increase in light intensity would be effective. Conversely, only the decrease a light intensity at dusk induces nocturnal emergence in shallow water. If changing light intensity does trigger emergence, then there are still depth differences in response to account for.

C. Migrations and Local Movements

Tagging experiments suggest that *Nephrops* does not undertake extensive migrations, though local movements occur. The results of various tagging experiments are summarized in Table II. These show that most recaptures by trawl were taken within 9 km (5 nautical miles) of the release site. This distance is less than the average length of a trawl tow. With more accurate positioning using

	Recaptures			Period of liberation (days)		Distance moved (km)		
Total	Male (%)	Female (%)	Both sexes (%)	Range	Mean	Range	Mean	Reference
90	—	—	4.5	Max 348	80% Caught within 60 days	0–35.2	8.9	Jensen (1965)
2	—	—	0.35	7–51	29	Negligible		Storrow (1912, 1913)
8	—	—	0.81	23–110	67	11.1–39.0	27.7	P. Warren (personal communication)
40	0.90	0	0.88	5–480	91	0–14.8	5.2	H. J. Thomas (unpublished)
16	0.77	1.75	0.80	3–337	73	0–13.0	4.1	H. J. Thomas (unpublished)
8	0.33	0	0.33	27–332	129	0.9–18.5	5.9	H. J. Thomas (unpublished)
9	0.80	0	0.80	11–79	44	0–10.2	4.8	H. J. Thomas (unpublished)
0	0	—	0	—	—	—	—	H. J. Thomas (unpublished)
60	20.7	5.5	14.7	8–347	119	0–1.0	0.25	C. J. Chapman (unpublished)

creels in Loch Torridon, 23 out of 60 recaptures (38%) showed no significant movement from the release site, and the average movement was only 0.25 km.

In all the experiments listed in Table II, the proportion of *Nephrops* recaptured was small. The reasons for this are uncertain since no control experiments were conducted to assess the likely survival rate of tagged *Nephrops* and the reliability of reported recaptures. Storrow (1913) considered that his tail-clipped animals were difficult to detect and Jensen (1965) found that 18 out of 90 (20%) marked animals that returned were sent from the market, having been missed by the fishermen. The mortality of trawl-caught *Nephrops*, is undoubtedly high after liberation, even without tagging (Symonds and Simpson, 1971). Creel-caught *Nephrops* would probably survive better than those caught by trawl, and this may have contributed to the higher proportion of returns in Loch Torridon. In P. Warren's experiment (unpublished) in the North Sea, tagged *Nephrops* were released outside the main fishing area in order to study movements into the fishery. His low return rate is indicative of little movement, even though the eight recaptured *Nephrops* had moved appreciably. This experiment is the first in which the animals were handled under dim lighting conditions to reduce the risk of damage to the eyes (P. Warren, personal communication).

Local movements of *Nephrops* have been studied by combining diving and tagging methods. Chapman and Rice (1971) tagged all the animals occupying burrows within a small area of sea bed and returned them to the same burrows. Although some individuals were recorded in the same burrow over a period of several days, others frequently changed their burrows. Large *Nephrops* often engaged in ritualized fighting for the possession of a burrow. The detailed movements of five *Nephrops* were recorded by attaching miniature ultrasonic transmitters and tracking them by means of a hydrophone array (Chapman *et al.*, 1975). In this study (but not the earlier one) tagging was conducted underwater to minimize handling and prevent damage to the eyes. The results suggested a sex difference in the degree of attachment to particular burrows. Two females made short excursions but always returned to the same burrow. Three males moved further afield, covering total distances of 24–71 m/day, never returning to the same burrow during the period of observation. Most exploited *Nephrops* grounds are likely to contain many empty burrows, and local movements between burrows probably occur frequently.

IV. POPULATION STRUCTURE AND LIFE CYCLE

Although direct methods of population assessment show promise for the future, only a limited use of these methods has been made so far (Section II, D). Much of our present knowledge of the ecology of *Nephrops* is therefore derived from the analysis of catch data from research and commercial fishing vessels. As

4. Ecology of Juvenile and Adult *Nephrops*

Nephrops can only be caught when they emerge from their burrows, it is obvious that the discussion in Section III is important for the interpretation of catch data.

A. Length and Sex Composition of Stocks

Information on the length and sex composition of *Nephrops* is collected in most countries by regular sampling of commercial catches and by research vessel surveys. There is a wealth of such data and only a brief summary of the important facts can be given here.

The length frequency distributions of catches obviously depend to a large extent on the mesh sizes and selection properties of the trawls used. Mesh sizes adopted by countries vary (Chapter 8 of this volume) and this makes it difficult to compare data from different stocks. The effect of mesh selection is illustrated in Fig. 9. This compares length frequency distributions obtained by photography and by trawl catches in the Sound of Jura (Chapman, 1979). These distributions have been combined to show the probable length distribution of the fully recruited population. As we shall see (Section IV, C), there is evidence that the juvenile *Nephrops*, though present on the grounds, does not emerge from its burrow during the first year. This accounts for the rising left-hand side of the curve in Fig. 9.

The length and sex composition of catches vary diurnally and seasonally. Andersen (1962) found that large *Nephrops* (≥ 52 mm CL) formed a higher proportion of day time trawl hauls compared to night hauls. Jensen (1965) and

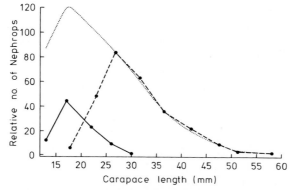

Fig. 9. Length frequency distributions of *Nephrops* (males and females combined) in the Sound of Jura determined by photography (●—●, 89 specimens) and by 70 mm mesh trawl (O- - - - -O, 290 specimens). The smooth dotted line (......) shows the probable length frequency distribution of the recruited population obtained by scaling up the photographic sample and combining the two sets of data. Numbers in 5 mm length groups plotted against the mean length within each group. [Modified from Chapman (1979), by permission of ICES.]

Farmer (1974a) reported similar observations. Chapman and Howard (1979) observed small but consistent differences in the mean size of *Nephrops* taken at different times in Scottish waters. The mean size was smallest during periods of peak catches at sunrise and sunset and was greatest during the day time. These results probably reflect variations in the emergence rhythm of *Nephrops* of different sizes (Section III).

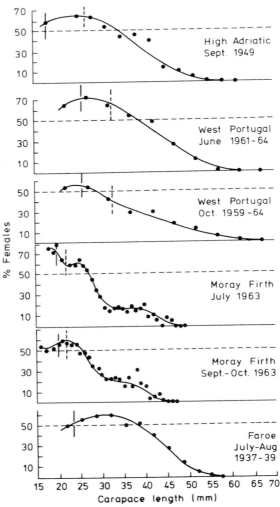

Fig. 10. The percentage of females in the total catch of *Nephrops* in each length group for different seasons and areas. The lowest (solid vertical line) and 50% (broken line) maturity lengths are shown. [From Figueiredo and Thomas (1976b), by permission of ICES.]

4. Ecology of Juvenile and Adult Nephrops

After reaching sexual maturity, the male *Nephrops* molts more frequently than the female, consequently growing faster. As a result, the males account for an increasing proportion of the larger size categories in the population and in the catch (Fig. 10). The recent increase in fishing effort in most countries (Chapter 8 of this volume) has brought about a general reduction in the mean size of *Nephrops* in the population and catch, with a corresponding increase in the proportion of females (Eiriksson, 1968).

Berried female *Nephrops* are largely absent from trawl catches (Section III, C) so that the sex ratio in the catch and the overall catch per unit of fishing effort varies seasonally according to the phase of the breeding cycle in different areas (Figueiredo and Thomas, 1967b; Farmer, 1975). Very few females are caught at any time of the year at Iceland, males accounting for 95% by weight of the landings (Eiriksson, 1979). In Scottish waters and in the Irish Sea, females are well represented in catches only during the summer months (June–August). Spawning takes place in August–September and thereafter females account for only 20–30% of the catch (Fig. 11B). However in the Farn Deep area of the North Sea, catches with a high proportion of females (40–50%) are taken during the egg bearing season of October–April (Fig. 11B) (Symonds, 1972b). In the

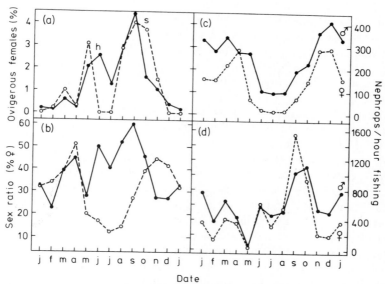

Fig. 11. Average seasonal fluctuation in trawl catches from the Firth of Forth during 1966–1970 (original data) and the Farn Deep during 1962–1967. (a) percentage of mature females carrying eggs and (b) percentage of mature, nonovigerous females in total catch (●—●, Firth of Forth; O- - - -O, Farn Deep). Mean catch/hour of males and nonovigerous females in (c) Farn Deep and in (d) the Firth of Forth. In (a) the two peaks in the graphs correspond to ripe eggs about to hatch (h) and newly spawned eggs (s). (Data for Farn Deep from Symonds, 1972b, by permission of H.M.S.O.)

Adriatic, the females form 60-70% of the catches from February to May and about 40% of the catch at other times. Interestingly, up to 65% of the captured females may be ovigerous (Karlovac, 1953) perhaps because of a high level of foraging activity in the warmer temperatures.

Clearly, the sex ratio in the catches is unlikely to represent the sex ratio in the population. The fact that the proportion of females approaches 50% at certain times of the year in most areas suggests that the true sex ratio is probably near 1:1, at least until sexual maturity is reached. Thereafter, the proportion of females may be higher than males because of the lower rate of exploitation on ovigerous females for most of the year.

B. Fecundity and Reproduction

Comprehensive accounts of the reproductive cycle in *Nephrops norvegicus* have been given by Farmer (1974b) for the Irish Sea and by Thomas (1964) for Scottish waters. The cycle is very similar in these two areas, and the following brief account is based on their observations.

Females reach maturity at a carapace length of 20-23 mm and males at about 26 mm (Fig. 10). The mature *Nephrops* of both sexes molt during the period from May to July, with a peak in May, and mating takes place while the female is "soft." The ovaries mature so that by August-September, fertilization has occurred and egg laying has taken place. The eggs are incubated on the pleopods until the following April-June, when the larvae hatch. The time for full egg development is about 8-9 months, and for most of this time the ovigerous females are mainly confined to their burrows. From studies showing the progressive development of the ovary during June-August, at least 90% of mature females would be expected to spawn each year. Almost all mature females carry spermatophores during the period prior to egg laying.

The chief geographical differences in the breeding cycle are summarized in Table III. The times of spawning and hatching and the length of the incubation period vary with latitude and are probably temperature dependent. Hatching begins earlier in the year off Portugal and in the Adriatic than in more northerly areas. Thomas and Figueiredo (1965) expressed doubts about the biennial spawning frequency at Faroe and in the Farn Deep area of the North Sea (Andersen, 1962; Symonds, 1972b). Andersen's conclusion was based on the relative frequency of newly molted, ovigerous and nonovigerous females in catches taken during July and August. Apart from one area (Skaale Fjord), where Andersen believed the breeding cycle to be retarded by lower temperatures than elsewhere, his data are not inconsistent with an annual spawning. In Skaale Fjord, hatching took place in August at the same time that newly berried females appeared in the catches. In this area it is quite likely that many females would not spawn every year. Elsewhere in the Faroes, hatching reached a peak in May, which would

TABLE III

Details of the Breeding Cycle of *Nephrops* in Different Areas

Area	Latitude	Egg laying (spawning)	Main period egg hatching	Average period of incubation (months)	Periodicity of spawning	Reference
Iceland	64°N	May–Jun	May–Jul	12–13	Biennial	Eiriksson (1970b)
Faroe	62°N	Jun–Aug	May–Aug	9–12	Biennial (?)	Andersen (1962)
Scotland	56–58°N	Aug–Nov	May–Jun	9	Annual	Thomas and Figueiredo (1965)
North Sea (Farn Deep)	55.5°N	Aug–Sep	May–Aug	10	Biennial (?)	Storrow (1912), Symonds (1972b)
Irish Sea	54°N	Aug–Sep	Apr–Jun	8–9	Annual	Farmer (1974b)
Adriatic	44°N	Jun–Jul	Jan–Feb	7	Annual	Karlovac (1953)
Portugal	38°N	Aug–Sep	Feb–Mar	6	Annual	Figueiredo and Barraca (1963)

leave ample time for molting and copulation to take place before the egg laying in August. In the absence of more data for other seasons, evidence regarding the frequency of spawning at the Faroes remains inconclusive.

Symonds (1972b) provided comprehensive data for the Farn Deep fishery, which is worth comparing to Scottish data from the nearby Firth of Forth (Fig. 1). Figure 11A shows that few ovigerous females were caught (maximum 4%), but it is clear that the timing of the breeding cycle is similar in the two areas. The percentage of ovigerous females in these catches shows two peaks, the first in May–June consisting of females carrying brown-colored eggs ready to hatch, and the second in September–October, comprising females with recently spawned green eggs. There is considerable overlap, however, and in July–August eggs at both stages of development are commonly found on different females in the same catch. This overlap between hatching and spawning seasons (Table III) probably arises simply because some females spawn much earlier than others and is not necessarily indicative of biennial spawning. During the winter months, catches from the Farn Deeps show marked increases in the catch per hour (Fig. 11C) and also in the proportion of nonovigerous females (Fig. 11B) in contrast to catches from the Firth of Forth (Fig. 11B, D). Thus, it is clear that a substantial proportion of mature females in the Farn Deep do not spawn every year. This is surprising, because examination of the ovaries showed that by July the majority of females were fully mature. Thus, indications are that mating and fertilization does not take place in a large proportion of the females. Figure 11C shows that from May to August the catch per unit effort of both sexes fell to a low level. Evidently *Nephrops* at this time are less available to the trawl in the Farn Deep area, presumably because they tend to remain in their burrows (Section III). If this is true, there may be less opportunity for mating, which must take place soon after the female has molted. This possibility could be checked by determining the proportion of females carrying spermatophores during July–August. The reason why *Nephrops* may be less available to capture during May–August is uncertain. One possibility, suggested by Symonds (1971), was that large concentrations of cod on the grounds at that time may have caused *Nephrops* to remain in their burrows.

Information on the fecundity of *Nephrops* from different areas is summarized in Table IV. Two sets of data are given relating the carapace length of the female to (1) the number of oocytes in the ovary and (2) the number of eggs on the pleopods. Initially, the number of eggs on the pleopods varies between 150 and 5000, but there is a progressive loss of eggs during the course of development. By the time of hatching, Figueiredo and Nunes (1965) estimated that about 70% of the initial number of eggs are lost during development. A lower figure of 40% egg loss was estimated by Morizur (1979) for *Nephrops* in the Bay of Biscay.

There appear to be geographical differences in fecundity. Thomas (1964) found significant differences between five locations around Scotland.

TABLE IV

Fecundity of *Nephrops*

Area	Equation[a]	Size range examined (mm)	Approx. range in number of eggs	Reference
A. Based on number of oocytes				
Scotland (overall)	$F = 0.526L^{2.350}$		—	
Firth of Forth	$F = 0.803L^{2.269}$		550–4600	
Moray Firth	$F = 0.408L^{2.450}$	20–45	450–4400	Thomas (1964)
Minch	$F = 0.808L^{2.204}$		250–3600	
Clyde	$F = 0.353L^{2.423}$		150–3200	
E. Shetland	$F = 0.128L^{2.758}$		400–4000	
Portugal	$\log F = 3.0557 \log L - 1.3717$	25–45	800–5000	Figueiredo and Nunes (1965)
Iceland	—	25–50	640–4400	Eiriksson (1970b)
France (Golfe de Gascogne)	$F = 0.108L^{2.797}$	23–41	750–3400	Fontaine and Warluzel (1969)
B. Based on number of eggs on pleopods				
Portugal				
Newly laid eggs, 0–16 weeks	$\log F = 3.0557 \log L - 1.6387$	25–45	430–2600	Figueiredo and Nunes (1965)
Before hatching, 20–24 weeks	$\log F = 3.0557 \log L - 1.8217$	25–45	280–1700	
Iceland (time after spawning not given)	—	25–50	500–2800	Eiriksson (1970b)
Irish Sea	$F = 78.204L - 1376.421$	22–35	340–1360	Farmer (1974b)

[a] L, carapace length (mm); F, number of eggs.

C. Juvenile Phase and Recruitment

Allowing for the egg losses during incubation, most ovigerous females would be expected to produce between 250 and 2000 pelagic larvae during each breeding season, depending on their size. There are three pelagic larval stages (spanning from 2 to 3 weeks) before metamorphosis to the juvenile or postlarval stage that settles on the sea bed (Figueiredo and Thomas, 1967a; Farmer, 1975). The few studies that have been made on the distribution of the larval stages suggest that they are most abundant over the adult beds. Aquarium studies show that the postlarval stage burrows in mud (Crnkovic, 1968).

At metamorphosis, juvenile *Nephrops* have a carapace length of 3–4 mm (Farmer, 1973; Figueiredo, 1975). Most of them probably settle on the sea bed soon after metamorphosis, although some may molt once or twice before settlement. This is suggested by the occurrence of juveniles, 5–7 mm CL, in catches made by a small mesh pelagic trawl in Scottish waters during July (S. Hay, personal communication). The juveniles grow rapidly and attain a mean length of 14 mm after one year, having molted up to ten times (Farmer, 1973). At a size of 11–15 mm CL, small numbers start appearing in commercial trawl catches, but prior to reaching this size few are caught even using small mesh nets. In the Irish Sea, Hillis (1972), using a bottom trawl with small mesh covers over the cod end and other parts of the net, caught a few juveniles of 4 mm CL in June–July and seven specimens of 7–9 mm CL in a total catch of 2200 larger individuals (13–22 mm CL) in October–November. By the following April the catches showed a prominent carapace length mode centered on 11 mm. Considering the large numbers of juveniles that must settle on the sea bed to give rise to the adult populations and the large numbers that can be caught in mid water, these catches of juveniles on the sea bed are very small indeed. These observations indicate that the juveniles inhabit burrows in the same areas as the adults and rarely emerge from their burrows during most of their first year, until they have reached a size of 10–15 mm CL. This would account for the length frequency distribution obtained by photography in Fig. 9, showing a gradual increase in the frequency of juveniles of 13 mm CL or greater. In these photographs, and in many others from different locations, the smallest *Nephrops* seen on the sea bed had a carapace length of 11 mm. In addition, if large numbers of juveniles were present on the mud surface, one would expect them to be found in the stomach contents of predatory fish (Section III). In Scottish waters, examination of the stomachs of large numbers of fish feeding on *Nephrops* have not produced a single juvenile (H. J. Thomas and G. Howard, personal communication). Conversely, Andersen (1962) found 33 juveniles, 5–8 mm CL, in stomaches of the ray, *Raja clavata*. Since juveniles of up to 7 mm can be caught in mid-water trawls, it is possible that some juveniles were eaten by rays at the time of settlement on the sea bed.

The burrows of juvenile *Nephrops* have seldom been found by divers in shallow water. Rice and Chapman (1971, 1974) found a few burrows and cast them in polyester resins. These casts give some clues about the possible mode of life of juvenile *Nephrops*. The most interesting feature of some of the casts is that they show small tunnels linked to larger ones, known to be used by adult *Nephrops*. A good example is illustrated in Fig. 5. This cast consists of three tunnel systems of different diameter, each occupied by different sized male *Nephrops*, measuring 41 mm CL (in the large tunnel between openings 1 and 8), 22 mm CL (tunnel between openings 0, 7, and 6), and 8 mm CL (multibranched tunnels marked by arrows). The latter tunnel has three vertical shafts leading to small circular openings on the mud surface. These ventilation shafts are unlikely to have been used for access which lends further support to the idea that juveniles do not normally emerge. This would imply that juvenile *Nephrops* must feed within the burrow. Crnkovic (1968) concluded that juveniles tunneled to feed, which could explain the complexity of their burrows (Fig. 5). The mud is very rich in organic material, particularly in the vicinity of *Nephrops* burrows (McIntyre, 1973). Hillis (1974) also reported two cases of interconnection between burrows occupied by *Nephrops* of different sizes, but no accurate size measurements of the occupants were given.

Rice and Chapman (1974) suggested two ways in which tunnel interconnections could arise. In areas of high density it would be possible for the burrows of two or more individuals to connect by chance. Alternatively, linkages may arise at the time the metamorphosed *Nephrops* settles on the sea bed. The authors have suggested that instead of risking the predation of fish by burrowing through the mud surface, the juveniles enter already existent adult burrows, and only then excavate their own tunnels through the walls of the adult burrow.

D. Growth

The processes of molting and growth of *Nephrops* are basically similar to those already described in Volume I. Saila and Marchessault (Chapter 6 of this volume) discuss various mathematical formulations of growth and give some equations for *Nephrops*.

Growth is a function of the molting frequency and the increase in length (or weight) at each molt. Growth increments in length have been measured from *Nephrops* molting in aquaria (Thomas, 1965b; Hillis, 1971a; Farmer, 1973; Figueiredo, 1975) and in cages at sea (Charuau, 1977). The results were generally presented as a graph of postmolt against premolt length (Hiatt diagram) or as percentage increase in length (growth factor) against premolt length. Neither of these methods are satisfactory, since the same parameter (premolt length) appears in both variables, virtually guaranteeing an association between them and masking any underlying biological relationship. A better way of examining the data is simply to plot the growth increment against premolt length. This has been done in Fig. 12. The results of Thomas (1965b), obtained under aquarium conditions, show that the molt increment tends to increase with premolt length in both sexes over the size range examined (Fig. 12A). Conversely, the results of Charuau (1977) reveal no correlation in males and a significant negative correlation in females (Fig. 12B). Whether these differences reflect geographical variation in growth rate or whether they were caused by the experimental conditions is uncertain. Charuau and Conan (1977) have suggested that differences in growth

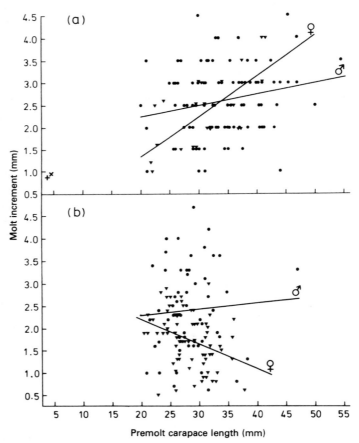

Fig. 12. Graphs showing the increase in length at molting plotted against premolt carapace length for male (●) and female (▼) *Nephrops*. (a) growth data from Thomas (1965b) obtained in aquarium conditions. Both regressions are statistically significant ($P<.05$). Data for postlarvae from Figueiredo (1975), x, mean of 27 observations; and Hillis (1971a), +, mean of 5 observations. (b) growth data from Charuau (1977) obtained in enclosures at sea. The regression for females is statistically significant ($P<.01$).

4. Ecology of Juvenile and Adult Nephrops

between the sexes could be reduced under aquarium conditions, when metabolic energy may be directed to growth rather than reproduction.

Information on the frequency of molting is scanty. Data from Thomas (1965b) show that the time interval between successive molts increases with size in male *Nephrops* kept in aquarium conditions. The average interval was 109 days (frequency 3–4 molts per year) at a carapace length of 20–30 mm, increasing to 334 days (approximately 1 molt per year) at a carapace length above 40 mm.

There appear to be definate seasons during which molting takes place. Charuau (1975) and Conan (1978) have shown that the proportion of *Nephrops* about to molt or recently molted ("soft") in the population off South Brittany is greatest during the spring and autumn, although the precise timing of these molting periods varies from year to year. It is clear that mature females molt only once per year during the spring, whereas mature males molt more frequently in the period from spring to autumn.

Annual growth rates have been estimated from the progression of prominent length modes in successive size–frequency distributions of trawl catches (Hillis, 1972; Farmer, 1973; Conan, 1978). While this method gives a satisfactory estimate of annual growth for the first few year classes, the modes corresponding to older age groups are very difficult to detect. For these larger *Nephrops*, an extensive tagging program employing tags that would remain in place through the stages of molting would appear to be the only way to measure their growth.

There is evidence that the mean size of *Nephrops* in a population is inversely related to its density. At one extreme in the Sound of Jura, there is a very high density of mainly juvenile *Nephrops* (Fig. 9; Chapman, 1979). At the other extreme, low density populations of predominantly large *Nephrops* are found in some Scottish sea lochs (Chapman and Rice, 1971). Supporting evidence from fishing data is difficult to interpret because so many factors influence the size and sex composition of catches. Variations in density may be caused by physical factors such as the nature of the substratum and its suitability for burrowing, or by patchy settlement of the postlarval stage. Following heavy postlarval settlement, it is possible that competition for available food would limit growth. This may account for the lower mean size of *Nephrops* in high density populations.

E. Mortality

Little information is available on the mortality rate of *Nephrops*, though it is undoubtedly high in commercially exploited populations. Based on the numerical strength of assumed age groups, the instantaneous rate of mortality Z was estimated to be about 1.0 (37% survival) in the Irish Sea (Anonymous, 1976) and 1.3 (27% survival) off northwest Spain between 1974 and 1976 (Fernández-Garcia, 1976). From cohort analyses of length composition data, using the method of Jones (1979), Eiriksson (1979) obtained estimates of the instantaneous rate of fishing mortality F from 0.22 to 0.62 in the Iceland fishery during 1962–1973. This fishery is interesting because males account for over 95% of the landings by weight and thus, mortality of the females due to fishing is very small.

No reliable information is available on the rate of natural mortality M, though it is generally thought to be high (Thomas, 1960). Predation by fish is undoubtedly one of the major natural causes of mortality, but as we have seen, the protection afforded by the burrow and the restriction of emergence to low light levels must considerably improve *Nephrops'* chances of survival. There is evidence that they tend to remain in their burrows for long periods of particular susceptibility, e.g., when newly molted, ovigerous, or juvenile.

The only known common parasite in *Nephrops* is the trematode, *Stichocotyle nephropis*. Their larval cysts are found attached to the outer wall of the lobster's hindgut. The distribution and the level of infestation in British waters were reported by MacKenzie (1963) and Symonds (1972a). This parasite occurs mainly in stocks to the west of the British Isles, the highest incidence being in the Sound of Jura where 20% of all *Nephrops* examined were infected. It is not known whether metabolism and/or growth is affected by the parasite. Although the growth rate of *Nephrops* in the Sound of Jura may be lower than elsewhere, it is unlikely that the relatively inactive larval stage of *Stichocotyle* can be wholly responsible for this. The high incidence of the parasite is more likely to be caused by the ease of transmission in such a dense population via the primarly host, *Raja clavata*.

V. CONCLUSIONS

The ecology of *Nephrops* is dominated by its burrowing behavior. Diurnal and seasonal rhythms of emergence, which frequently differ between the sexes and with size, are generally reflected in trawl catches. For this reason, the behavior aspects have been discussed in some detail in this chapter, so that the material will prove useful in the interpretation of catch data. Many countries carry out regular trawling surveys to assess the distribution and abundance of *Nephrops* populations from year to year. Although Andersen (1962) postulated a relationship between the activity of *Nephrops* and light intensity, he is the only author to have applied appropriate corrections to his catches. In planning trawling surveys to estimate abundance, it is important to ensure that hauls are timed to coincide with the period of emergence.

Hitherto, little attention has been given to population dynamics of *Nephrops* populations (hence the absence of a separate section devoted to the subject in this volume). For the majority of populations there is a general lack of basic information on recruitment, growth, and natural mortality, as well as on the selectivity of fishing gear. Jones (1979) has carried out a preliminary assessment of the *Neph-*

rops population in the Firth of Forth in order to determine the effect of changes in fishing effort and mesh size on yields. It was assumed that recruitment was constant, and a range of values for the other parameters was used in the assessments. These gave a range of estimates of stock size and density which were then compared with other values derived independently from underwater television observations (Chapman, 1979). In this way it was possible to set narrower limits to the growth and mortality parameters. Thus, he was able to conclude that long-term benefit would be derived from a reduction in fishing effort or an increase in mesh size (from 70 to 90 mm), providing that the instantaneous rate of natural mortality did not exceed 0.2.

Tagging experiments have been conducted with only limited success. The low recapture rate (Table II) is probably indicative of high mortality after release. With care in handling and adherence to precautions that avoid damage to the eyes, there is no reason why tagging methods should not be more successful in the future in providing much needed information on growth and mortality rates.

REFERENCES

Andersen, F. S. (1962). The Norway lobster in Faroe waters. *Medd. Danm. Fisk. Havunders.* [N. S.] **3,** 265-326.
Anonymous (1976). Report of the working group on *Nephrops* stocks. *ICES Coop. Res. Rep.* No. 55, pp. 1-42.
Aréchiga, H., and Atkinson, R. J. A. (1975). The eye and some effects of light on locomotor activity in *Nephrops norvegicus. Mar. Biol.* **32,** 63-76.
Atkinson, R. J. A. (1974a). Behavioural ecology of the mud-burrowing crab *Goneplax rhomboides. Mar. Biol.* **25,** 239-252.
Atkinson, R. J. A. (1974b). Spatial distribution of *Nephrops* burrows. *Estuarine Coastal Mar. Sci.* **2,** 171-176.
Atkinson, R. J. A., and Naylor, E. (1973). Activity rhythms in some burrowing decapods. *Helgol. Wiss. Meeresunters.* **24,** 192-201.
Atkinson, R. J. A., and Naylor, E. (1976). An endogenous activity rhythm and the rhythmicity of catches of *Nephrops norvegicus* (L.). *J. Exp. Mar. Biol. Ecol.* **25,** 95-108.
Bagge, O., and Münch-Petersen, S. (1979). Some possible factors governing the catchability of Norway lobster in the Kattegat. *Rapp. P.-V. Reun., Cons. Perm. Int. Explor. Mer* **175,** 143-146.
Chapman, C. J. (1979). Some observations on populations of the Norway lobster, *Nephrops norvegicus* (L.) using diving, television, and photography. *Rapp. P.-V. Reun., Cons. Perm. Int. Explor. Mer* **175,** 127-133.
Chapman, C. J., and Howard, F. G. (1979). Field observations on the emergence rhythm of the Norway lobster, *Nephrops norvegicus* (L.) using different methods. *Mar. Biol.* **51,** 157-165.
Chapman, C. J., and Rice, A. L. (1971). Some direct observations on the ecology and behaviour of the Norway lobster, *Nephrops norvegicus. Mar. Biol.* **10,** 321-329.
Chapman, C. J., Priestley, R., and Robertson, H. (1972). Observations on the diurnal activity of the Norway lobster, *Nephrops norvegicus* (L.). *ICES, C.M.* **K:20,** 1-18.
Chapman, C. J., Johnstone, A. D. F., and Rice, A. L. (1975). The behaviour and ecology of the

Norway lobster, *Nephrops norvegicus* (L.). *Proc. 9th Eur. Mar. Biol. Symp.*, pp. 59-74. Aberdeen University Press.

Charuau, A. (1975). Croissance de la langoustine sur les fonds due sud-Bretagne. *ICES, C.M.* **K:11**, 1-5.

Charuau, A. (1977). Essai de détermination du taux d'accroissement à la mue de la langoustine dans le milieu naturel. *ICES, C.M.* **K:25**, 1-3.

Charuau, A., and Conan, C. (1977). Growth at moult of Norway lobsters: Methods of assessment, sexual dimorphism and geographic trends. *ICES, C.M.* **K:34**, 1-10.

Conan, G. Y. (1978). Life history, growth, production and biomass modelling of *Emerita analoga, Nephrops norvegicus*, and *Homarus vulgaris* (Crustacea, Decapoda). Ph.D. thesis. University of California, San Diego, 1-349.

Crnkovic, D. (1968). Some observations regarding the burrows of juvenile *Nephrops norvegicus* (L.). *Rapp. P.-V. Reun., Cons. Int. Explor. Sci. Mer Mediterr.* **19**, 171-172.

Dybern, B. I., and Höisaeter, T. (1965). The burrows of *Nephrops norvegicus* (L.). *Sarsia* **21**, 49-55.

Eiriksson, H. (1968). The effect of fishing on catch per effort, size, and sex-ratio of the Norway lobster (*Nephrops norvegicus* L.) in Icelandic waters during the years 1962-1967. *ICES, C.M.* **K:4**, 1-7.

Eiriksson, H. (1970a). The *Nephrops* fishery of Iceland 1958-1969. *ICES, C.M.* **K:5**, 1-5.

Eiriksson, H. (1970b). On the breeding cycle and fecundity of the Norway lobster at South-west Iceland. *ICES, C.M.* **K:6**, 1-3.

Eiriksson, H. (1979). A study of the Icelandic *Nephrops* fishery with emphasis on stock assessments. *Rapp. P.-V. Reun., Cons. Perm. Int. Explor. Mer* **175**, 270-279.

Farmer, A. S. D. (1973). Age and growth in *Nephrops norvegicus* (Decapoda: Nephropidae). *Mar. Biol.* **23**, 315-325.

Farmer, A. S. D. (1974a). Field assessments of diurnal activity in Irish Sea populations of the Norway lobster, *Nephrops norvegicus* (L.) (Decapoda: Nephropidae). *Estuarine Coastal Mar. Sci.* **2**, 37-47.

Farmer, A. S. D. (1974b). Reproduction in *Nephrops norvegicus* (Decapoda: Nephropidae). *J. Zool.* **174**, 161-183.

Farmer, A. S. D. (1975). Synopsis of biological data on the Norway lobster, *Nephrops norvegicus* (Linnaeus, 1758). *FAO Fish. Synopsis* No. 112, pp. 1-97.

Fernández-Garcia, A. (1976). Data on the Norway lobster population of Galicia (NW Spain). *ICES, C.M.* **K:29**, 1-11.

Figueiredo, M. J. (1975). Some studies on the growth of early post-larvae of *Nephrops norvegicus* (L.) reared from the egg. *ICES, C.M.* **K:16**, 1-9.

Figueiredo, M. J., and Barraca, I. F. (1963). Contribuicão para o conhecimento da pesca e da biologia do Lagostim (*Nephrops norvegicus* L.) na costa postuguesa. *Notas Estud. Inst. Biol. Mar. Lisb.* No. 28, pp. 1-44.

Figueiredo, M. J., and Nunes, M. C. (1965). The fecundity of the Norway lobster, *Nephrops norvegicus* (L.) in Portuguese waters. *ICES, C.M.* Doc. 34, pp. 1-5.

Figueiredo, M. J., and Thomas, H. J. (1967a). *Nephrops norvegicus* (Linneaus, 1758). Leach—a review. *Oceanogr. Mar. Biol.* **5**, 371-407.

Figueiredo, M. J., and Thomas, H. J. (1967b). On the biology of the Norway lobster, *Nephrops norvegicus* (L.). *J. Cons. Perm. Int. Explor. Mer.* **31**, 89-101.

Fontaine, B., and Warluzel, M. (1969). Biologie de la Langoustine du Golfe de Gascogne *Nephrops norvegicus* (L.). *Rev. Trav. Inst. Peches Marit.* **33**, 223-246.

Gibson, F. A. (1967). Stocks of *Nephrops norvegicus* off the south coast of Ireland. *Ir. Fish. Invest., Ser. B* **1** (1), 1-11.

Hammond, R. D., and Naylor, E. (1977a). Effects of dusk and dawn on locomotor activity rhythms in the Norway lobster *Nephrops norvegicus*. *Mar. Biol.* **39**, 253–260.

Hammond, R. D., and Naylor, E. (1977b). Laboratory study of the emergence pattern of *Nephrops* from burrows in habitat mud. *Mar. Behav. Physiol.* **4**, 183–186.

Hillis, J. P. (1971a). Growth studies in *Nephrops*. *ICES, C.M.* **K:2**, 1–3.

Hillis, J. P. (1971b). Effects of light on *Nephrops* catches. *ICES, C.M.* **K:3**, 1–5.

Hillis, J. P. (1972). Juvenile *Nephrops* caught in the Irish Sea. *Nature (London)* **238**, 280–281.

Hillis, J. P. (1974). A diving study on Dublin bay prawns, *Nephrops norvegicus* (L.) and their burrows off the east coast of Ireland. *Ir. Fish Invest., Ser. B* No. 12, pp. 1–9.

Högland, H., and Dybern, B. I. (1965). Diurnal and seasonal variations in the catch-composition of *Nephrops norvegicus* (L.) at the Swedish west coast. *ICES, C.M.* **K:146**, 1–6.

Howard, A. E., and Bennett, D. B. (1979). The substrate preference and burrowing behaviour of juvenile lobsters (*Homarus gammarus* (L.)). *J. Nat. Hist.* **13**, 433–438.

Jensen, A. J. C. (1965). *Nephrops* in the Skagerak and Kattegat (length, growth, tagging experiments and changes in stock and fishery yield). *Rapp. P.-V. Reun., Cons. Perm. Int. Explor. Mer* **156**, 150–154.

Jones, R. (1979). An analysis of a *Nephrops* stock using length composition data. *Rapp. P.-V. Cons., Perm. Int. Explor. Mer* **175**, 259–269.

Karlovac, O. (1953). An ecological study of *Nephrops norvegicus* (L.) of the High Adriatic. *Izvj. Inst. Oceanogr. Ribarst.* **5**, 1–51.

Kitching, J. A., Ebling, F. J., Gamble, J. C., Hoare, R., McLeod, A. A. Q. R., and Norton, T. A. (1976). The ecology of Lough Ine. XIX. Seasonal changes in the western trough. *J. Anim. Ecol.* **45**, 731–758.

Loew, E. R. (1976). Light and photoreceptor degeneration in the Norway lobster, *Nephrops norvegicus* (L.). *Proc. R. Soc. London, Ser. B* **193**, 31–44.

McIntyre, A. D. (1973). Meiobenthos. *Proc. Challenger Soc.* **4**, 1–9.

McIntyre, A. D., and Johnston, R. (1975). Effects of nutrient enrichment from sewage in the sea. *In* "Discharge of Sewage from Sea Outfalls" (A. L. H. Gameson, ed.), pp. 1–16. Pergamon, Oxford.

MacKenzie, K. (1963). *Stichocotyle nephropis* Cunningham, 1887 (Trematoda) in Scottish waters. *Ann. Mag. Nat. Hist.* [13] **6**, 505–6.

Morizur, Y. (1979). Evaluation de la perte d'oeufs lors de l'incubation chez *Nephrops norvegicus* dans la région de Sud-Bretagne. *ICES, C.M.* **K:45**, 1–9.

Naylor, E., and Atkinson, R. J. A. (1976). Rhythmic behaviour of *Nephrops* and some other marine crustaceans. *Perspect. Exp. Biol.* **1**, 135–143.

Rae, B. B. (1967). The food of cod in the North Sea and on West of Scotland grounds. *Mar. Res.* No. 1, pp. 1–68.

Rice, A. L., and Chapman, C. J. (1971). Observations on the burrows and burrowing behaviour of two mud-dwelling decapod crustaceans, *Nephrops norvegicus* and *Goneplax rhomboides*. *Mar. Biol.* **10**, 330–342.

Rice, A. L., and Chapman, C. J. (1974). Scampi. *Sea Front.* **20**, 258–268.

Rice, A. L., and Johnstone, A. D. F. (1972). The burrowing behaviour of the gobiid fish *Lesueurigobius friesii* (Collett). *Z. Tierpsychol.* **30**, 431–438.

Simpson, A. C. (1965). Variations in the catches of *Nephrops norvegicus* at different times of day and night. *Rapp. P.-V. Reun., Cons. Perm. Int. Explor. Mer* **156**, 186–189.

Storrow, B. (1912). The prawn (Norway lobster, *Nephrops norvegicus*), and the prawn fishery of North Shields. *Rep. Dove Mar. Lab.* **1**, 10–31.

Storrow, B. (1913). The prawn (Norway lobster, *Nephrops norvegicus*), and the prawn fishery of North Shields. *Rep. Dove Mar. Lab.* **2**, 9–12.

Symonds, D. J. (1971). The *Nephrops* fisheries of England and Wales. *ICES, C.M.* **K:18,** 1-4.

Symonds, D. J. (1972a). Infestation of *Nephrops norvegicus* (L.) by *Stichocotyle nephropis* Cunningham in British waters. *J. Nat. Hist.* **6,** 423-426.

Symonds, D. J. (1972b). The fishery for the Norway lobster, *Nephrops norvegicus* (L.) off the north-east coast of England. *Fish. Invest. (London), Ser. 2* **27,** 1-35.

Symonds, D. J., and Simpson, A. C. (1971). The survival of small *Nephrops* returned to the sea during commercial fishing. *J. Cons. Perm. Int. Explor. Mer* **34,** 89-98.

Thomas, H. J. (1960). *Nephrops*. VI. Some problems associated with regulation of the Norway lobster (*Nephrops norvegicus* L.). *ICES, C.M.* Doc. 181, pp. 1-6.

Thomas, H. J. (1964). The spawning and fecundity of the Norway lobsters (*Nephrops norvegicus* L.) around the Scottish coast. *J. Cons., Cons. Int. Explor. Mer* **29,** 221-229.

Thomas, H. J. (1965a). The white-fish communities associated with *Nephrops norvegicus* (L.) and the by-catch of white fish in the Norway lobster fishery, together with notes on Norway lobster predators. *Rapp. P.-V. Reun., Cons. Perm. Int. Explor. Mer* **156,** 155-160.

Thomas, H. J. (1965b). The growth of Norway lobsters in aquaria. *Rapp. P.-V. Reun., Cons. Perm. Int. Explor. Mer* **156,** 209-216.

Thomas, H. J., and Davidson, C. (1962). The food of the Norway lobster, *Nephrops norvegicus* (L.). *Mar. Res.* No. 3, pp. 1-15.

Thomas, H. J., and Figueiredo, M. J. (1965). Seasonal variations in the catch composition of the Norway lobster, *Nephrops norvegicus* (L.) around Scotland. *J. Cons. Perm. Int. Explor. Mer* **30,** 75-85.

Part II
MANAGEMENT

Introduction

D. A. HANCOCK

The restriction of the subject matter of this book to "lobsters" may initially appear to confine the field of interest to a very narrow base. However, on closer inspection it becomes clear that the lobster group contains species exhibiting highly diversified behavior that influences the conduct of fisheries dependent upon them and provides variety in their population dynamics, a proper understanding of which is required for the various management techniques preferred by the countries concerned.

The genera and species of lobster have in common the crustacean mode of discontinuous growth by ecdysis, separate sexes, high fecundity, incubation of attached eggs and early larvae after either internal or external fertilization, extended periods of free larval development, and so on. They share important difficulties of assessment, particularly in relation to growth and aging. Some of these difficulties have been satisfactorily overcome with recent innovations (e.g., a tag designed to survive ecdysis), but for others, particularly age determination, technical problems remain. In Chapter 8 of this volume, Dow includes aging among a long list of biological parameters for which American lobster fisheries lack information. The problem was discussed in a recent review (Hancock, 1977) where it was concluded that little progress has been made toward the identification of individual age in crustacea.

The biology and behavior of the various species of lobsters have been examined minutely in Volume I. Of particular importance to the operation and management of a fishery are the size, shape, and structure of the individuals of a species, their density, abundance and population age structure, their habitat requirements, their short-term behavioral responses to diurnal stimuli, their longer-term behavior (particularly their migrations), and their interrelationships with fishing gear (i.e., trawl and pot selection, susceptibility to entrapment, etc.).

For example, *Nephrops norvegicus* lives territorially in burrows, emerging periodically to provide daily and seasonal fluctuations in trawl catches. Activity is inversely correlated with light intensity, although other variables also influence catch rates (Farmer, 1975). As with other lobster species, when traps are used, the best catches are taken during the night. Species that do not burrow (e.g., spiny lobsters, clawed lobsters) are usually cryptic in habit, favoring crevices between rocks and under reefs, where they can normally be taken only in traps or by divers. The exception to this is during periods where migrations take place over bottoms suitable for trawling. For example, *Panulirus ornatus* is taken by trawling in the Gulf of Papua and *Palinurus gilchristi* in deep water off the southern coast of Africa. Animals will be particularly vulnerable to capture by trawling during migrations across sand. In deference to this, *Panulirus argus* which takes part in such spectacular migrations, (Chpt. 7, Vol. I), and *Jasus edwardsii* in New Zealand may no longer be trawled by law.

In passing, it is worth referring to the fact that the choice of legal method of capture is very often the outcome of heated debate between amateur and professional users of the resource, as well as between traditional trap fishermen, divers, and trawl fishermen.

Data obtained from commercial or experimental fishing must be examined for bias in order to pool and compare results. Size composition of the catch may be affected by the action of the gear, e.g., selection by meshes of trawls, by meshes, escape gaps and entrance apertures of traps, and by differences in response to bait by sex and size, including hierarchical interactions between individuals. The latter effects, which are quantitatively referred to as the "catchability" parameter, are likely to occur in both the short term and on a seasonal basis, under the influence of environmental variables such as temperature (Morgan, 1974b), or as a result of some event in the sexual or growth cycle. Factors causing bias to capture may be identified or quantified by comparing with catches obtained by other means, e.g., by diving or by tagging experiments (Morgan, 1974a). Other information required to standardize units of fishing effort for yield assessment purposes includes relative efficiency of different baits, comparisons of different gears used in the same fishery, adjustments needed when more than one species is caught by the same gear, and effects of extended pot immersion or soak time.

The inability to determine age directly has put greater emphasis on the need for cohort identification, though this is usually only practical for the early years, e.g., in *Nephrops* (Nicholson, 1979). While the increment added at each molt is fairly readily measured by tagging experiments, the frequency of molting is less easy to quantify directly. Some success has been achieved by analyzing tagging data on an annual ("anniversary") basis, though this can be wasteful of data unless recaptures can be restricted to essential times. The problem of representing discontinuous growth data by a suitable quantitative description is underlined

Introduction

in Chapter 6 by Saila and Marchesseault. Clearly the purpose for which growth data are to be employed must determine how they are analyzed. For purely descriptive purposes or for "short cut" assessments predicting the effect of varying the minimum legal size on yield (Hancock, 1975), the precise form of the growth model is of much less consequence than when it needs to be constrained to fit the assumptions involved by an encompassing yield model. For example, the isometric growth model of von Bertalanffy is a requirement of the standard Beverton and Holt yield equation, though it can be modified to incorporate other descriptions of growth. Formulation of a suitable crustacean growth model is clearly an area which will continue to command attention.

Management

Lobsters include some of the most individually valuable and popular of fished species and have been in great demand for many years on world markets. Lobster stocks are therefore much less likely to be in the categories of un- and underexploited than of fully- and over-exploited. The Food and Agriculture Organization (1970), in a review of its Indicative World Plan commented, "the (rather reliable) quantitative data for all forms of lobster-like crustaceans reflect the high esteem with which these animals are held in world markets . . . the modest size of the increase in production [51% compared with nearer to 100% for crabs, prawns, and traditionally caught scalefish] is probably just a measure of the high intensity of exploitation, and of management, to which stocks of 'lobsters' are already subjected." Measures to control exploitation, collectively referred to as fisheries management, may range from individual regulations based on empirical reasoning, to comprehensive management policies involving in-depth studies of the population dynamics of the species with a view to some quantitative model to describe the fishery. The sophistication of the approach will be determined to a great extent by the relative importance and value of the fishery. Simple regulations may have their origins in sociological, practical, economic, and political, as well as biological considerations. For example, in 1897 the minimum size of *P. cygnus* was introduced in Western Australia, in order to control the size for market acceptability. This minimum size happens to be close to the average size at first maturity. It is interesting to note how often a minimum size established years ago appears from more careful analysis (e.g., Morgan, 1977) to have been correctly chosen. It then becomes important to distinguish whether this apparent success stems from the stabilization of fishing effort at a level appropriate to the long established minimum legal size, or whether the estimates of parameters used for the assessment are sufficiently reliable to have enough confidence in predicted yields to recommend a change.

The highly irregular shape of lobsters, compared with fish, has led to practical difficulties in achieving minimum legal size based on the structure of the gear.* The shape of *Nephrops* in particular has been responsible for a selection range by trawl meshes which is too wide to provide a reliable means for control. Neither is it always easy to adjust escape gap size or shape in lobster traps to ensure the escape of undersized animals while at the same time satisfying fishermen that all of their legal catch will be retained by the gear. Nevertheless, escape gaps are being used successfully in various lobster and spiny lobster fisheries around the world.

Stock and Recruitment

The value of various measures that maximize spawning, including a minimum legal size chosen to allow breeding before first capture, the prohibition of capture from breeding areas or during the breeding season (closed areas and season, respectively), and prohibited landing of spawning females, has never been questioned until recently. However, considerable attention is now being focused on the desirability of such restrictions.

The relationship between stock and recruitment, which may be of fundamental importance to successful management has, apart from the estimation of natural mortality, proved to be one of the most difficult to quantify. Earlier fisheries models were based on the assumption of constant annual recruitment, but the identification of many examples at variance with this led to a detailed review of the subject at a special symposium in 1970 (Parrish, 1973). At that time there were few data available for invertebrates, particularly for the Crustacea (Hancock, 1973). Morgan (this volume, Chapter 5) gives one recent example in which the recruitment of settling larvae of a palinurid is correlated with, but not linearly related to, abundance of the spawning stock, approximating a Ricker model. In this species, however, survival of the recruits into the fishery is controlled by a number of density-dependent factors (Chittleborough and Phillips, 1975). Consequently, there is an apparent lack of correlation between the abundance of spawning stock and recruits into the fishable stock.

It is not surprising therefore that one of the most difficult concepts to win conviction, particularly from fishermen, revolves about the benefits or otherwise of protecting breeding stocks. This stems from an historical indoctrination of the belief of a positive correlation between spawners and recruits. The possibilities and suggestions for controls to protect "motherhood" are endless, e.g., prohibition on landing or (by closed seasons or areas) lifting of "berried" females, protection of mated females (observed by spermathecae attached to palinurids),

---

*Bennett (Chapter 9) details the problem of retention of juvenile finfish by trawl meshes small enough to effectively capture *Nephrops*.

Introduction

protection of *all* females (as with the Alaskan king crab), protection of the largest females because they produce most eggs, and protection of the largest males needed to service the large females. It has to be explained that at each such level of control the immediate catch would be reduced, almost always without adequate scientific justification. It was the lack of such evidence which led the United Kingdom to abandon the universal prohibition of taking berried females, in favor of local controls. Other countries, e.g., Canada (Wilder, 1965) and Australia, have retained berried lobster regulations as a precautionary measure, in the absence of any real evidence of their usefulness. However, it would be a confident manager who could abandon all such restrictions, particularly in a fishery where fishing effort has long since exceeded the level required to achieve maximum sustainable yield and where mathematical imprecision cannot distinguish with certainty whether the total catch is being maintained (Beverton and Holt prediction) or has begun to decline (Schaefer prediction).

Fisheries Models and Management Objectives

Nowadays, choosing the most appropriate size at first capture to ensure maximum yield from a fishery will usually involve at least some reference to a yield assessment model. Such models are well described in the chapters which follow. The stock production model of Schaefer has been a popular choice for management, but it has the limitation that it cannot be used to determine the optimum age at first capture, nor can it be applied to series of data during which age at first capture has been varied. Moreover the model requires a series of catch and effort data over a period when the level of fishing effort has varied significantly. For a fishery in which levels of fishing effort have not varied sufficiently for such a practical approach, a more theoretical description is necessary. This will have the added advantage of providing a basis for predicting the effects of varying a number of parameters, including size at first capture, as well as fishing effort. Notable among such "dynamic pool" models are those of Beverton and Holt and of Ricker, with their respective extensions.

Such catch equations have been in vogue for many years (since Baranov, 1918; Russell, 1931; Graham, 1935), but the idea of managing fisheries by maximum sustainable yield (MSY) first really began to catch on about 30 years ago (Larkin, 1977). Although the concept of MSY became increasingly fashionable, its value has recently been challenged (Larkin, 1977) so that it is perhaps timely to question more closely the purpose and data requirements of management. It is still true to say that few lobster fisheries possess data of the quality and continuity necessary to make a sufficiently precise estimation of MSY for confident management. The few examples that exist have been cited in the following chapters by Morgan and by Dow.

Despite Larkin's conclusion that from a biological standpoint the concept of MSY is simply not sufficient (i.e., it cannot encompass mixed species, it is likely to lead to instability because of the progressive tendency towards less quantitatively and qualitatively successful spawners, etc.) he *does* accept MSY as a valuable rough index of production potential. However, once attained, "it should be expected that it may not be sustained." Misgivings about MSY led naturally to the consideration of the *optimum* sustainable yield, particularly in recreational fisheries where economic criteria are more difficult to define. This concept seeks to rationalize the commercial and recreational exploitation of a stock. Optimum sustainable yield has been defined by Roedel (cited by Larkin) as "a deliberate melding of biological, economic, social and political values designed to produce the maximum benefit to society from stocks that are sought for human use, taking into account the effect of harvesting on dependent, or associated species." The reconciliation of a commercial fishery that depends on quantity rather than size with an amateur fishery that has traditionally sought fewer individuals of record breaking dimensions, is not an easy task. However, aiming toward a total catch that is somewhat less than the predicted MSY results in two important benefits, i.e., the catch more closely approximates the theoretical maximum economic yield, and reduces the possibility of overshooting MSY, which is usually difficult to rectify and could involve biological danger.

A working group of the International Council for the Exploration of the Sea (1976) proposed the identification of a total allowable catch (TAC) which should be modified against the objectives for optimum fishing (a) to maintain the spawning stock size within the defined range and (b) to keep the fishable biomass above the agreed minimum level. From all these considerations it should be possible to think of the yield from "rational" exploitation as somewhere between maximum sustainable yield and maximum economic yield, i.e., the level of exploitation which is at a maximum in terms of public responsibility for a common property resource while achieving an acceptable economic return.

It may be salutary to examine how these somewhat lofty ideals have been met by management practices in the lobster fisheries of the world. Many examples are given in the following chapters of the application of isolated regulations. The individual success of these is difficult to assess without effective management research to integrate them into the broader understanding provided by a yield model. Dow (Chapter 8) concludes that "resource management has rarely been exercised effectively in regulating the clawed lobster fisheries." He gives as an example, a management program that was designed to stabilize catches in the lobster fishery of the State of Maine by adjusting annual fishing effort to sea temperature trends, which, however, did not prevent a decline in landings.

Management by a quota system has met with mixed success in South African rock lobster fisheries. Forecasting of quotas is likely to be unreliable, particularly with short-lived species, and as with management based on temperature predic-

Introduction

tions, can be misinterpreted as a guarantee of catch. Failure of catches to reach quota levels can have embarrassing consequences to administrators. The opposite could be due to inefficient utilization of the resource.

According to Larkin (1977) "the best way of reconciling the MSY and economic religions has been held to be the limitation of entry into a commercial fishery; if there is a continued regulation of the *number* of fishing units and their fishing power, then at least MSY can be taken inexpensively." In common with his general view on commercial fisheries, lobster fisheries are not characterized by regulation of entry. Despite the general support of limited entry as a system of management by a Working Group on *Homarus* (International Council for the Exploration of the Sea, 1975), it appears as though recognition of the traditional right of an individual to take fish, at least in the United Kingdom, will hamper any attempts to control the number of fishing licenses. Such traditionalism is evidently not uncommon in determining management policies in lobster fisheries (Dow, Chapter 8).

Where applied, management by limited entry has apparently enjoyed considerable success (e.g., in the Western Australian spiny lobster fishery, described in detail by Bowen in Chapter 7). As Bowen concludes, limitation of entry produces its own problems, but to its protagonists these are more acceptable than serious overexploitation, though the latter must not be assumed to be completely ruled out by limited entry. As Morgan shows, even in a fishery in which the number of boats and the number of fishing units is fixed, there is still considerable potential for greater effective effort. It is then, or better still in anticipation of this, managers look hopefully at a "buy-back" scheme (i.e., the removal of boats from the fishing fleet by purchase using Government or Industry funds) such as that practiced in the Canadian salmon fishery, and more recently in a Canadian lobster fishery (Dow, Chapter 8). Reduction of effort by this means should not, at least theoretically, lead to a reduction in total catch in the long term, but it will improve the economics of the individual fisherman and reduce the need for biological concern.

Whatever the chosen method, management of this valuable resource places a great responsibility on the scientist and the administrator, a far cry from the days when *H. americanus* was caught for agricultural fertilizer and cod fish bait (Dow, Chapter 8) and *N. norvegicus* was the "perk" of scalefish trawlermen.

These introductory paragraphs have so far not made reference to the important topic of aquaculture, which may provide a practical means of supplementing declining natural stocks or a method of farming a species of little previous commercial usage. The economic success of aquaculture will depend on most of the parameters requiring evaluation in the natural stock—growth rates, natural mortality, success of recruitment, marketable size, and quality. These and the many various problems of aquaculture in lobsters have been dealt with in the comprehensive review by Van Olst, Carlberg, and Hughes.

REFERENCES

Baranov, F. I. (1918). On the question of the biological basis of fisheries. *Nauchno-Issled. Ikhtiol. Inst. Iz.* **1**(1), 81-128.

Chittleborough, R. G., and Phillips, B. F. (1975). Fluctuations of year-class strength and recruitment in the western rock lobster, *Panulirus longipes* (Milne-Edwards). *Aust. J. Mar. Freshwater Res.* **26**, 317-328.

Farmer, A. S. D. (1975). Synopsis of biological data on the Norway lobster. *FAO Fish. Synopsis* **112** (FIRS/S112), 1-97.

Food and Agriculture Organization (1970). The fish resources of the ocean (edit. J. A. Gulland). *FAO Fish. Tech. Pap.* **97** (FIRS/T97), 1-425.

Graham, M. (1935). Modern theory of exploiting a fishery, and application to North Sea trawling. *J. Cons., Cons. Int. Explor. Mer* **10**, 264-274.

Hancock, D. A. (1973). The relationship between stock and recruitment in exploited invertebrates. *Rapp. P.-V. Réun., Cons. Int. Explor. Mer* **164**, 113-131.

Hancock, D. A. (1975). The Yorkshire fishery for edible crabs (*Cancer pagurus*): Assessment of the effects of changes in the minimum legal size. *Fish. Invest. (London), Ser. 2* **27**(8), 1-11.

Hancock, D. A. (1977). Population ecology and growth. *Circ.—CSIRO, Div. Fish. Oceanogr. (Aust.)* **7**, 279-286.

International Council for the Exploration of the Sea (1975). Report of the working group on *Homarus* stocks. *ICES CM* **K:38**, 1-19 (mimeo).

International Council for the Exploration of the Sea (1976). Report of the *ad hoc* meeting on the provision of advice on the biological basis for fisheries management. *ICES CM, Gen* **3**, 1-16 (mimeo).

Larkin, P. A. (1977). An epitaph for the concept of maximum sustained yield. *Trans. Am. Fish. Soc.* **106**(1), 1-11.

Morgan, G. R. (1974a). Aspects of the population dynamics of the western rock lobster, *Panulirus cygnus* George. I. Estimation of population density. *Aust. J. Mar. Freshwater Res.* **25**, 235-248.

Morgan, G. R. (1974b). Aspects of the population dynamics of the western rock lobster, *Panulirus cygnus* George. II. Seasonal changes in the catchability coefficient. *Aust. J. Mar. Freshwater Res.* **25**, 249-259.

Morgan, G. R. (1977). Aspects of the population dynamics of the western rock lobster and their role in management. Ph.D. Thesis, University of Western Australia.

Nicholson, M. D. (1979). The use of length frequency distributions for age determination of *Nephrops norvegicus* (L.). *Rapp. P.-V. Reun., Cons. Int. Explor. Mer* **175**, 176-181.

Parrish, B. B., (1973). Fish stocks and recruitment. *Rapp. P.-V. Reun., Cons. Int. Explor. Mer* **164**, 1-372.

Russell, F. S. (1931). Some theoretical considerations on the "overfishing" problem. *J. Cons., Cons. Int. Explor. Mer* **6**, 3-27.

Wilder, D. G. (1965). Lobster conservation in Canada. *Rapp. P.-V. Reun., Cons. Int. Explor. Mer* **156**, 21-29.

Chapter 5

Population Dynamics of Spiny Lobsters

G. R. MORGAN

I.	Introduction	189
II.	Population Parameters	191
	A. Delineation of Stocks	191
	B. The Measurement of Abundance	193
	C. Growth	193
	D. Reproductive Dynamics	197
	E. Mortality Rates	200
III.	Population Dynamics Models	204
	A. Surplus Yield Models	205
	B. Dynamic Pool Models	206
IV.	Conclusions	211
	References	213

I. INTRODUCTION

The spiny lobsters of the family Palinuridae are found in all major tropical and temperate oceans of the world and support subsistence or large-scale fisheries in most areas. The better known and more widespread species are listed in Table I, where they are separated into three broad geographic categories. As is common with other groups, the Palinuridae in tropical waters tend to be characterized by a relatively large number of species but a low density of individuals within each species. This is reflected in the fact that of the 12 species listed as tropical in Table I, only three (*Panulirus argus, P. polyphagus,* and *P. laevicauda*) support major fisheries. Temperate waters, however, are characterized by a low number

TABLE I

The More Common Species of Palinuridae and Their Fisheries[a]

Species	Areas of a significant fishery	Approximate 1976 catch (ton)
Tropical		
Panulirus guttatus	None	—
Panulirus longipes (2 subspecies)	None	—
Panulirus polyphagus	Thailand, India, Pakistan, South East Asia	3700
Panulirus versicolor	None	—
Panulirus ornatus	New Guinea, East Africa	540
Panulirus penicillatus	Reunion, Pacific Islands, Galapagos	400
Panulirus homarus (3 subspecies)	East Africa, Indonesia	400
Panulirus argus	Caribbean, Brazil	22,800
Panulirus laevicauda	Brazil	3000
Panulirus regius	Northwest Africa, Portugal	450
Panulirus gracilis	Equador, Panama	270
Panulirus echinatus	None	—
Subtropical		
Panulirus marginatus	Hawaii	10
Panulirus stimpsoni	Hong Kong	10
Panulirus japonicus	Japan, South China Sea	1200
Panulirus cygnus	Western Australia	8900
Panulirus inflatus	West coast of Mexico, Guatemala	1500
Panulirus pascuensis	Easter Islands	5
Palinurus delagoae	Southeast Africa	60
Jasus verreauxi	Eastern Australia, New Zealand	125
Palinurus mauritanicus	Mauritania, West Africa	150
Panulirus interruptus	California	120
Palinurus charlestoni	Cape Verde Islands	5
Temperate		
Palinurus elephas	France, Spain, U.K., Italy	1500
Palinurus gilchristi	South coast of Africa	970
Jasus lalandii	Southwest Africa	6200
Jasus novaehollandiae	South and Southeast Australia	3500
Jasus paulensis	St. Paul and New Amsterdam Islands	900
Jasus edwardsii	New Zealand	3700
Jasus tristani	Tristan da Cunha	5
Jasus frontalis	Juan Fernandez	48

[a] The data for this Table have been collected by the author from a number of sources. They do not necessarily coincide with the data published by FAO for 1976.

of species, but a relatively high density of individuals so that of the eight species listed as temperate-water species in Table I, six support major fisheries and the other two (*Jasus frontalis* and *J. tristani*) supply important local fisheries.

The total world catch of spiny lobsters amounted to 60.7 thousand metric tons in 1976 (Food and Agriculture Organization, 1977). The Carribean, the southeast Atlantic, and the eastern Indian Ocean contributed 42.7 thousand metric tons or 70.0% of the total. An increasing awareness of the need to manage these fisheries has, in recent years, provided an impetus to the study of the ecology and population dynamics of the adult stock in several areas. Because of this motivation, the aspects that have received attention have usually been those that are directly concerned with the estimation of yield curves, the establishment of optimum sizes at first capture, and other fisheries-orientated problems. These aspects include studies of migration patterns, growth rates, natural mortality rates, reproduction, and measurements of abundance, although the priorities given to these various studies are usually related to the particular management needs of the fishery in question. Various methods have been employed in the measurement of these population parameters within the Palinuridae, including examination of length frequency distributions, tagging studies, and experimental fishing. The commercial fisheries based on several species have also proved valuable in providing data, particularly on mortality rates. These methods, their limitations, their problems, and the information gained from them will be discussed below. The more commercially orientated data (e.g., fishing gear design) have necessarily been omitted.

II. POPULATION PARAMETERS

A. Delineation of Stocks

Studies of the dynamics of any population generally rely on the inherent assumption that the population behaves as a single entity containing no subgroups within the population that would have significantly different population characteristics. The choice of what to regard as a "unit stock" is not an easy one. In many instances it relies on the classifier's ability to detect differences in the population characteristics of the species in question. Hence the choice of the unit stock may change as a body of information is built up and as the population characteristics become measured with greater precision.

Separation of a species into stocks can usually be based on two criteria. First, a stock may be completely separated from other stocks of the same species throughout its life cycle. This results in a completely self-contained stock that may differ genetically from other stocks, although the genetic differences are not great enough for it to be classified by taxonomists as a separate species. Second, a stock may be genetically coherent because of mixing during part of the life cycle. However, different parts of the stock may have vastly different population parameters such as mortality rates, growth rates, etc., so that they cannot be reasonably treated as a single stock.

In the Palinuridae, neither of these aspects has received adequate attention for any species. Morgan (1977), after noting similarities in catch per unit effort data from various parts of the commercial fishery, concluded that *Panulirus cygnus* probably forms a genetically coherent stock along the western Australian coast. However, regional differences in some population parameters (e.g., growth rates, size at first maturity, etc.) might make it advantageous to separate stocks on a longitudinal basis. Sims and Ingle (1966) considered the population of *P. argus* in the Florida–Carribean area to be part of one genetic stock, although Berry (1974) was able to distinguish two self-maintaining stocks (which he regarded as subspecies) of *P. homarus* in the southeast Africa–Madagascar region. Johnson (1960a,b, 1971, 1974) established that larval separation due to prevailing hydrography was responsible for maintaining the geographically adjacent species of *P. interruptus* and *P. inflatus* as separate genetic units in California.

A promising approach to the problem of genetic interchange within populations of spiny lobsters is being developed by Robert Menzies and his associates at the Nova University in Florida. Using polyacrylamide gel electrophoresis, they are examining esterase systems in adult populations of *P. argus* from various geographical regions. Results to date seem to suggest a detectable genetic difference between populations of *P. argus* in Central America and those of the Florida coast (R. Menzies, personal communication). These findings suggest genetic heterogeneity, which is contrary to what one might expect if larval dispersion was complete throughout the range of *P. argus*. The possibility of self-sustaining stocks of spiny lobsters should therefore not be precluded—particularly in species, such as *P. argus,* that inhabit large geographic areas.

However, in most instances the long pelagic larval stages of the Palinuridae appear to ensure adequate genetic interchange over wide areas (subject to local hydrographic conditions). Thus, separation of stocks within the Palinuridae seems to rely on the ability to detect differences in population parameters throughout the range of the species and not on the identification of genetically separated subunits of the species. In the Palinuridae, variations in growth rates between localities within the same species are quite common and have been demonstrated in wild adult *P. cygnus* (Morgan, 1977); *P. argus* (Munro, 1974; Dawson, 1949; Buesa Mas, 1965; Smith, 1958), *P. japonicus* (Kubo and Hattori, 1947; Nakamura, 1940), *P. interruptus* (Lindberg, 1955; Mitchell *et al.,* 1969), and *Jasus lalandii* (Newman, 1972; Newman and Pollock, 1974). Other population parameters which have been shown to vary in a nonrandom way between localities are (1) the size at first maturity in *P. cygnus* (Morgan and Barker, 1974, 1975, 1976) and *P. argus* (Sutcliffe, 1957) and (2) the time of spawning in *J. lalandii* (Newman, 1972) and *P. cygnus* (George, 1958a).

Some of the observed differences in the values of the population parameters measured for various species can be attributed to differences in techniques. However, there is sufficient evidence to suggest that species of the Palinuridae inhabiting large geographical areas will show nonrandom variations in at least

5. Population Dynamics of Spiny Lobsters

some of these parameters. The failure to recognize and measure these variations has precluded the identification of separate stocks within most species. This in turn has confused the interpretation of measured population parameters, particularly when performed by different workers.

B. The Measurement of Abundance

Few attempts have been made to measure the absolute abundance of wild adult populations of spiny lobsters, although relative abundances have been estimated from information on catch rates in most species that support commercial fisheries, e.g., *P. argus* (Munro, 1974), *P. cygnus* (Bowen and Chittleborough, 1966), *J. lalandii* (Heydorn et al., 1968), and *J. edwardsii* (Street, 1970). However, comparisons between species cannot be made on the basis of catch rate data because of the various types of fishing gear used in its measurement.

Peacock (1974) measured the absolute density of *P. argus* in the tropical Barbuda Lagoon using tagging techniques and found the mean density of catchable rock lobsters to be 3.9–7.0 animals/ha. This is a much lower figure than that obtained by Morgan (1974a) for *P. cygnus*. Morgan measured densities (again using tagging techniques) of between 329 and 2065 animals/ha at the Abrolhos Islands, in western Australia. However, these densities of *P. cygnus* varied seasonally, with the lowest densities being recorded in late spring (i.e., November) and the highest being recorded in midsummer (i.e., January). The lower density of spiny lobsters in the tropical versus the more temperate situation is apparently a common feature in the comparative distribution and abundance of spiny lobsters. The difference is undoubtedly due to the generally lower productivity of tropical waters, which results in a high species diversity but a low population density (Chekunova, 1972).

Besides the technological difficulties of finding a suitable tag for spiny lobsters, estimation of abundance by tagging studies may be subject to the usual problems associated with mark-recapture studies. These include tagging mortality, random distribution of tagged and untagged animals, and migration. Morgan (1974a) has discussed some of these problems in relation to *P. cygnus*. In particular, he demonstrated that tagged animals had a higher probability of recapture than untagged animals. Such factors should definitely be taken into account when measuring absolute abundance by mark-recapture methods.

C. Growth

Of all the parameters of adult palinurid populations, the growth rate of individuals is probably the aspect that has been most intensively studied; despite this effort, complete descriptions of growth of spiny lobsters are surprisingly rare. This apparent anomaly has arisen from the difficulty in separating the two components of the growth process in spiny lobsters, i.e., the molt increment and the molt frequency. Since most studies have relied on the recovery of tagged animals of a known size or length when released, the observed growth between release and recapture have consisted of the unmeasured contributions from these two growth processes.

The few attempts that have been made to separate molt increment and molt frequency in wild populations have usually involved estimating the molt frequency by determining the proportion of recovered tagged animals which had molted within various time periods (e.g., Munro, 1974). These data are then used to calculate an intermolt period and an annual molt frequency. This method, however, makes several assumptions, among which is that molt frequency is assumed constant throughout the year. This is certainly not the case in *P. cygnus* (George, 1958a; Chittleborough, 1975), *J. lalandii* (Patterson, 1969), *P. japonicus* (Hattori and Oisi, 1899; Nakamura, 1940), *P. argus* (Travis, 1954), and probably most other species of the Palinuridae. In most of the cases studied, the length of the intermolt period has been shown to be inversely related to sea water temperature (e.g., Smale, 1978, for *P. homarus*). Kurata (1962) found this to be a common feature of most crustacean growth patterns.

The length of the intermolt period also increases with increasing size in all species of the Palinuridae and probably in all species of crustacea. In a useful assessment of the growth of shrimps, crabs, and lobsters, Mauchline (1977) established a common feature of the intermolt period of the species he studied; i.e., he determined that the logarithm of the intermolt period is linearly related to carapace length (CL). The relevance of this observation will be discussed later in this Section.

Although the intermolt period increases with size in all species studied, work on wild populations, as well as observations of captured adult animals in aquaria, have indicated that the intermolt periods of mature animals are between 60 and 90 days for the tropical *P. argus* (Peacock, 1974; Munro, 1974) and *P. homarus* (Berry, 1971; Smale, 1978), 100–180 days for the more temperate *P. cygnus* (Morgan, 1977) and *P. interruptus* (Lindberg, 1955), and 180–360 days for the colder water *J. lalandii* (Patterson, 1969; Heydorn, 1969b), *J. novaehollandiae* (Fiedler, 1964), and *J. tristani* (Pollock and Roscoe, 1977).

The other component of the growth process in the Palinuridae, the molt increment, has also been measured in wild populations by separation of the molt frequency and molt increment in capture–recapture studies. This method obviously relies on the ability to measure the molt frequency accurately which (as noted earlier) is not usually possible. However, the measurement of the molt increment has been possible for some species, because the majority of the population molts at a specific time of the year (e.g., the "whites" molt of *P. cygnus*, George, 1958b) or when the intermolt period is long enough so that most recaptures are made before the animals have an opportunity to molt more than once.

5. Population Dynamics of Spiny Lobsters

Field studies of molt increments have also, in most instances, been supported by observations on captive animals, although the stress imposed by inadequate holding facilities can have profound influences on both molt increment and molt frequency (e.g., Chittleborough 1975, 1976; Lindberg, 1955).

Chittleborough (1976) has shown that in *P. cygnus* the molt increment of aquarium held animals increased with age until the second year of postlarval life, remained steady during the third year, and decreased thereafter. In the wild population, Chittleborough (1976) also demonstrated that where there was an extreme of either abundant food or a severe shortage, there was no difference in molt increment between sexes. When there was a moderate shortage of food, however, juvenile females appeared to compete less successfully than males of the same age, resulting in different molt increments between the sexes. Morgan (1977) has shown that in wild adult *P. cygnus,* the molt increment decreased with increasing size, with adult males having a larger molt increment than females of the same size. Pollock and Roscoe (1977) also showed a decrease in molt increment with increasing size for large (75–110 mm CL) *J. tristani* at Tristan da Cunha.

This decrease in molt increment with increasing size has been shown by R. Winstanley (personal communication) to be a nonlinear relationship in adult *J. novaehollandiae* from Tasmania. This is in accordance with the findings of Mauchline (1977), who expressed the molt increment of several species of crustacea as a "growth factor," i.e., the percentage increase in some linear measure of body size during a molt. He showed that the logarithm of the growth factor decreased linearly with carapace length. It may easily be shown that such a relationship generally leads to a nonlinear increase in molt increment with size, followed by a nonlinear decrease. Moreover, the curve of molt increment versus carapace length may be symmetrical or skewed in either direction.

Because of the relationship between molt increment and size, comparisons of molt increment between species are difficult. However, Peacock (1974) estimated molt increments of between 5.6 and 8.5 mm in *P. argus* of 50–100 mm CL, Chittleborough (1976) measured molt increments of between 2.9 and 5.4 mm in juvenile *P. cygnus* between 3+ and 5+ years of age (approximately 40–80 mm CL), and Pollock and Roscoe (1977) measured increments for *J. tristani* of 2.8–5.5 mm for animals between 73 and 109 mm CL.

The measurement of the growth rate of individual animals is important in establishing a mean growth curve applicable to the population as a whole. It is this "population" growth curve that forms a vital part of the description of the population dynamics of the species in question, as well as forming part of the yield assessment models for several species of the Palinuridae. In the cases studied (see later), the most adequate description of the growth curve has been provided by the von Bertalanffy (1938) growth function, although variation around the curve is generally high. This high degree of variation may be attributable to the plasticity that both the molt increment and, more particularly, the molt frequency exhibit in their response to environmental pressures such as food supply, temperature, etc. These responses have been discussed in detail by Chittleborough (1976) for *P. cygnus* and his conclusions are generally similar to those reached for other species of the Palinuridae that have been studied (e.g., Smale, 1978, for *P. homarus*). This high degree of variation, of course, makes the precise definition of the growth curve difficult, which in turn has contributed to reasonably imprecise yield models for fisheries based on palinurid species (e.g., Munro, 1974; Morgan, 1977). Munro (1974) concluded that the von Bertalanffy (1938) growth function best fitted his growth rate data on *P. argus*. Moreover, he noted that the theoretical growth curve was in close agreement with previous experimental observations of Sutcliffe (1957), Ting (1973), and Peacock (1974). Morgan (1977) also found that the von Bertalanffy (1938) growth function provided the best description of growth in adult *P. cygnus*, while Chittleborough (1976) was also able to adequately represent the growth rate of captive juvenile *P. cygnus* by the von Bertalanffy (1938) curve. However, Saila et al. (1979) detected significant differences between an empirical growth model (based on molt frequency and molt increment data) and the von Bertalanffy curve for *J. edwardsii* in New Zealand.

Mauchline's (1977) linear relationships between the logarithm of the intermolt period and carapace length and the logarithm of the growth factor and carapace length may be used in many cases to provide complete descriptions of growth for species of palinurids, although in some instances (e.g., *J. edwardsii*, J. L. McCoy personal communication) the intermolt period does not increase with size *ad infinitum*. The data requirements for this approach are minimal, but must include accurate assessments of the intermolt period, as well as growth factors of a few successive molts or of molts of a few animals representing the size range of the species.

As an example, Fig. 1 depicts the growth of *P. cygnus* based on molt frequency and molt increment data of Chittleborough (1976), Morgan (1977), Phillips et al. (1977), and unpublished data of G. R. Morgan. By combining the relationships between intermolt period and carapace length and the growth factor and carapace length, a growth curve is produced that (apart from not reaching an asymptote) closely approximates a von Bertalanffy type curve. Thus, it is not surprising that in the species studied, the von Bertalanffy curve usually provides an adequate description of observed growth data.

The adequacy of the von Bertalanffy (1938) growth function as a realistic description of the growth process in palinurid populations, however, depends on the validity of several inherent assumptions. The most important of these is that the growth rate is assumed constant throughout any one year, so that there are no within-season variations. This is obviously not the case in the Palinuridae, where the molt frequency varies seasonally and mass moltings occur. The effects of these seasonal changes in growth rate on the form of the growth curve have not

5. Population Dynamics of Spiny Lobsters

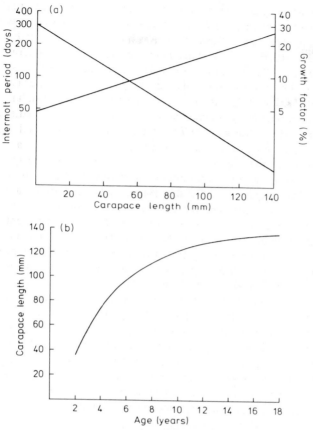

Fig. 1. (a) The relationships between the length of the intermolt period and size and the percentage molt increment and size for *Panulirus cygnus*. (b) The growth curve which results from a combination of these two relationships.

been examined in any species of the Palinuridae, and it may well be that a modified form of the von Bertalanffy (1938) growth function which takes these changes into account will provide the most adequate representation of palinurid growth, although the observed high degree of variability in the growth rates of most species will tend to make the recognition of seasonal changes difficult.

D. Reproductive Dynamics

The sexes are morphologically distinct in all species of the Palinuridae with various secondary sexual characteristics being well developed (George and Morgan, 1979). Various sex ratios of the adult stock have been reported ranging from 59.2% males in *P. argus* (Munro, 1974) to 10% males during some months of the year in *P. cygnus* (Morgan and Barker, 1974). The estimation of sex ratios from trap caught animals, however, is inevitably confused by seasonal variations in availability due to molting and reproductive patterns. Thus, it is unlikely that some extreme sex ratios recorded represent the actual sex ratio in the population, which is probably close to unity in most species (e.g., Feliciano, 1958, for *P. argus*), although differential growth rates of males and females will usually result in a preponderance of males in the larger size categories (e.g., Munro, 1974, for *P. argus;* Smale, 1978, for *P. homarus*).

1. Size at First Maturity

Estimates of the size at first maturity within the Palinuridae have often included data on the size of the smallest mature animal, the size of the largest immature animal, and the size cohort at which a certain proportion have mature gonads (50 and 90% are commonly used). External secondary sexual features of females have also been used to indicate maturity; these include the presence of fibrillar pleopods or viable eggs on the abdomen. Neither internal nor external maturity indicators for males, however, have been regarded as satisfactory by most workers and, in fact, testes examinations may produce results indicating gonad maturity at a much smaller size than that of physical maturity, i.e., capability to perform the mating act (Fielder, 1964; Heydorn, 1965). Information on the size at first physical maturity of males is therefore generally lacking, although aquarium studies (e.g., Berry, 1971) have indicated that males reach maturity at a similar or slightly larger size relative to female size at maturity. George and Morgan (1979) have estimated the size at first physical maturity of male and female *P. versicolor* by using the relative change in length of the legs. They obtained values of 66 mm CL as the size at first maturity of females and 72 mm CL as the size at first maturity of males.

Berry (1971) considered the "majority of females" of *P. homarus* to be mature at 54 mm CL, whereas males were physically capable of reproduction at a carapace length of between 50 and 60 mm. Heydorn (1969a) estimated the size at first maturity of female *P. homarus* from the Transkei to be 50 mm, De Bruin (1962) estimated the Ceylon population to attain sexual maturity between 55 and 59 mm, and George (1963) suggested that the East Aden population reached sexual maturity between 60 and 70 mm. In *P. argus,* Smith (1948) reported mature females of 45 mm CL, but the smallest sizes reported later are somewhat larger: 54 mm CL in British Honduras (Food and Agriculture Organization, 1968), 57 mm CL in Cuba (Buesa Mas and Mota-Alves, 1971), and 65 mm CL in Venezuelan waters (Cobo de Barany *et al.,* 1972). In other parts of the range of *P. argus,* the size at first maturity has been measured in the 80–90 mm CL range (Davis, 1975; Creaser, 1950; Sutcliffe, 1952; Feliciano, 1958; Dawson, 1949; Peacock, 1974; Munro, 1974), although in many cases a failure to define

5. Population Dynamics of Spiny Lobsters

what has been measured to establish the size at first maturity makes comparisons hazardous.

In *P. cygnus*, Morgan and Barker (1974, 1975, 1976) have shown that the size at first maturity (i.e., the smallest carapace length at which 50% have mated) and the mean size of breeding females both vary significantly throughout the range of the species. They found that the size at first maturity ranged from $\simeq$ 80 mm in northern areas of the fishery to $\simeq$ 95 mm in southern areas. At the subtropical Abrolhos Islands, sizes at first maturity were even smaller, the majority of females reaching maturity at a length less than the legal minimum length of 76 mm CL (George, 1958a,b).

Similar geographical variations in size at first maturity within the distribution of *P. japonicus* have been observed, sizes at first maturity ranging from 45 (Nakamura, 1940) to 55 mm (Kinosita, 1931).

Jasus edwardsii in New Zealand also shows geographical variations in the size at first maturity. The 50% maturity level of females is reached at 80 mm CL in the Fiordland, 110 mm CL at Stewart Island, and 120 mm CL at Moeraki (Street, 1969). It was suggested that these variations result from differences in water temperature between the three localities, so that the smaller size at first maturity corresponding to areas of relatively higher water temperatures. In addition, Street (1969) also noted that in areas where the size at first maturity was largest, a greater maximum length was reached. Heydorn (1964) reported sizes at first maturity of female *J. lalandii* in South Africa of 70 mm CL while Lindberg (1955) estimated that female *P. interruptus* in California reached sexual maturity at a total length of 21.5 cm (about 78 mm CL). B. T. Hepper (personal communication) has estimated the size at first maturity of female *Palinurus elephas* (previously *P. vulgaris*) off the coast of Cornwall to be 125 mm CL.

The attainment of sexual maturity within the Palinuridae appears to occur at a single molt, termed the "maturity molt" (Fielder, 1964; Lindberg, 1955). The size at which this occurs is variable both within species and between species. Tropical species generally appear to reach sexual maturity at a relatively smaller size than subtropical or temperate species, and temperature effects may also account for geographical variations in the size at first maturity within a single species.

2. Fecundity

Fecundity is high in all species studies and usually ranges up to some 700,000 eggs per spawning. Lindberg (1955) and Allen (1916) reported a range in fecundity of between 50,000 and 800,000 eggs per female in *P. interruptus*, depending on size. Similar ranges in fecundity have been measured in *P. argus* (Mota-Alves and Bezerra, 1968), *P. laevicauda* (Paiva and Silva, 1962), *P. cygnus* (Morgan, 1972), *P. japonicus* (Ino, 1950), and *P. homarus* (Berry, 1971). Members of the genus *Jasus*, however, appear to be less fecund than members of the genus, *Panulirus* of equivalent size; e.g., *J. novaehollandiae* and *J. lalandii*, both of 100 mm CL produce 236,000 eggs (Hickman, 1946) and 180,000 eggs (Heydorn, 1964), respectively, in contrast to *P. cygnus* of 100 mm CL, which produces 460,000 eggs (Morgan, 1972). *P. elephas* also appears to be an animal of relatively low fecundity; B. T. Hepper (personal communication) reports 110 mm CL specimens carrying only 87,000 eggs. The total number of eggs carried by mature animals of the genus *Jasus* can, of course, be as numerous as members of the genus *Panulirus*, since *Jasus* generally tends to reach larger maximum sizes. An extreme case is found in *J. verreauxi*, which is probably the largest species of spiny lobster. Specimens of this animal that reach a length of 235 mm CL can carry over 2,000,000 eggs (Kensler, 1967). The relationship between carapace length and number of eggs carried is generally a linear one (Morgan, 1972). In *P. cygnus*, Morgan (1972) showed that fecundity does not vary over the geographical range of the species, and little egg loss occurs during incubation. Infertile eggs comprised a mean of 4.5% of the total carried, but the percentage was not related to the size of the animal.

3. Number of Spawnings

Several members of the Palinuridae have demonstrated the ability to spawn more than once each year in the wild. These include *P. argus* (Sutcliffe, 1952), *P. japonicus* (Ino, 1950; Okada *et al.*, 1947), and *P. homarus* (Berry, 1971b). *Panulirus cygnus* can be induced to spawn repetitively in the aquarium (Chittleborough, 1974) and probably does so in the northern part of its range in the wild. All other temperate and perhaps subtropical members of the Palinuridae seem to be confined to one spawning each year.

E. Mortality Rates

The estimation of both natural and fishing mortality rates in palinurids usually relies on data from the operations of a commercial fishery. Since not all populations or species of the Palinuridae are commercially exploited, and since very few of the exploited species have adequate commercial statistics available, the information on mortality rates is necessarily sparse. The problems are compounded by the inability to determine age for any of the palinurid species, so that the method of comparing the abundance of successive year classes (commonly used in studies of fish populations) cannot be used. All of the available methods of mortality rate estimation therefore rely on some knowledge of the length distribution of the species in question, rather than the age.

1. Total Mortality

In an exploited population, the total instantaneous mortality rate will consist of contributions from both the fishing mortality rate (i.e., the losses due to fishing) and the natural mortality rate (i.e., the losses from all other causes). Since the usual population analysis method of comparing the abundances of successive year classes cannot be used for the estimation of the total mortality rate Z in the Palinuridae, recourse has been made to other forms of commercial statistics.

Bowen and Chittleborough (1966) used the decline in catch per unit of effort during a season to calculate values of Z for *P. cygnus* at the Abrolhos Islands of western Australia. They arrived at annual Z values of between 0.80 and 2.40. Using the same method, Vranckx (1973) calculated Z values of between 0.30 and 1.19 for *J. paulensis* at St. Paul and New Amsterdam Islands. The Bowen and Chittleborough (1966) method, however, suffers from the important disadvantage of assuming that the catch per unit effort is proportional to the abundance of spiny lobsters at all times throughout the fishing season. Morgan (1974b) has shown that in *P. cygnus* this is certainly not the case. Catch per unit of effort is significantly influenced by changes in catchability during a year, as well as changes in abundance. Since these changes in *P. cygnus* are related to molting activity and hydrological conditions, it is likely that similar seasonal changes occur in other species of palinurids. Use of catch per unit effort data, particularly within a fishing season, will therefore be unlikely to provide realistic estimates of Z for exploited spiny lobster populations.

The use of length composition data has probably been the most fruitful approach in the estimation of total mortality in the Palinuridae. In general, as the total mortality increases, the length composition and mean size of the population will change. Under steady state conditions, length and mean size are influenced by both the total mortality rate and the growth rate. Beverton and Holt (1956) derived an expression for the steady state relationship between Z and the average length of the population for animals growing according to the von Bertalanffy (1938) growth curve.

$$Z = \frac{K(L_\infty - \bar{l})}{\bar{l} - l_c}$$

where K and L_∞ are the von Bertalanffy (1938) growth parameters, $\bar{l}$ is the average length of fish in the exploited population (or if the fishery is not selective for size, the average length of fish in the catch), and l_c is the length at entry into the exploited phase.

Munro (1974) used this relationship to calculate values of Z for *P. argus* in the Caribbean and found values ranging from 0.50 to 1.52, depending on the degree of exploitation of the populations. The disadvantage of using this method for spiny lobsters arises from their decreasing vulnerability to capture (by baited traps) with increasing size (e.g., Morgan, 1979; Newman, 1972). Thus, the estimated average length of the exploited population will be biased if sampling is carried out by trapping. However, alternative sampling methods (e.g., diving) should overcome this problem.

If more comprehensive data exists on the size composition of an exploited population, then comparisons of catches taken in size classes can be used to calculate values of Z over the size range of the exploited population. Thus, the numbers of animals captured (C_t) in any time period, t_1 to t_2, is a function of the instantaneous fishing mortality rate F and the number of animals present N_t.

$$C_t = \int_{t_1}^{t_2} FN_t \, dt$$

$$= \int_{t_1}^{t_2} FN_0 e^{-Zt} \, dt$$

$$= \frac{-FN_0}{Z} (e^{-Zt})\Big|_{t_1}^{t_2}$$

$$= \frac{FN_0}{Z} \exp(-Zt_1) \{1 - \exp[-Z(t_2 - t_1)]\}$$

Similarly, the catch taken during a second interval of time, t_1^1 to t_2^1, is given by

$$C_t^1 = \frac{FN_0}{Z} \exp(-Zt_1^1) \{1 - \exp[-Z(t_2^1 - t_1^1)]\}$$

If $T = t_1^1 - t_1$, and Z and F are assumed equal in the two time intervals, then the ratio of the catches taken in the two intervals is (after some rearrangement of terms)

$$\frac{C_t^1}{C_t} = \exp(-ZT) \left\{ \frac{1 - \exp[-Z(t_2^1 - t_1^1)]}{1 - \exp[-Z(t_2 - t_1)]} \right\}$$

The time periods $(t_2^1 - t_1^1)$ and $(t_2 - t_1)$ can now be expressed in terms of the von Bertalanffy (1938) growth parameters, since $t_1 = t_0 - (1/K) \ln[1 - (l_1/L_\infty)]$. Therefore

$$t_2 - t_1 = \frac{1}{K} \ln\left(\frac{L_\infty - l_1^1}{L_\infty - l_2^1}\right)$$

and

$$\frac{C_t^1}{C_t} = \exp(-ZT) \left[\frac{1 - \exp(-Z\{1/K[\ln(L_\infty - l_1^1)/(L_\infty - l_2^1)]\})}{1 - \exp(-Z\{1/K[\ln(L_\infty - l_1)/(L_\infty - l_2)]\})} \right]$$

The ratio of the catches in each time (or length) interval can therefore be expressed as a function of the instantaneous total mortality coefficient Z and the von Bertalanffy (1938) growth parameters. However, the equation cannot be

5. Population Dynamics of Spiny Lobsters

rearranged to express Z as a simple function of the catch ratios, but must be solved using an iterative technqiue such as the Newton–Raphson method.

Using this method, Morgan (1977) calculated values of Z over the exploited size range of *P. cygnus*, demonstrating that Z decreases with increasing size, presumably as a result of decreasing vulnerability. The overall value of Z for this fishery in recent years was in the range 0.87–1.01. Variable recruitment influenced the calculation of Z for the smaller size classes, but appeared to have little influence on the larger size classes.

2. Fishing Mortality

The separation of total mortality into its components of fishing and natural mortality may be achieved (1) by examining the changes in Z over a wide range of effort, thereby estimating M (i.e., Z when fishing effort is zero), or (2) by estimating F directly from tagging studies. The first approach requires good data not only on Z, but also on the effort expended in the fishery over a number of years. These data are generally not available in palinurid fisheries, possibly the only exception being the fishery for *P. cygnus* in western Australia. There, this approach has provided estimates for F of between 0.64 and 0.78 for recent years (Morgan, 1977).

Tagging studies have been used in several species to estimate the fishing mortality rate, including *J. lalandii* in South Africa (Newman, 1972) and *P. cygnus* in western Australia (Morgan, 1977). However, many tagging studies of palinurids have been more directed toward growth and movement studies. The method of estimating fishing mortality from tagging studies has usually involved the methods of Ricker (1975), in which $\ln N_R$ is plotted against r [where N_R is the number of animals captured in the time interval between rT and $(r + 1)T$, and T is the length of the time interval in which the tag returns are grouped]. This results in a straight line with a slope of $-ZT$ and an intercept on the y axis of $\ln[FN_0/Z (1 - e^{-ZT})]$. Thus, both the total mortality and the fishing mortality may be estimated.

Such tagging studies may be subject to a variety of biases that have been fully discussed by Ricker (1975). These biases should be taken into account in the evaluation of the results of any studies involving tagging. Morgan (1974a), for example, showed that tagged *P. cygnus* are more likely to be recaptured than untagged animals, which would lead to overestimates of both Z and F using the Ricker (1975) analysis. Loss of tags and nonreporting of recaptures are other sources of bias that may have particular relevance to spiny lobster fisheries.

3. Natural Mortality

Munro (1974), using data on the average carapace length of the population together with data on the growth parameters, utilized the method of Beverton and

Holt (1956) to estimate total and natural mortality rates of various populations of *P. argus*. He calculated annual values of the instantaneous natural mortality coefficient M of 0.52 for an unexploited stock, 0.23 for a moderately exploited stock, and 0.14 for a heavily exploited stock. Moreover, he suggested that the natural mortality resulted very largely from predation, so that

$$M = gP + M_0$$

where g is the mortality generated in the prey species by one unit of biomass of predators, P is the biomass of predators, and M_0 is the mortality caused by other factors (which is probably negligible).

In *P. cygnus*, Bowen and Chittleborough (1966), using commercial catch and effort statistics, calculated values of M for the commercially fished stock at the Abrolhos Islands to range from 0.222 to 0.781 for a 4½-month period, equivalent to 0.592 to 2.083 on an annual basis. Vranckx (1973) used this method in estimating population parameters for *J. paulensis* at St. Paul and New Amsterdam Islands and obtained values of M (annual) ranging from 0.20 to 1.04 for a number of years. Again using commercial catch records, Morgan (1977) calculated an annual value of M for commercial sized *P. cygnus* of 0.226 by comparing the catches taken in various discrete length groups for a number of years and relating these to the fishing effort expended in each year.

Apart from the observations of Buesa Mas (1965) on *P. argus*, no work has been done on the variation in M with size or age in any adult population of the Palinuridae, although this aspect is likely to be important. However, the natural mortality coefficient M has been shown to be density dependent in adult *P. cygnus* (Morgan, 1974a) as well as juvenile *P. cygnus* (Chittleborough, 1970). The results of Munro (1974), cited earlier, could be interpreted as indicating a similar density dependence in *P. argus*.

III. POPULATION DYNAMICS MODELS

The study of population dynamics models that are particularly applicable to exploited palinurid populations is, at the moment, a study of potential usefulness rather than one of case histories. In only a few instances has sufficient information on both the animal and the fishery been available for attempts to be made at describing how the long-term equilibrium yield might change with fishing effort. The attempts that have been made have invariably used models originating from the study of fish populations. Although this does not invalidate their use with spiny lobsters per se, the data requirements are usually more orientated toward information that is readily available for fish populations. Problems arise when the use of the model requires data that are very difficult or impossible to obtain for a lobster population. For example, since age cannot be determined for any of the species of the Palinuridae, the age distribution models used for fish populations are not applicable. Unfortunately, the more applicable models that use length rather than age composition for population studies have received scant attention from palinurid population dynamicists.

The types of models that have application in the study of spiny lobster populations may be divided into two broad types, empirical models and conceptual models. Empirical models, which involve fitting relationships to observed data on yield and one or more variables (e.g., fishing effort and water temperature), have not been used in spiny lobster fisheries and will not be discussed here. Comments on these models in connection with the population dynamics of clawed lobsters have been made in Chapter 6 (this volume) and will have the same relevance in application to spiny lobster fisheries.

Conceptual models involve the formulation of hypotheses that explain how a particular population will react to changes in fishing effort, in terms of changing biomass. When data are available, the hypotheses are tested by comparison with observed data. The conceptual models may be divided into two types, surplus yield models and dynamic pool models.

A. Surplus Yield Models

Surplus yield models, which rely only on catch and associated fishing effort data for their formulation, are fully discussed with reference to clawed lobsters by Saila and Marchesseault (this Volume, Chapter 6). Their comments on the models' usefulness in assessing stocks of lobsters apply equally well to palinurid stocks.

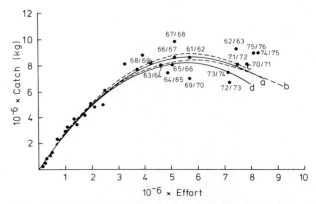

Fig. 2. Surplus yield models fitted to observed catch and effective effort data for the fishery of *Panulirus cygnus*. (a) Schaefer (1957), (b) GENPROD (Pella and Tomlinson, 1969), (c) PRODFIT (Fox, 1975), and (d) delayed Recruitment (Marchesseault *et al.*, 1976).

Surplus yield models have been used to provide descriptions of three palinurid fisheries, i.e., *J. novaehollandiae* in Tasmania (J. Bradbury, personal communication), *J. edwardsii* in New Zealand (Saila *et al.*, 1979), and *P. cygnus* in western Australia (Morgan, 1977). In each case the models have provided surprisingly good fits to the observed transitional data on catch and effort. Figure 2 illustrates the application of various surplus yield models to the *P. cygnus* fishery in western Australia and also demonstrates a common problem in the evaluation of conceptual models, i.e., insufficient data being available over large ranges of fishing effort.

B. Dynamic Pool Models

Population dynamics models that attempt to relate the obtainable yield of a fishery to the biological aspects of the species (e.g. growth and mortality rate) have had a long history, beginning with the formulations of Baranov (1918) and Thompson and Bell (1934). At the present time, the two most widely used models are those of Beverton and Holt (1957) and Ricker (1975). The Ricker method is slightly more general because it is able to cope with age-dependent mortality and variations in growth rate over the life span of the animal. Both models, however, use essentially the same relationship in expressing the yield in weight for any fishing mortality rate as a simple function of (1) the number of recruits entering the fishable stock; (2) the fishing and natural mortality rates; (3) the ages of first capture and recruitment; and (4) the von Bertalanffy (1938) growth parameters K, W_∞, and t_0. All of these can usually be measured for a given population, except for the number of recruits, which is a generally unknown fact. To overcome this problem, the precise number of recruits is usually disregarded, and the yield is expressed as the yield per recruit. This, however, introduces problems when the number of recruits varies from year to year (discussed in this section).

1. Yield per Recruit Analyses

Yield per recruit anaylses have been carried out on only three species of the Palinuridae because *P. cygnus* in western Australia, *P. argus* in the Caribbean, and *J. edwardsii* in New Zealand are probably the only species for which there is sufficient knowledge on both the population parameters of the animal and the fishery for such analyses to be attempted. Munro's (1974) assessment of the *P. argus* fishery on the South Jamaica shelf made use of a modified form of the Beverton and Holt yield equation that used the incomplete beta function to express the yield per recruit as a function of the ratios M/K, the exploitation rate E, and a quantity c (which equals l_c/L_∞, where l_c is the length at recruitment or first capture and L_∞ is the maximum length attained). Although this modification of the Beverton and Holt equation greatly simplifies the calculations involved, it

5. Population Dynamics of Spiny Lobsters

also introduces some approximations (e.g., an isometric length–weight relationship). A more crucial disadvantage, however, which is applicable to the Beverton and Holt equation in general, is that it cannot take into account changes in the natural mortality rate or the fishing mortality rate over the size range of the exploited population. As mentioned earlier, this is likely to be important for calculations describing spiny lobster populations.

Despite these shortcomings, the Beverton and Holt yield equation has provided a useful assessment of the *P. argus* fishery of the South Jamaica shelf. Using the relationships shown in Fig. 3, Munro (1974) concluded that any increase in fishing effort would result in a decrease in the yield per recruit. Moreover, he concluded that any increase in the length at first capture (up to 125 mm) would result in significant improvements in the yield per recruit.

Saila et al. (1979) applied two versions of a yield per recruit model to the fishery for *J. edwardsii* in New Zealand. These versions differed only in the type of growth function incorporated into the model, and they provided qualitatively similar results. Both indicated that the maximum yield per recruit could be achieved at substantially lower levels of fishing mortality than that prevailing in the fishery. Also, no strong evidence for changing the existing minimum size at first capture was evident from the model results.

The assessment of the fishery for *P. cygnus* in western Australia (Morgan, 1977) using the Ricker (1975) method provided quite different conclusions from

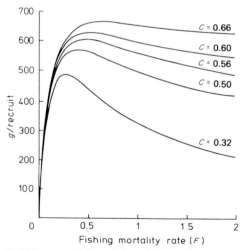

Fig. 3. Theoretical yield per recruit from the stock of *Panulirus argus* on the South Jamaican shelf ($M/K = 1.0$, $F = 0.27$, $l_r = 60$ mm CL, $c = 0.32$) compared with potential yield per recruit if recruitment were at a mean carapace length of 95 mm ($c = 0.50$), 106 mm ($c = 0.56$), 114 mm ($c = 0.60$), or 125 mm ($c = 0.66$). (From Munro, 1974, with permission.)

those reached for *P. argus* in the Caribbean. This is to be expected, since the two species have different growth rates, mortality rates, and lengths at first capture. On the basis of the application of the Ricker method, it was concluded that the yield per recruit would change very little with increasing fishing effort, but that increases or decreases in the size at first capture would result in a decline in the yield per recruit at the present level of fishing effort.

It is important to recognize that the reliability of conclusions based on a Beverton and Holt type of analysis are closely associated with the accuracy and precision with which the input parameters of mortality, growth, and length at first capture are measured. Also, yield equations are not equally sensitive to inaccuracies in the estimation of these individual parameters. Thus, it can easily be demonstrated that the Ricker type model has a much greater sensitivity to erroneous estimations of L_∞ and M than to similar inaccuracies of t_0 and K. However, errors in more than one parameter may be compounded so that the final error in the yield per recruit is greater than the sum of the errors produced by inaccuracies in the estimation of the individual parameters. In Table II the yield per recruit in a *P. cygnus* fishery is shown to vary with changing values of the input parameters, for three different values of the relative fishing mortality rate.

TABLE II

Sensitivity Analysis for the Western Rock Lobster, *P. cygnus*.[a]

Percentage change in parameters				Relative F		
L_∞	K	t_0	M	0.5	1.0	1.5
0	0	0	0	0	0	0
0	0	0	+20	−10.5	−7.5	−5.5
0	0	+20	0	+.3	+.4	+.5
0	0	+20	+20	−10.2	−7.0	−4.9
0	+20	0	0	+.2	+.9	+1.2
0	+20	0	+20	−10.7	−7.1	−4.8
0	+20	+20	0	+.9	+1.7	+2.0
0	+20	+20	+20	−10.1	−6.3	−4.0
+20	0	0	0	+28.4	+26.9	+24.1
+20	0	0	+20	+11.9	+14.2	+14.3
+20	0	+20	0	+29.8	+28.1	+25.3
+20	0	+20	+20	+13.0	+15.3	+5.6
+20	+20	0	0	+30.5	+31.0	+28.9
+20	+20	0	+20	+13.9	+17.6	+18.3
+20	+20	+20	0	+32.6	+32.9	+30.8
+20	+20	+20	+20	+15.7	+19.3	+20.0

[a] Table shows percentage change in yield per recruit for 20% increases in the various parameters of the Ricker application of the Beverton and Holt yield equation at three different values of relative fishing mortality.

Although the quantitative changes in yield per recruit shown in Table II will vary according to the initial values of the parameters, the trends that are shown will nevertheless remain unaltered.

2. Stock and Recruitment

The yield per recruit relationship discussed previously will be of the same form as the yield versus fishing mortality relationship, provided that recruitment to the fishery is constant. In a discussion of factors affecting recruitment to the *P. cygnus* fishery, Chittleborough and Phillips (1975) concluded that if the initial settlement of puerulus larvae on the shallow water reefs is adequate, then the density-dependent mortality during the juvenile phase of the life cycle will result in a significantly smaller range of variation in recruitment to the fishery in contrast to the range of variation that occurs in recruitment of puerulus larvae. However, in some years (e.g., 1968/69) the abundance of settling puerulus larvae may be inadequate, i.e., below the holding capacity of the shallow water reefs, so that poor recruitment to the fishery will result, as was the case in 1973/74. So although density-dependent mortality during the juvenile phase will ensure reasonably constant recruitment to the fishery over a wide range of initial year-class strengths, conditions may occur which will result in lowered recruitment to the fishery. One factor that has obvious impact on the relative strength of a particular year class is the number of mature animals that produce it. Information concerning the relationship between abundance of the parent stock and subsequent recruitment of puerulus larvae is only available for the western rock lobster, *P. cygnus*. But since the larval life cycle of most species of palinurids is similar, there does not seem any reason to suggest that other species will depart significantly from the strategy adopted by *P. cygnus* in the maintenance of its abundance. The abundance of the spawning stock in *P. cygnus* has been measured from commercial records as the catch per pot lift of ovigerous females (these data not available from one part of the fishery). The settlement of the subsequent puerulus larvae some nine months later has been measured by Phillips (see Chittleborough and Phillips, 1975). The relationship between the two is shown in Fig. 4, together with a Ricker (1958) stock-recruitment relationship which has been fitted to the data. This relationship is of the form $R = ASe^{-BS}$, where R is the recruitment, S is the stock size, A is the coefficient of density dependent mortality, and B is the coefficient of density independent mortality. The Ricker relationship appears to provide a surprisingly good fit to the data, especially since most stock-recruitment relationships are characterized by a high degree of variability. It is unfortunate that the data also represents a single time series rather than points being randomly distributed over the range of spawning stock sizes. For despite the apparent good fit, environmental or other influences cannot be discounted as possible explanations of the changes in the levels of puerulus settlement.

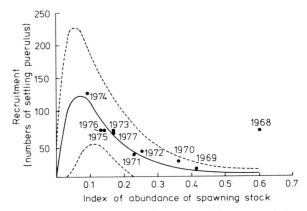

Fig. 4. Relationships between the index of abundance of spawning stock and the resulting numbers of settling puerulus found in artificial collectors for the western rock lobster, *Panulirus cygnus*. Broken lines represent 95% confidence limits. Data on numbers of settling puerulus from Phillips (personal communication).

Although the settlement of the puerulus larvae of *P. cygnus* varies from year to year, it is usually the density-dependent effects on growth and mortality between the puerulus and the juvenile stages that are of importance in maintaining steady recruitment to the fishery. However, if the settlement of puerulus larvae is small enough, a reduced recruitment to the fishery can occur. The assumption of more or less constant recruitment to the fishery (and to the spawning stock) appears to be realistic within the spawning stock size range so far encountered in *P. cygnus*. Thus, the total yield-fishing effort relationship may be regarded as being of the same form as the yield per recruit relationship, at least up to the present levels of fishing mortality. However, as fishing mortality continues to increase, it may be expected that the abundance of spawning females will decline to a point where recruitment is affected.

The fishery for the western rock lobster, *P. cygnus*, is characterized by having a legal minimum size of 76 mm CL which is below the size that females of most areas become mature (Morgan and Barker, 1974). Thus, any increased fishing pressure reduces their abundance. However, in areas or fisheries of other palinurids, where the legal minimum size (or the size at first capture) is above that of first spawning, the problem of reduction in the abundance of the spawning stock will not be as important, i.e., recruitment to the fishery should not be affected to such a great extent by increases in fishing pressure.

3. Unlimited versus Limited Resource Models

An essential feature of the simplest dynamic pool yield models is that they are unlimited resource models. As the fishing mortality rate is reduced, these models usually predict relatively large increases in stock biomass as well as increased catch per unit effort. However, in palinurid fisheries where data are available, there has been no trend in catch per unit effort at low fishing mortality rates that might indicate the occurrence of this phenomenon. In fact, for *P. cygnus* the relationship between catch per unit effort and fishing mortality rate is well represented by a linear relationship (Morgan, 1977). This rather circumstantial evidence suggests that there are not unlimited resources available for the growth of at least some palinurid populations, but that some resource (e.g., food and shelter) is limiting the population in the unexploited state. As a stock is exploited, this constant resource concept predicts that the population is at first able to respond (e.g., by increasing recruitment or growth or by reducing natural mortality) so that all the available resource continues to be fully or nearly fully utilized. Thus for a time biomass should decline only slowly. Only when the population response becomes inadequate should the stock start to decline rapidly.

An indication of this type of population response has been observed by Chittleborough (1979), who suggested that the growth rate of *P. cygnus* had increased with increasing fishing intensity, resulting in increasing sizes of first maturity. Examination of the length frequency distribution of the commercial catch over a number of years provides some support for this hypothesis of increased growth rate.

Since dynamic pool models are generally utilized in their constant parameter form, resource-limited population responses will invalidate their conclusions, particularly at low levels of fishing mortality rate. Surplus yield models, however, which can take these responses into account and are essentially resource-limited models may provide more realistic descriptions of palinurid populations, until such responses can be measured and incorporated into dynamic pool type models.

IV. CONCLUSIONS

Information available on the population dynamics and ecology of adult palinurids ranges from reasonably complete data on most aspects (e.g., in *P. argus*, *P. cygnus*, and *J. lalandii*) to an almost total lack of information (e.g., *P. echinatus*, *P. rissoni*, and *P. pascuensis*). The completeness of information is usually related to the importance of the commercial fishery (if any) on the species in question, and in some cases (e.g., *P. argus* and *P. cygnus*) data are complete enough to allow fisheries yield models to be constructed for the species. Despite the lack of information on many species, however, certain generalizations can be made regarding the major aspects of the population dynamics and ecology of the palinurids which may be of assistance in planning work on these lesser known species.

A geographical gradation in the distribution and abundance of palinurids as one proceeds from tropical to subtropical to temperate waters has already been noted. A similar geographical gradation may be seen in the growth rate of palinurids. Within the species studied, the molt frequency is highest in tropical species and lowest in temperate species, but there is no discernible trend in molt increment. This high molt frequency in tropical species leads to a high annual growth rate and generally a smaller maximum size than in temperate species. Similarly, size at first maturity is generally smaller in tropical palinurids than in temperate species, and temperate species seem restricted to a single spawning each year, in contrast to the repetitive and even continuous spawning that is common among tropical species.

The variability exhibited by members of the Palinuridae in the various parameters of their population is evident not only between species, but also within a single species. In the few instances where studies have been made, geographic variation in growth rates and size at first maturity have been detected. In previous studies, inadequate attention has been given to this geographical variation and the related problem of the identification of "unit stocks". If the failure to recognize the importance and possible magnitude of these problems persists, then apparent inconsistencies in measured population parameters and yield models of the group will continue.

Since the present knowledge of the population dynamics of the Palinuridae is very incomplete, it would appear to be generally unprofitable to use dynamic pool yield models that require good information on current growth and mortality parameters, as well as how they might have changed with exploitation. Thorough studies of the mechanisms affecting recruitment to the exploitable stock are also necessary for confident predictions to be made of future yields. In fact, there is probably no species (including the much studied fish stocks of the North Sea and north Atlantic) to which a Beverton and Holt yield model could be applied to estimate future total yields with even a reasonable degree of confidence.

Conversely, surplus yield models require a minimum of data, are biologically reasonable for spiny lobsters, and more importantly for the species that have been studied, appear to provide a reasonable description of observed catch and effort data. However, these are not without their problems. For example, there is a great deal of uncertainty as to whether the assumptions of the models hold at low stock biomass levels. They probably do not for spiny lobsters, since reduced recruitment at very low spawning stock levels will probably lead to low rates of change of population size, in contrast to the very high rates that the models predict. As fishing mortality increases, the accuracy with which a surplus yield model is able to predict yields probably diminishes rapidly. Surplus yield models are also inflexible because they use data and predict yields for fisheries only under one set of conditions. Thus, they cannot take into account the effects of changing size at first capture, which is easily handled by the dynamic pool models. Some assessment of the effects of changing size at first capture can, however, be gleaned from the application of techniques such as those of Allen (1953) or Gulland (1961), for which minimal data is required.

Since the possibilities of aging spiny lobsters are remote, it is somewhat disappointing that more attention has not been given to yield models based on the length rather than the age structure of the stock. Jones (1974) has developed a length-related cohort analysis technique, which has been used successfully on several crustacean species. More developments along these lines would be welcome. Such techniques, however, eventually need to refer to a time base. Therefore, it is necessary either to be able to convert lengths to ages by using an appropriate growth equation, or to determine the rate of growth over the lengths being considered. The detailed study of growth rates and their variation in spiny lobsters is therefore likely to become increasingly important in the formulation of yield models for palinurid fisheries.

REFERENCES

Allen, B. M. (1916). Notes on the spiny lobster (*Panulirus interruptus*) of the California coast. *Univ. Calif., Berkeley, Publ. Zool.* **16**, 139–152.

Allen, K. R. (1953). A method for computing the optimum size limit for a fishery. *Nature (London)* **172**, 210.

Baranov, F. I. (1918). On the question of the biological basis of fisheries. *Nauchw. Issled. Ikhtiol. Inst. Izv.* **1** (1), 81–128. (In Russian)

Berry, P. F. (1971). The biology of the spiny lobster, *Panulirus homarus* (Linnaeus) off the east coast of southern Africa. *S. Afr. Assoc. Mar. Biol. Res., Invest. Rep.* **28**, 1–76.

Berry, P. F. (1974). A revision of the *Panulirus homarus* group of spiny lobsters (Decapoda, Palinuridae). *Crustaceana* **27**, 31–42.

Beverton, R. J. H., and Holt, S. J. (1956). A review of methods for estimating mortality rates in exploited populations with particular reference to sources of bias in catch sampling. *Rapp. P.-V. Reun., Cons. Int. Explor. Mer* **140**, 67–83.

Beverton, R. J. H., and Holt, S. J. (1957). On the dynamics of exploited fish populations. *Fish. Invest. (London), Ser. 2* **19**, 1–391.

Bowen, B. K., and Chittleborough, R. G. (1966). Preliminary assessments of stocks of the Western Australian crayfish, *Panulirus cygnus* George. *Aust. J. Mar. Freshwater Res.* **17**, 93–121.

Buesa Mas, R. J. (1965). Biology and fishing of spiny lobsters, *Panulirus argus* (Latreille). In "Soviet-Cuban Fishery Research" (A. S. Bogdonov, ed.) [transl. from Russian by Israel Program for Scientific Translations, Jerusalem, 1969 (TT69-49016), 62–77].

Buesa Mas, R. J., and Mota-Alves, M. I. (1971). Escala de Colores para el estudio del circlo reproductor de la langosta *Panulirus argus* (Latreille) en el area del mar Caribe. *FAO Fish. Rep.* **71**(2), 9–12.

Chekunova, V. I. (1972). Geographical distribution of spiny lobsters and ecological factors determining their commercially important concentrations. *Tr. Vses. Nauchno-Issled. Inst. Morsk. Rybn. Khoz. Okeanogr.* **77**(2), 110–19 (transl. into English by NMFS, Foreign Fish. Transl. TT73-50035, 11 pp.).

Chittleborough, R. G. (1970). Studies on recruitment in the western Australian rock lobster

Panulirus longipes cygnus George: Density and natural mortality of juveniles. *Aust. J. Mar. Freshwater Res.* **21**, 131-148.

Chittleborough, R. G. (1974). Western rock lobster reared to maturity. *Aust. J. Mar. Freshwater Res.* **25**(2), 227-234.

Chittleborough, R. G. (1975). Environmental factors affecting growth and survival of juvenile western rock lobsters, *Panulirus longipes* (Milne-Edwards). *Aust. J. Mar. Freshwater Res.* **26**, 177-196.

Chittleborough, R. G. (1976). Growth of juvenile *Panulirus longipes cygnus* George on coastal reefs compared with those reared under optimal environmental conditions. *Aust. J. Mar. Freshwater Res.* **27**,(2), 279-296.

Chittleborough, R. G. (1979). Natural regulation of the population of *Panulirus longipes cygnus* George and responses to fishing pressure. *Rapp. P.-V. Reun., Cons. Int. Explor. Mer* **175**.

Chittleborough, R. G., and Phillips, B. F. (1975). Fluctuations in year-class strength and recruitment in the western rock lobster *Panulirus longipes* (Milne-Edwards). *Aust. J. Mar. Freshwater Res.* **26**, 317-328.

Cobo de Barany, T., Ewald, J., and Cadima, E. (1972). La pesca de la langosta en el archiplelago de Los Roques, Venezuela. *Inf. Tech. Proy. Invest. Desarollo Pesquero, Venez.* **43**, 1-34.

Creaser, E. P. (1950). Repetition of egg laying and number of eggs of the Bermuda spiny lobster. *Gulf. Caribb. Fish. Inst., Miami Univ., Proc.* **2**, 30-31.

Davis, G. E. (1975). Minimum size of mature spiny lobsters, *Panulirus argus*, at Dry Tortugas, Florida. *Trans. Am. Fish. Soc.* **104**(4), 675-76.

Dawson, C. F. (1949). Florida Crawfish Research. *Gulf Caribb. Fish. Inst., Miami Univ., Proc.* **1**, 21-28.

De Bruin, G. H. P. (1962). Spiny lobsters of Ceylon. *Bull. Fish. Res. Stn., Ceylon* **14**, 1-28.

Feliciano, C. (1958). The lobster fishery of Puerto Rico. *Gulf Caribb. Fish. Inst., Miami Univ., Proc.* **10**, 147-156.

Fielder, D. R. (1964). The spiny lobster *Jasus lalandii* (H. Milne-Edwards) in South Australia. II Reproduction. *Aust. J. Mar. Freshwater Res.* **15**, 133-144.

Food and Agriculture Organization (1968). Report to the government of British Honduras on investigations into marine fishery management; research and development policy for spiny lobster fisheries. *Rep. FAO/UNDP (TA) Rep.* **2481**, 1-95.

Food and Agriculture Organization (1977). Yearbook of fisheries statistics - catches and landings. Vol. 42, 1-320.

Fox, W. W. (1975). Fitting the generalized stock production model by least squares and equilibrium approximation. *Fish. Bull.* **73**(1), 23-36.

George, R. W. (1958a). The biology of the western Australian commercial crayfish, *Panulirus longipes*. Ph.D Thesis, University of Western Australia.

George, R. W. (1958b). The status of the white crayfish in western Australia. *Aust. J. Mar. Freshwater Res.* **9**(4), 537-545.

George, R. W. (1963). Report to the Government on Aden on crawfish resources of Eastern Aden Protectorate. *FAO/EPTA Rep.* **1696**, 1-23.

George, R. W., and Morgan, G. R. (1979). Linear growth stages in the rock lobster, *Panulirus versicolor*, as a method for determining size at first physical maturity. *Rapp. P.-V. Reun., Cons. Int. Explor. Mer. 175.*

Gulland, J. A. (1961). The estimation of the effect on catches of changes in gear selectivity. *Cons., J. Cons. Int. Explor. Mer.* **26**, 204-214.

Hattori, T., and Oisi, Y. (1899). On the rearing of *Panulirus japonicus* 1. *J. Imp. Fish. Inst. (Jpn.)* **1**(2), 67-79.

Heydorn, A. E. F. (1964). Notes on the reproductive biology and size limit of South African rock lobsters. Part 2. *S. Afr. Ship. News Fish. Ind. Rev.* **19**(6), 93-105.

5. Population Dynamics of Spiny Lobsters

Heydorn, A. E. F. (1965). The rock lobster of the South African west coast *Jasus lalandii* (H. Milne-Edwards). 1. Notes on the reproductive biology and the determination of minimum size limits for commercial catches. *Invest. Rep., Div. Sea Fish. S. Afr.* **53**, 1-32.

Heydorn, A. E. F. (1969a). Notes on the biology of *Panulirus homarus* and on length/weight relationships of *Jasus lalandii*. *Invest. Rep., Div. Sea Fish. S. Afr.* **69**, 1-26.

Heydorn, A. E. F. (1969b). The rock lobster of the South African west coast, *Jasus lalandii* (H. Milne-Edwards). 2. Population studies, behaviour, reproduction, moulting, growth, and migration. *Invest. Rep., Div. Sea Fish. S. Afr.* **71**, 1-52.

Heydorn, A. E. F., Newman, G. G., and Rossouw, G. S. (1968). Trends in the abundance of west coast rock lobster, *Jasus lalandii* (Milne-Edwards). *Fish. Bull. S. Afr.* **5**, 1-10.

Hickman, V. V. (1946). Notes on the Tasmanian marine crayfish, *Jasus lalandii* (Milne-Edwards). *Pap. Proc. R. Soc. Tasmania* 1945, 27-28.

Ino, T. (1950). Observations on the spawning cycle of Ise-ebi, *Panulirus japonicus* (v. Siebold). *Bull. Jpn. Soc. Sci. Fish.* **15**(11), 725-727.

Johnson, M. W. (1960a). The offshore drift of larvae of the Californian spiny lobster, *Panulirus interruptus*. Symposium on the Changing Pacific Ocean in 1957 and 1958. *Calif. Coop. Oceanogr. Fish., Invest. Rep.* **7**.

Johnson, M. W. (1960b). Production and distribution of larvae of the spiny lobster *Panulirus interruptus* (Randall) with records of *P. gracilis* Streets. *Bull. Scripps. Inst. Oceanogr.* **7**(6), 413-462.

Johnson, M. W. (1971). The palinurid and scyllarid lobster larvae of the tropical eastern Pacific and their distribution as related to the prevailing hydrography. *Bull. Scripps. Inst. Oceanogr.* **19**, 1-36.

Johnson, M. W. (1974). On the dispersal of lobster larvae into the eastern Pacific barrier (Decapoda, Palinuridae). *Fish. Bull.* **72**(3), 639-647.

Jones, R. (1974). Assessing the Long Term Effects of Changes in Fishing Effort and Mesh Size from Length Composition Data. *Cons. Int. Explor. Mer, Document* CM1974/F:33, 13 pp. (mimeo)

Kensler, C. B. (1967). Fecundity in the marine spiny lobster, *Jasus verreuaxi* (H. Milne-Edwards) (Crustacea: Decapoda:Palinuridae). *N.Z. J. Mar. Freshwater Res.* **1**(2), 143-155.

Kinosita, T. (1931). Biological minimum size of *Panulirus japonicus*. *Rep. Fish. Exp. Stn., Wakayama Prefect.* **28**, 1-22.

Kubo, I., and Hattori, S. (1947). An analysis of tagging results of *Panulirus japonicus* (v. Siebold), with special reference to rate of growth and ecdysis. *Bull. Jpn. Soc. Sci. Fish.* **12**(2), 108-110.

Kurata, H. (1962). Studies on the age and growth of Crustacea. *Bull. Hokkaido Reg. Fish. Res. Lab.* **24**, 1-115.

Lindberg, R. G. (1955). Growth, population dynamics, and field behaviour in the spiny lobster, *Panulirus interruptus*. *Univ. Calif., Berkeley, Publ. Zool.* **59**, 157-248.

Marchesseault, G. O., Saila, S. B., and Palm, W. J. (1976). Delayed recruitment models and their application to the American lobster (*Homarus americanus*) fishery. *J. Fish. Res. Board Can.* **33**, 1779-1787.

Mauchline, J. (1977). Growth of shrimps, crabs, and lobsters—an assessment. *J. Cons., Cons. Int. Explor. Mer* **37**(2), 162-169.

Mitchell, C. T., Turner, C. H., and Strachan, A. R. (1969). Observations on the biology and behaviour of the California spiny lobster, *Panulirus interruptus* (Randall). *Calif. Fish. Game* **53**(2), 121-131.

Morgan, G. R. (1972). Fecundity in the western rock lobster, *Panulirus longipes cygnus* (George) (Crustacea:Decapoda:Palinuridae). *Aust. J. Mar. Freshwater Res.* **23**, 133-141.

Morgan, G. R. (1974a). Aspects of the population dynamics of the western rock lobster, *Panulirus cygnus* George. I. Estimation of population density. *Aust. J. Mar. Freshwater Res.* **25**, 235-248.

Morgan, G. R. (1974b). Aspects of the population dynamics of the western rock lobster, *Panulirus cygnus* George. II. Seasonal changes in the catchability coefficient. *Aust. J. Mar. Freshwater Res.* **25**, 249-259.

Morgan, G. R. (1977). Aspects of the population dynamics of the western rock lobster and their role in management. Ph.D. Thesis, University of Western Australia.

Morgan, G. R. (1979). Trap response and the measurement of effort in the fishery for the western rock lobster. *Rapp. P.-V. Reun., Cons. Int. Explor. Mer.* **175**.

Morgan, G. R., and Barker, E. H. (1974). The western rock lobster fishery, 1972-73. *West. Aust. Dep. Fish., Wildl. Rep.* **15**, 1-22.

Morgan, G. R., and Barker, E. H. (1975). The western rock lobster fishery, 1973-74. *West. Aust. Dep. Fish., Wildl. Rep.* **19**, 1-22.

Morgan, G. R., and Barker, E. H. (1976). The western rock lobster fishery, 1971-72. *West. Aust. Dep. Fish., Wildl. Rep.* **22**, 1-22.

Mota-Alves, M. I., and Bezerra, R. O. F. (1968). Sobre o numero de ovos da lagosta *Panulirus argus* (Latreille). *Arq. Est. Biol. Mar. Univ. Ceara* **8**(1), 33-35.

Munro, J. L. (1974). The biology, ecology, exploitation, and management of Caribbean reef fishes. Scientific Report of the ODA/UNI fisheries ecology research project, 1962-1973. Part VI. The biology, ecology, and bionomics of Caribbean reef fishes—Crustaceans (Spiny lobsters and crabs). *Univ. West Indies Zool. Dep., Res. Rep.* **3**, 1-57.

Nakamura, S. (1940). Ecological studies on the spiny lobster, *Panulirus japonicus* (v. Siebold), with special reference to its conservation. *J. Imp. Fish. Inst. (Jpn.)* **34**, 101-113.

Newman, G. G. (1972). Assessment and management of some fish stocks in Southern Africa. Ph.D Thesis, University of Stellenbosch.

Newman, G. G., and Pollock, D. E. (1974). Growth of the rock lobster, *Jasus lalandii,* and its relationship to benthos. *Mar. Biol.* **24**(4), 339-345.

Okada, Y. *et al.* (1947). Study of *Panulirus japonicus* (v. Siebold). III. Number of broods. *Dobutsugakau Zasshi* **57**(3), 27-28.

Paiva, M. P., and Silva, A. B. (1962). Sobre o numero de ovos da lagosta *Panulirus laevicauda* (Latreille). *Arq. Est. Biol. Mar. Univ. Ceara* **2**(1), 17-19.

Patterson, N. F. (1969). The moulting frequency in captive adult Cape rock lobsters, *Jasus lalandii* (H. Milne-Edwards). *S. Afr. J. Sci.* **65**(3), 72-74.

Peacock, N. A. (1974). A study of the spiny lobster fishery of Antigua and Barbuda. *Gulf. Caribb. Fish. Inst., Univ. Miami, Proc.* **26**, 117-130.

Pella, J. J., and Tomlinson, P. K. (1969). A generalized stock production model. *Bull. Int.-Am. Trop. Tuna Comm.* **13**, 421-496.

Phillips, B. F., Campbell, N. A., and Rea, W. A. (1977). Laboratory growth of early juveniles of the western rock lobster, *Panulirus longipes cygnus. Mar. Biol.* **39**, 31-39.

Pollock, D. E., and Roscoe, M. J. (1977). The growth at moulting of crawfish *Jasus tristani* at Tristan da Cunha, South Atlantic. *J. Cons., Cons. Int. Explor. Mer* **37**(2), 144-146.

Ricker, W. E. (1958). Handbook of computations for biological statistics of fish populations. *Bull. Fish. Res. Board Can.* **119**, 1-300.

Ricker, W. E. (1975). Computation and interpretation of biological Statistics of fish populations. *Bull. Fish. Res. Board Can.* **191**, 1-382.

Saila, S. B., Annala, J. H., McKoy, J.L. and Booth, J. D. (1979). Application of Yield Models to the New Zealand Fishery for Rock Lobster *Jasus edwardsii* (Hutton). *N.Z. J. Mar. Freshwater Res.* **13**(1), 1-11.

Schaefer, M. B. (1957). A study of the dynamics of the fishery for yellowfin tuna in the eastern tropical Pacific Ocean. *Bull. Int.-Am. Trop. Tuna Comm.* **2**(6), 245-285.

Sims, H. W., and Ingle, R. W. (1966). Caribbean recruitment of Florida's spiny lobster populations. *Q. J. Fla. Acad. Sci.* **29**(3), 207-242.

Smale, M. J. (1978). Migration, growth, and feeding in the Natal rock lobster, *Panulirus homarus* (Linneaus). *S. Afr. Assoc. Mar. Biol. Res., Invest. Rep.* **47,** 1–56.

Smith, F. G. W. (1948). The spiny lobster industry of the Caribbean and Florida. *Caribb. Res. Counc. Fish. Ser.* **3,** 1–49.

Smith, F. G. W. (1958). The Spiny Lobster Industry of the Caribbean. Fla. St. Bd. Conserv., Educ. Ser. 11, Mar. Lab., Univ. Miami, 1–34.

Street, R. J. (1969). The New Zealand crayfish, *Jasus edwardsii* (Hutton). *N.Z. Mar. Dep. Fish., Tech. Rep.* **30,** 1–53.

Street, R. J. (1970). New Zealand rock lobster *Jasus edwardsii* (Hutton) South Island Fisheries. *N.Z. Mar. Dep. Fish., Tech. Rep.* **54,** 1–58.

Sutcliffe, W. H. (1952). Some observations of the breeding and migration of the Bermuda spiny lobster, *Panulirus argus*. *Gulf. Caribb. Fish. Inst., Miami Univ., Proc.* **4,** 64–68.

Sutcliffe, W. H. (1957). Observations on the growth rate of immature Bermuda spiny lobster *Panulirus argus*. *Ecology* **38,** 526–529.

Thompson, W. F., and Bell, F. H. (1934). Biological statistics of the Pacific halibut fishery. (2) Effect of changes in intensity upon total yield and yield per unit of gear. *Rep. Int. Fish. (Pac. Halibut) Comm.* **8,** 1–49.

Ting, R. Y. (1973). Investigation of the resource potential of the spiny lobster (*Panulirus argus* Latreille) in Puerto Rico. *Contrib. Agropecu. Pesq./Agric. Fish. Contrib.* **5**(2), 1–18.

Travis, D. F. (1954). The moulting cycle of the spiny lobster, *Panulirus argus* Latreille. I. Moulting and growth in laboratory-maintained individuals. *Biol. Bull. (Woods Hole, Mass.)* **107,** 433–450.

von Bertalanffy, L. (1938). A quantitative theory of organic growth. *Hum. Biol.* **10**(2), 181–213.

Vranckx, R. (1973). Evolution du stock de langoustes sur les fonds de peche des îles Saint Paul et Nouvelle-Amsterdam de 1962 à 1970. *Bull. Mus. Natl. Hist., Nat.* (France) (3e Sér.) (Ecol. Gén.) **155,** 193–203.

Chapter 6

Population Dynamics of Clawed Lobsters

S. B. SAILA AND G. MARCHESSEAULT

I. Introduction	219
A. Species Involved	219
B. Description of Fisheries	221
II. Vital Statistics	224
A. Stock Identification	224
B. Growth Summary	226
C. Mortality	228
D. Fecundity and Sex Ratios	228
III. Population Dynamics—Models	229
A. Empirical Models	229
B. Conceptual Models	233
References	239

I. INTRODUCTION

A. Species Involved

The purpose of this review is to bring together some of the available information on the fisheries and the population dynamics of the clawed lobsters, *Homarus americanus*, *Homarus gammarus*, and *Nephrops norvegicus*. Although all three are Atlantic marine boreal species, the American lobster, *H. americanus*, is restricted to its geographic distribution in the northwest Atlantic Ocean continental shelf area and its margins. The common lobster, *H. gammarus*, and the Norway lobster, *N. norvegicus*, both inhabit the continental shelves of the northeastern Atlantic Ocean and the Mediterranean Sea. The latter two species occupy discrete areas where suitable substrates are found. According to Havinga (1929), the Norway lobster is found on muddy substrates, but the common lobster inhabits a more diverse sedimentary environment and is frequently found in the vicinity of rocky areas.

The annual yield of the clawed lobsters is partitioned by species for 1975 as follows: (Food and Agriculture Organization, 1977)

Species	Annual Yield (metric tons)
H. americanus	30,387
H. gammarus	1,834
N. norvegicus	40,950
Total	73,171

It is clear from this table that the fishery for the common lobster, *H. gammarus*, is extremely small, in contrast to that of the other two species.

Although the three species listed are all members of the family Nephropidae, the nature of the fishery for *N. norvegicus* is distinct from the others. The most common gear used in this fishery is a trawl net, in contrast to the stationary traps or creels that are primarily used to catch the two *Homarus* species. The nature of the fishery may influence the valid estimation of certain vital statistics. For *N. norvegicus*, mesh selectivity, fishing power of the vessels, and the relationship of the behavioral ecology of the animal to the gear must be carefully considered. It is known, for example, that trawl catches on any particular trawling ground show marked daily (Chapman *et al.*, 1975) and seasonal fluctuations (Symonds, 1972).

The extremely high value of the American lobster (annual Canadian and United States landings recently valued in excess of $80 million) has been a stimulus for considerable research effort in both the United States and Canada. However, resolution of several problems relating to the population dynamics of the species has been dependent on the development of a suitable tag that would be retained during ecdysis. Such tags are now generally available because of the pioneering efforts of Scarratt and Elson (1965). For the clawed lobsters, the analysis of mark and recapture data not only increases our knowledge on movements, but also provides information that is necessary for the estimation of growth curves and that is useful in determining mortality coefficients and population sizes (Jones, 1976). In addition to the information gained from tagging studies, there have been some recent developments relating to improved management models. These include delayed recruitment models and empirical prediction models. This review will address some of these newer developments. Although much of the latter work has related to *H. americanus*, it is believed to be applicable to *H. gammarus* with relatively little modification. However, these management models may have less direct application to the *Nephrops*, fishery for previously mentioned reasons (e.g., habitat, distribution, and fishing methods) and the life history of the species is considerably shorter than that of the others.

B. Description of Fisheries

1. United States (Homarus americanus)

The American lobster, *H. americanus,* is fished off the coasts of eleven states in the northeastern and middle Atlantic states. The area fished extends from Maine (45°N) to North Carolina (35°N). Landings in the United States are on the order of 13,000 metric tons annually (28.7 million pounds). This amounts to somewhat less than one-half the total catch, the remainder of which is taken by Canada.

Both an inshore and an offshore fishery operate in the United States. The inshore fishery is limited to a depth of about 40 m and operates from Maine to Delaware. The lobster trap usually used inshore consists of a rectangular frame approximately 0.69–1.22 m (2¼–4 ft) in length and 0.46 m (1½ ft) in width. It is covered by wooden laths, and netting is used to construct entrance funnels as well as compartments within the trap. It is weighted with cement or other heavy material and tied singly or in a series to a buoy line, which serves as a marker. The size of the fishing vessel is small, ranging from skiffs to motor boats about 15 m in length. The vessels and gear used in the United States are generally similar to those used in Canada.

The offshore fishery is located in the vicinity of the continental slope and shelf areas, at depths ranging from 100 m to more than 600 m. It extends from the southeastern edge of Georges Bank to the vicinity of Virginia and North Carolina at the southern extremity. The offshore fishery developed during the early 1960s, primarily as an otter trawl fishery. It rapidly expanded and reached a peak in landings of about 4000 tons (9 million pounds) and since then has declined to slightly more than the amount harvested from the inshore fishery. The traps are fished in a buoyed series.

The five major lobster-producing states are Maine, Massachusetts, Rhode Island, New York, and New Jersey. The fishery in Maine is exclusively an inshore fishery. During the early years of lobstering, Maine landings exceeded three-fourths of the total United States yield. The catch from Maine has remained fairly stable at about 7727 metric tons (17 million pounds) for the past decade.

In general there is some evidence for increased fishing effort and declining catches in the inshore lobster fishery of the United States. Concern about the state of lobster stocks resulted in efforts to develop a unified lobster management program, which was initiated in 1972. Considerable progress has been made since then in gaining information on local stocks and in developing preliminary stock assessments. Five distinct management areas have been identified as containing separate stock complexes or as having unique management problems. The management areas include both offshore and inshore grounds.

The fishery for offshore lobsters is concentrated near the heads of the canyons at the edge of the continental shelf from April to June. Subsequently (from July to September) the fishery is active on the shelf at depths less than 160 meters, and in the late fall, the operation withdraws to even shallower water.

The inshore fishery for the American lobster in the United States is active throughout the year, but the fishing effort is usually restricted by adverse weather conditions during the winter months. There are no seasonal restrictions in the United States, with the exception of Monhegan Island off the coast of Maine, where there is a 6-month season. The rules and regulations concerning minimum size limits for the lobster fishery vary by state, ranging from 78-81 mm ($3^1/_{16}$-$3^3/_{16}$ inches) in carapace length. In addition, the penalties for taking berried females and short lobsters are variable by state, as are the licensing fees and buoy restrictions.

2. Canada (Homarus americanus)

Since its inception more than a hundred years ago, the Canadian fishery for *H. americanus* has been largely a small boat operation. The most common type of boat in general use is an open motor boat up to 15 m in length, operated by one or two men. Lobsters are taken in conventional lobster traps. The usual form is a half cylinder, 0.76-1.2 m (2½-4 ft) long, composed of a wooden frame covered by laths and netting. The lobster traps may be set singly or several to a buoyed trawl line and are baited with fresh or salt fish. In contrast to the United States, the traditional inshore fishery of Canada is subject to season lengths as well as size limits. These vary among the ten lobster districts into which the Maritime Provinces of Canada have been divided (DeWolf, 1974). In 1968, the first of a series of regulations was introduced to limit the number of licences issued and the number of lobster traps fished per boat. This program for clawed lobsters is unique to Canada.

The inshore fishery of Canada takes place mainly within the 40 m contour, as it does in the United States, and a Canadian offshore fishery has developed on the Georges and Browns Banks since 1971. Lobster landings in the Maritime Provinces have fluctuated somewhat in the past few decades, ranging from 13,600 to 19,500 metric tons (30-43 million pounds). The most productive fishing areas appear to be southeastern Nova Scotia, Prince Edward Island, and the Northumberland Strait. However, lobsters are landed in all but two of the coastal districts of the Maritime Provinces, though in recent years some of the landings have declined (e.g., those of the Northumberland Strait).

3. The European Fishery (Homarus gammarus)

The fishery for the common lobster is pursued by at least 13 northern European countries. The major producing countries in recent years have been Scotland,

England-Wales, and France. Landings for the entire European fishery have been on the order of 2000 metric tons for several years.

Homarus gammarus looks very much like its close relative, the American lobster. However, it is a bit more slender and its basic color pattern is bluish dorsally and mottled white on the ventral surface.

The European fishery operates from lobster boats which vary in length from 4 to 18 m. The lobster traps (called creels in Scotland) are made of a wooden frame covered with netting. They are usually smaller than the average North American lobster trap. However, they are set similarly, either singly or in trawls of up to 40 traps. The maximum depth fished appears to be about 25 m. At present there are no closed seasons in Europe. The catch varies seasonally in a manner somewhat analagous to the United States catch, i.e., the greatest landings occurring in late summer and fall. Most of the European lobster-producing countries have adopted minimum size limits, based on a total length in the 200-230 mm. (8-9 inch) range. Southern Ireland has adopted a minimum carapace measure of 83 mm (3¼ inches). This is larger than any existing size limit in North America for *H. americanus*. The protection of berried females is accomplished by restricting sales in some countries. No gear or limited entry restrictions are known to exist for this species.

4. The European Fishery (Nephrops norvegicus)

Farmer (1975) has summarized much of the available information in a recent synopsis of biological data. Much of the information which follows is condensed from this important document.

The potential yield of *N. norvegicus* in the northeastern Atlantic has been estimated at about 60,000 metric tons annually (Gulland, 1971), in contrast with the actual annual yield of about 41,000 metric tons. The production for the Mediterranean areas has remained relatively constant over the past 15 years at approximately 2600 metric tons. No specific potential production estimates are available for this region, but it is believed that the species is heavily exploited here.

Fishing for *N. norvegicus* is conducted in the Mediterranean by 6 countries and in the northeast Atlantic by at least 14 countries. The fishery occurs over a suitable muddy or clay bottom on the continental shelf of both of these locations. The fishing grounds are usually close to the coast, but some may be as far as 200 km from shore. The areas of highest abundance in Europe are the east and west coasts of Scotland, the North Sea, the Atlantic coast of France, the Icelandic waters, the Skagerrak and Kattegat, the Atlantic coast of Spain, and the Mediterranean and Adriatic Seas.

The most common gear used in the European fishery is a prawn trawl. However, traps or creels are also used. The mesh sizes vary by geographic areas, though a size of 70 mm seems to be common. The rigging of the trawls is also quite variable. There are some restrictions on the sizes of boats that can be used in certain coastal waters, and the trawlers used for this species usually range about 12–30 m in length.

The pattern of landings for *N. norvegicus* appears to follow that of the other clawed lobsters, i.e., peak landings being made in the summer and fall.

The most common unit of effort employed for *Nephrops* statistics is the catch or landing, expressed in kilograms per hour. Mention has already been made of the problems which might be encountered with this kind of measure of effort. In addition, mesh selection for *N. norvegicus* is not a simple relationship depending only on mesh size. Not only do the animals tend to cling to the meshes and to each other, but their behavior appears to vary with the type of twine material used.

Regulatory measures for the Norway lobster vary by country. They include limitations on the size of fishing boats, closed seasons, protection of berried females, and minimum size regulations.

II. VITAL STATISTICS

A. Stock Identification

In attempting to understand and measure causes of fluctuations in the abundance of a species, it is important to establish the number and identity of any existing subsets of the species. This is crucial because many management plans are based on vital statistics (e.g., growth rates and mortality rates) that are assumed to be constant. Subsets of a population may have a characteristic distribution in space and time as well as having unique growth and mortality coefficients. Those subsets possessing unique vital statistics should be recognized and treated separately when applying population dynamics models.

1. Homarus americanus

Inshore populations of this species consist of local groups. Wilder (1963) has clearly demonstrated that the home range and territory of American lobsters in Canadian inshore areas is restricted to a radius of about 16 km (10 miles) or less. Although there appear to be differences in the average size of lobsters caught from various inshore fishing grounds, these differences may reflect gear selectivity, variable exploitation rates, or ingress from offshore stocks (Cooper, 1970; Uzmann *et al.*, 1977). However, it is generally believed that many of the inshore groups are reproductively isolated.

A multivariate analysis of morphologic measurements obtained from geographically separated samples of lobsters was made by Saila and Flowers (1969). The results of this analysis indicated profile differences between the inshore and offshore samples when matched by size and sex. Berried females displaced from

6. Population Dynamics of Clawed Lobsters

an offshore site (Veatch Canyon) tended to move toward the general area of first capture after the eggs had been shed (Saila and Flowers, 1968). These data were used to infer the existence of spearate inshore and offshore stocks. Parasite studies by Uzmann (1970) further supported these conclusions, but Tracey et al. (1975), using genetic differentiation analyses, found relatively little genetic variability in eight samples of inshore and offshore lobsters. The results of subsequent extensive tagging studies (Uzmann et al., 1977; Lund et al., 1973) have indicated that there are mixtures of inshore stocks with migrant lobsters from offshore areas during the summer months. For the most part, this mixing seems to take place in the fishing grounds of southern New England and Montauk, New York, and it involves the movement toward shore of offshore stocks.

Available evidence concerning stock identity suggests that inshore and offshore stocks are separate, although seasonal mixing has been observed regularly. The level of differences among the inshore groups remains to be further resolved. The observed seasonal movement (dispersion) pattern of the offshore stock is significantly greater than that observed for inshore stocks or for any other clawed lobster species. It must be recognized, however, that the tagged inshore animals have been smaller than the tagged offshore lobsters. The relationship between size and movement requires further study, since some evidence for large-scale movements of the bigger inshore lobsters has been reported (Dow, 1974).

2. *Homarus gammarus*

Available information from mark and recapture studies demonstrates that *H. gammarus* is not migratory (Gibson, 1967). The greatest distance traveled by an individual prior to recapture (in these studies) was reportedly about 8 km. This suggests that the identity of *H. gammarus* stocks may be similar to the inshore stocks of *H. americanus*. In both instances it is presumed that the population consists of several local groups which may be reproductively isolated.

3. *Nephrops norvegicus*

Recent observations by underwater television, photography, and ultrasonic tagging (Chapman et al., 1975) confirm previous findings that the *N. norvegicus* species probably consists of several local stocks that are reproductively isolated. Additional evidence for this statement includes observed differences in body proportions between the inhabitants of different fishing grounds, and variable incidence of the parasite *Stichocotyle nephropsis* (Farmer, 1975).

4. Population and Biomass Data

The reader is referred to Chapters 2-4 of this volume for descriptions of available biomass estimates and related information concerning the ecology of juvenile and adult lobsters.

B. Growth Summary

The material presented herein will be confined to a brief discussion of growth data as it applies to population dynamics models and a listing of some growth equations.

Frequently, growth models are expressed as univariate functions of time. As such, they are often based upon a differential equation expressing the rate of change in size as a function of size. Models of this type have been successfully used to predict size at age for individual animals that grow continuously. They have also predicted the average size at age for entire populations of continuously growing animals. The von Bertalanffy growth curve (a decaying exponential type of function) is an especially popular model because it seems to describe the growth of many aquatic animals fairly accurately, and it can be derived from a model that gives some physiological significance to the parameters. In addition, the parameters of the von Bertalanffy growth curve are incorporated into the widely used Beverton–Holt yield equation.

McCaughran and Powell (1977) have outlined the problems of using classical fisheries growth models to describe crustacean growth; (1) crustaceans do not grow in length continuously, but only at ecdysis, and (2) the large variability of size at age may not be adequately described by an average growth function for the population. In addition, they developed a stochastic growth equation to represent growth of the carapace length of the Alaska king crab. Mauchline (1977) made an analytical assessment of crustacean growth, which included data on *H. americanus* and *H. gammarus*. The authors of both of these papers clearly recognized that the frequency of molting and the magnitude of the molt increment as functions of size must be incorporated into a growth model for it to be realistic. However, they differed in the importance each attached to variability of size at age.

A preliminary examination of growth data of the clawed lobsters suggests that there may be substantial differences between the shapes of the growth curves described by Mauchline (1977) for the American lobster and the curve that Thomas (1973) obtained by fitting a von Bertalanffy function to empirical data. The von Bertalanffy growth curve for growth in terms of weight (the usual input for yield-per-recruit models) has a slight reverse curvature at the beginning and an inflection point at 8/27 of asymptotic weight, under the assumption that a cubic relation exists between length and weight. Conversely, Mauchline's empirical model describes relatively faster growth and larger increments at the earlier stages of growth. Both models are quite sensitive to changes in growth rate. However, the differences in the growth curve for the same species based on different growth models will substantially affect inferences derived from yield-per-recruit models. It is suggested that empirical growth functions be incorporated into yield-per-recruit models for crustacea where possible.

6. Population Dynamics of Clawed Lobsters

The following list briefly summarizes the available growth equations of the decaying exponential type that have been used to describe the growth in carapace length (mm) for clawed lobsters.

Homarus americanus
  Massachusetts (both sexes) $l_t = 253[1-e^{-0.063(t-0.548)}]$ Lobster Scientific Committee (unpublished)
  Maine (both sexes) $l_t = 267[1-e^{-0.048(t-0.772)}]$ Thomas (1973)

Homarus gammarus
  females $l_t = 217[1-e^{-0.094(t-0.344)}]$ Gibson (1967)

Nephrops novegicus
  males $l_t = 116.91[1-e^{-0.08(t+0.44)}]$ Conan (1975)
  females $l_t = 56.04[1-e^{-0.18(t+0.44)}]$ Conan (1975)

The assumption that the von Bertalanffy function is a reasonable approximation for crustacean growth is a principal concern in lobster population dynamics. Discontinuous growth in crustaceans coupled with the variability in size at age make the application of a classical, continuous growth model difficult. Russell and Borden (1980) have voiced arguments similar to those suggested by Mauchline (1977) and McCaughran and Powell (1977) in response to the inadequacy of the continuous von Bertalanffy growth function. Russell and Borden have demonstrated that a strong relationship exists between the intermolt period and the postmolt weight gain, agreeing with the relationships reported by Hewett (1974). In addition, they have used these results to argue that it should be possible to develop a quantitative expression for lobster growth without having to rely on continuous growth models. Recently Bayley (1977) presented a method by which the applicability of the von Bertalanffy growth functions could be determined, given adequate length–weight data.

Bennett (1976) presented a crustacean yield model (*Cancer pagurus*) that uses weight at age data in lieu of the constant growth assumptions of the von Bertalanffy model to avoid the constraint of isometric growth. He determined that changes in growth had a greater relative influence on the model's yield per recruit predictions than did changes in fishing or natural mortaltiy. The findings serve to underscore the need to reliably describe crustacean growth.

In spite of potential inadequacies, the von Bertalanffy growth equation has been used widely to describe the dynamics of lobster populations. The estimation of parameters for the growth equations has been complicated by early and late season molt groups that require the careful examination of size frequency data to obtain values for t_0 and K (See Russell and Borden, 1980). Estimates of fishing and total mortality may be made for lobster populations using cohort analysis (Pope, 1972) with length frequency data (Jones, 1974). When these parameters are incorporated into the Beverton–Holt type of dynamic pool model (Thomas, 1973), estimates of F vary between sexes, as do the growth constants. As a result, Y/R must be estimated for both sexes, and extensive knowledge of seasonal behavior must be obtained before various effort-based management strategies may be effectively implemented.

C. Mortality

1. Homarus americanus

Thomas (1973) has reviewed studies on *H. americanus* and has found values of M (the instantaneous natural mortality coefficient) ranging from 0.04 to 0.08. The general concensus is that F (the instantaneous fishing mortality coefficient) is quite high, with values in excess of 1.0 for the inshore stocks throughout much of their range. Thomas (1973) has reported an F value of 2.30 for the 1966-1970 period in Maine.

2. Homarus gammarus

Total mortality estimates of 46-58% were obtained for *H. gammarus* off the Irish coast (Gibson, 1967). These data are for males; no comparable figures are available for females. It is suggested that the natural mortality rate for *H. gammarus* may be similar to that for the American lobster.

3. Nephrops norvegicus

No comparable data on mortality rates were found.

D. Fecundity and Sex Ratios

1. Homarus americanus

Saila and Flowers (1965) performed a simulation study of sex ratios and regulation effects with the American lobster. The model of stock composition was constructed to permit evaluation of stock changes under varying conditions of exploitation, natural mortality, and regulation. The effects of the decreased molt frequency of mature females on sex ratios and the regulations for protecting berried females (e.g., the double-gauge regulation) were examined. The results of the model studies suggested that the regulation involving protection of large animals has no biological significance, because of the small numbers involved under most conditions of exploitation. Conversely, protection of berried females clearly maintained higher proportions of females in the stocks. The average fecundity of model lobster stocks enjoying protection of their berried females was consistently higher than those with no protection.

Saila *et al.* (1969) have derived the following equation for the carapace length-fecundity relationship:

$$\log Y \text{ (fecundity)} = -1.6017 + 2.8647 \log X \text{ (carapace length, mm)}$$

It should be noted, however, that egg loss occurs during incubation (Perkins, 1971) and was estimated to be about 36% from the period of extrusion (October) to that of hatching (June). Obviously, it is desirable to obtain fecundity estimates from animals near hatching time.

2. *Homarus gammarus*

No data on the size-fecundity relationship was found. However, Thomas (1955) demonstrated stable sex ratios of model stocks when berried females were protected.

3. *Nephrops norvegicus*

Farmer (1975) has summarized available information for this species. Its fecundity is considerably lower than that of the other two clawed lobsters. The number of eggs produced ranges from about 1400 to 4100, and there is a progressive loss of eggs from the pleopods during incubation. This loss is estimated at about ten % per month. The period of incubation from egg laying to egg hatching is variable according to temperature, ranging from 6 to 10 months.

III. POPULATION DYNAMICS—MODELS

A. Empirical Models

Trends in the yields of many commercial fisheries have been associated with trends in components of environmental variability. Sutcliffe *et al.* (1977) have pointed out that temperature may not be the most important environmental variable, but it has the longest continuous record of observations and also serves as an indicator of local oceanographic climate. The influence of temperature on the early life history stages of the American lobster has been documented by Templeman (1936a,b), Scarratt (1964, 1973), and Caddy (1976), among others. A lobster's susceptibility to capture has been shown to be affected by temperature (Paloheimo, 1963), and movement has been related to temperature by McLeese and Wilder (1958), Cooper and Uzmann (1971), and Uzmann *et al.* (1977).

1. *Simple Regression and Correlation Models*

Based on the apparent relationship between stock abundance and various temperature dependent mechanisms, several attempts have been made to demonstrate associations between temperature variability and variations in the yield of the American lobster. Dow (1969) has shown a consistent association between sea surface temperature, inshore lobster abundance, and catch fluctuations—in Maine and in other parts of the species range. Dow *et al.* (1975) demonstrated a squared multiple correlation coefficient (R^2) value of 0.933 among catch, effort, and sea surface temperature. The partial correlation between yield and the other predictor variables was 0.440. In addition, Dow (1977) has calculated product-moment correlation coefficients (r values) between Maine's mean annual sea surface temperatures and Maine and European lobster landings 4, 5, and 6 years later. The correlations were statistically significant for both data sets. Other temperature series were correlated with American and European lobster landings, and again significant correlations were found.

Sutcliffe (1973) also found positive correlations between the monthly discharge of the St. Lawrence River and the annual regional catch of American lobster. These correlations were lagged for periods of 8 and 9 years, and the author suggested some predictive value in the method.

Flowers and Saila (1972) derived predictive yield equations by multiple regression, using present and lagged temperatures as independent variables. Stepwise multiple regression techniques were used to add variables to the regression in descending order of importance, based on the reduction in the total sum of squares. The multiple regression equation used is of the form

$$Y = B_0 + \sum_{i=1}^{N} B_i X_i \qquad (1)$$

where Y is the dependent variable, X_i are independent variables, and B_0 and B_i are constants.

The choice of lagged temperature variables by stepwise multiple regression permitted a testing of the effects of variability in catch in relation to temperature and the development of optimum lags. From this work a prediction equation for Maine was found to be:

$$Y_0 = -1280.965 + 14740.687 \log(T_{-6,7,8}) = 345.823\, T_{w0} \qquad (2)$$

where Y_0 is the present year's yield (metric tons), $T_{-6,7,8} = (T_{-6} + T_{-7} + T_{-8})$, the sum of mean annual temperatures for 6, 7, and 8 years previous to T_0, and T_{w0} is the mean temperature for the month of January, February, and March (present year).

Equation (2) can be used to predict the present year's catch at the end of March for the Maine fishery. Even more precise predictions are possible for the Nova Scotia fishery, if bottom temperature data is added to the regression analysis.

2. *Autoregressive-Moving Average Model*

This methodology was first described in a comprehensive manner by Box and Jenkins (1970). One of the early applications was by Poole (1972) in studying the population change in an experimental population of *Daphnia magna*. More recently, Boudréault *et al.* (1977) have applied an autoregressive-moving average model to annual American lobster landings (1912–1974) in the Magdalen

6. Population Dynamics of Clawed Lobsters

Islands region of the Gulf of St. Lawrence. They were able to explain about 42% of the total variance by a first-order autoregressive model. Since this approach has not been frequently used to date, a brief explanation follows.

A useful class of probability models for time series examinations are linear random processes, which are sufficiently flexible to describe many time series. The time series considered is assumed to be discrete with observations at unit time intervals. The time series may be designated by z_t, where the index t progresses from 1 to n through a series of n observations. Discrete linear random processes are described by relating an "output" random process z_t to an "input" random process a_t.

$$z_t = (\phi_1 Z_{t-1} + \ldots + \phi z_{t-p}) + a_t - (\theta_1 a_{t-1} + \ldots + \theta_q a_{t-q}) \qquad (3)$$

where z_t is the observation deviation from the mean occurring at time t, ϕ_1, $\phi_2, \ldots \phi_p$ are the autoregressive parameters, $\theta_1, \theta_2, \ldots \theta_q$ are the moving-average parameters, and a_t is the independent, normally distributed random variable with mean zero and variance σ^2.

This is an autoregressive moving-average (ARMA) process of order (p,q). Special cases of this model, i.e., when $q = 0$ or $p = 0$, are called autoregressive [AR (p)] or moving-average [MA (q)] models.

Given a set of n observations from a time series (assumed stationary in this case), the procedure for finding an adequate ARMA (p,q) model to represent the time series can be described as an iterative process involving three steps; (1) model identification as AR(p), MA(q), ARMA(p, q), or even ARMA(p, d, q) (the d refers to differencings used to remove the nonstationarity of the mean function if it is nonstationary), (2) model parameter estimation, and (3) checking the candidate model by an analysis of residuals.

The first step is accomplished by examining the autocorrelation (ρ_k) and partial autocorrelation (ρ'_k). Box and Jenkins (1970) produced a matrix to help in identifying a candidate model. After the candidate model is identified it is necessary to estimate the parameters. The procedure for arriving at maximum likelihood estimates is very involved. The model used by Boudréault et al. (1977) was an AR(p) model, which is linear in its parameters. Other models must be handled by numerial techniques to estimate parameters. The final step is diagnostic (i.e., to check the model), and it is usually done by examining sample autocorrelations.

3. Multiple Regression on Principal Components

DePont and Boudréault (1976) developed a multiple regression prediction model for the annual landings of American lobster for the same time series of landings mentioned in Section III,A,2. The model is a linear model with three independent variables; (1) the previous year's landings, (2) the mean December temperature 8½ years ago, and (3) the winter temperature 3½ years ago. The winter temperature is the average for the period from November 15 to December 15. The unique feature of this model is that its coefficients were estimated from a regression on principal components of the independent variables, which were initially chosen from an autocorrelation analysis of annual landings and lagged correlation analyses of monthly temperatures with landings. The coefficients of this model (obtained by a regression on principal components) accounted for 85% of the total variance. The model was used to predict next year's landings, and it provides a trend for two additional years.

4. Polynomial Distributed Lag Model

Orach-Meza and Saila (1978) have applied another technique, the Almon lag or polynomial distributed lag technique, to predict the landings of American lobster for Maine. The data base consisted of a time series of catch, effort, and temperature data for the period of 1933–1974. The polynomial distributed lag was considered appropriate for the analysis because it was reasoned that the yield of a fishery for a long-lived animal, such as the American lobster, does not react fully or immediately to environmental variables and fishing effort. Instead, it is assumed that the yields respond in a gradual manner to the values of the relevant lagged variables. The coefficients of the lagged independent variables are weighted in such a manner as to be approximated by a polynomial expression.

The distributed lag model can be formulated as

$$Y_t = \sum_{j=1}^{N} \sum_{k=0}^{L} b_k X_{t-k,j} + e_t \qquad (4)$$

where Y_t is the lobster yield at time t ($t = 1, 2, \ldots, n$), $X_{t-k,j}$ is the jth independent variable ($j = 1, 2, \ldots, N$) at the kth lag on the lagged variable ($k = 0, 1, \ldots, L$), b_k is a lag coefficient, the respective values of which are assumed to be points on a polynomial, and e_t is a random disturbance term at time t assumed to be normally distributed with a zero mean and unit standard error.

The degrees of the polynomial P and the lengths of the lags L are determined from a knowledge of biological criteria or from statistical criteria. Statistical criteria can include examination of the autocorrelation function. In the case where $P < L$, it can be assumed that $L + 1$ values of the lag coefficients lie on a polynomial of degree P (Almon, 1965). The basic equation [Eq. (4)] can be reduced by linear combinations and substitutions to

$$Y_t = \sum_{j=1}^{N} \sum_{i=0}^{P} a_{i,j} Z_{i,j,t} + u_t \qquad (5)$$

where $Z_{i,j,t} = \sum_{k=0}^{L} k^i X_{t-k,j}$ are linear combinations of the jth environment variable at the ith degree of the polynomial ($i = 0,1,2, \ldots, P$), i.e., the Z variables are moving averages of the original variables, and $a_{i,j}$ are linear estimates of the coefficients of the Z variables. Estimates of the lag coefficients can therefore be obtained from $a_{i,j}$ by the following power series:

$$b_{k,j} \sum_{i=0}^{P} a_{i,j} k^i \qquad (6)$$

If $P = L$, the polynomial lag technique and ordinary multiple regression are identical, and Eq. (4) is equivalent to Eq. (5).

The major advantage of using the polynomial distributed lag model rather than a direct multiple regression method to describe lagged relationships is that if the parameters of the lag regressors can be constrained by a set of linear combinations to be points on a polynomial of specified degree, then the distributed lag can be estimated by the standard linear regression method without problems of multicollinearity in the regressor variables. It has also been shown that if the disturbance terms are independently and identically distributed, with mean zero and unit standard error, the resulting multiple regression estimates will be unbiased, consistent, and efficient. They will lead to valid tests concerning the lag coefficients (Dhrymes, 1971; Schmidt and Waud, 1973).

The practical utility of the distributed lag technique for fisheries forecasting was demonstrated by Orach-Meza and Saila (1978). This model permits predictions of recruitment into the legal supply of lobsters of up to five years in advance with reasonable confidence. Obviously, this kind of predictive capability is an important goal in resource management.

5. Summary

It is evident that significant developments in the application of empirical models to the lobster fishery have recently taken place. These include careful tests of the validity of lag coefficients and improved predictive capabilities. There is no reason to believe that these models, devoloped initially for *H. americanus*, cannot be applied to the other species of clawed lobsters for which suitable time series exist. Indeed, better time series data already exist for finned fishes, so that these models are equally applicable to them.

B. Conceptual Models

Attempts to model the dynamics of lobster populations have generally used the same techniques that are commonly applied to fishery science. These techniques may be characteristically associated with the two generalized conceptual approaches that describe yield from a fishery. One approach considers abundance change in terms of production over time and is characterized by the surplus-yield type of model. This approach does not distinguish between recruitment, physiological growth, and natural mortality as factors contributing to overall changes in abundance, but considers only their resultant effect as a single function of the population size. The second approach describes production in terms of detailed, species-specific growth and mortality rates and is commonly referred to as the dynamic pool type of model. The conceptual bases and applicability of the surplus yield models are elaborated in the next section. The dynamic pool models were considered by Morgan in chapter 5.

1. Surplus-Yield Models

The surplus-yield approach to modeling is perhaps best exemplified by the work of Schaefer (1954, 1957). The model has its greatest applicability to fisheries for which elemental rates of growth and mortality are not easily obtained. The approach derives from the logistic law of population growth, describing a sigmoid curve on which the rate of biomass increase is maximized at one-half the maximum equilibrium population level. Under equilibrium conditions, the catch is euqivalent to the annual production, which is a function of stock size. Biomass is estimated directly from the weight of the catch and the age composition of the stock is assumed to remain constant over time.

In the absence of fishing mortality, the Schaefer model assumes the form

$$\frac{dP}{dt} = bP \left(\frac{K-P}{K} \right) \tag{7}$$

where P is the biomass of fish in the catchable stock, K is the maximum population abundance (carrying capacity), and b is a constant equivalent to the rate of natural increase. Letting $a = b/K$, one is left with the generalized form of the logistic equation

$$\frac{dP}{dt} = bP - aP^2 \tag{8}$$

The generalized Schaefer model, which explicitly considers fishing mortality, is given by the expression

$$\frac{dP}{dt} = bP - aP^2 - qfP \tag{9}$$

where f is applied effort and q is the coefficient of catchability. At equilibrium, yield may be expressed as

$$Y_0 = qf_0 P_0 = bP_0 - aP_0^2 \tag{10}$$

or in terms of applied effort as

$$Y_0 = \frac{qf_0}{a}(b - qf_0) \tag{11}$$

where the subscript 0 identifies the equilibrium condition.

6. Population Dynamics of Clawed Lobsters

The value in the Schaefer approach is that when adequate catch and effort data are available to serve as an index of abundance, it becomes possible to establish maximum equilibrium yield and optimum applied effort, given certain simplifying assumptions about growth and age composition. In practice, the catchability of lobsters varies widely throughout the year, and the estimation of annual CPUE as an index of abundance may be biased by the distribution of effort. It has been suggested that a more useful index of abundance may be calculated from catch and effort data specific to a period of characteristic behavior, in which catchability (and environment) may be assumed to be reasonably constant from year to year (Thomas, 1973).

Variations of the surplus yield model have been presented by Pella and Tomlinson (1969) and Fox (1970). The latter formulations have been offered in response to the symmetrical yield-effort relationship imposed by the Schaefer model.

As noted previously, the Schaefer surplus-yield model considers the effect of natural mortality only as an implicit part of an overall growth function. Clearly then, the model assumes no explicit stock-recruitment relationship, the existence of which has been discussed recently by Hancock (1973). As such, a time lag between spawning and recruitment is considered to have no effect on growth in the population. This assumption may be acceptable when the model is being used to describe the dynamics of a species that is characteristically fast growing and enters the fishery at an early age. Here physiological growth and recruitment may be combined as an instantaneous function of population size. When modeling the lobster fisheries, however, the time delay between spawning and recruitment becomes an important consideration, particularly when the fishery relies heavily on the recruiting year class.

A variation of the Schaefer model that explicitly considers the time delay between spawning and recruitment was proposed by Marchesseault et al. (1976). The model, termed delay differential in form, considers the impact of spawning stock size on recruitment to the fishery several years later. In doing so, the model suggests the existence of a stock-recruitment relationship. While the form of the stock-recruitment relationship for lobster is in debate, it is assumed that at high levels of exploitation (with depleted stock abundance) a relationship exists. For simplicity, the relationship is taken to be proportional in the delay model. The general form of the delayed recruitment model is

$$\frac{dP}{dt} = bP(t) - a_1 P(t)^2 + a_2 P(t-w) - q f(t) P(t) \qquad (12)$$

In this form, the maximum equilibrium yield (MEY) and optimum applied effort are

$$Y_0^* = (b + a_2)^2/4a_1 \quad \text{and} \quad f_0^* = (b + a_2)/2q \qquad (13)$$

and equilibrium yield as a function of effort assumes the form

$$Y_0 = \frac{qf_0}{a_1}(b + a_2 - qf_0) \qquad (14)$$

Where catch per unit effort (CPUE) data is used as an index of abundance, it can be shown that Eq. (13) and (14) may be equally expressed by dividing the right side of each expression by q.

Marchesseault *et al.* (1976) compared the Schaefer and the delayed recruitment models in an analysis of the inshore lobster (*Homarus americanus*) fishery. Because there is typically a 5-7 year delay between spawning and recruitment for lobster in New England waters, a management model that considers recruitment as a fundamental contributor to population dynamics is particularly appropriate. Accounting for recruitment in terms of a linear function of the spawning population is likely to fall short of describing the functional recruitment relationship actually operative in the lobster fishery, especially in light of the demonstrated influence of temperature (Flowers and Saila, 1972; Dow, 1977; and others noted previously in the discussion of empirical models). However, the motivation for this approach to modeling is to provide a management tool that may be easily applied, utilizing readily available catch and effort data.

The application of the delayed recruitment model to the Rhode Island inshore lobster fishery demonstrated that predictions of maximum equilibrium yield are more conservative relative to the Schaefer model. Moreover, the predicted optimum population level is shown to be higher and the optimum level of applied effort is lower than would be predicted using the Schaefer model. Generated yield versus effort curves are shown in Fig. 1. It may be concluded that, given

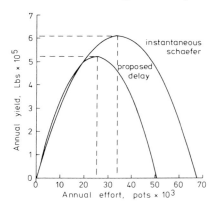

Fig. 1. Predicted values of annual yield with annual effort for the Rhode Island inshore lobster fishery, given by the delay and instantaneous Schaefer models. (After Marchesseault *et al.*, 1976.)

6. Population Dynamics of Clawed Lobsters 237

the dependence of the lobster fishery upon annual recruitment, management using the Schaefer model may lead to overfishing and continued depletion of existing stocks.

The effects of random fluctuations in recruitment on the dynamics and management analysis of the lobster fishery may be examined using a stochastic, delayed recruitment surplus yield model. Under the assumption of long-term average environmental conditions and optimum exploitation, mean annual yields from the fishery are expected to correspond closely to MEY. It is evident, however, that environmental fluctuations, particularly those influencing recruitment, may create instability and uncertainty in any calculations involving population dynamics. Sissenwine (1976) and Doubleday (1976) used similar approaches to study the effect of random elements on the time dependent solutions of the Schaefer surplus yield model for hypothetical fisheries. The authors independently demonstrated that the variability in the predicted mean catch resulting from random environmental inputs increased in relation to the fishing effort applied. Additionally, the results indicated that the long-term average yield taken by the fishery is likely to be below deterministic MEY under optimum exploitation conditions.

The delayed recruitment surplus-yield model just described provides a basis for examining the effect of fluctuations in recruitment on the dynamics of the lobster fishery. The stochastic representation of this model is given by Marchesseault (in preparation) using CPUE as an index of population abundance N

$$\frac{dN}{dt} = bn(t) - a_{12}\eta^2(t) + a_2\eta(t-w) - qf(t)\eta(t) \tag{15}$$

where $= N + (\sigma_N\zeta)$, σ_N = the standard deviation of the variable, and ζ is a standard random normal deviate, $\zeta \simeq N(0,1)$. The technique used to incorporate random effects into the simulation of the model differs conceptually from that used by Beddington and May (1977) for the Schaefer model. In a manner similar to Sissenwine (1976), the random environmental input is held constant over the time interval of one year, instead of being allowed to fluctuate continually as though the model reflected a continuous time process. It is not conceptually consistent with Schaefer-type models to describe processes that occur on a time interval of less than one year.

Because the dynamics of an exploited lobster population are dependent on annual recruitment, it is instructive to formulate the model with only the delayed recruitment variable $[N(t-w)]$ being subject to random fluctuations. As a result of stochastic simulations, random variability in recruitment is shown to increasingly translate into year to year yield fluctuations over time as the level of applied effort is increased. The standard error associated with estimates of mean yield is shown to increase with levels of effort above f_0 and decrease with applied effort below f_0. With yield fluctuations and standard errors of estimate increasing for

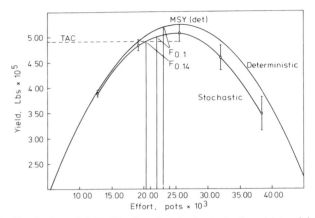

Fig. 2. Predicted values of yield with effort given by the stochastic and deterministic versions of the delayed recruitment surplus yield model. Yield values corresponding to $F_{0.1}$ are shown for both models, and yield corresponding to TAC is given by the stochastic model analysis.

overexploitation conditions, it is clear that random fluctuations in recruitment may result in instability and uncertainty associated with yield predictions at low stock levels.

Yield versus effort curves are shown in Fig. 2 for the deterministic and stochastic delayed recruitment models. The level of yield corresponding to MEY predicted by the delay model is shown to be below that predicted by the deterministic model. Under the assumption that it is desirable to limit annual yield so that the probability of overfishing is minimized, it may be reasonable to establish total allowable catch (TAC) at a level consistent with the lower 95% confidence interval of the mean MEY prediction. Figure 2 shows the latter yield constraint in conjunction with yield associated with $F_{0.1}$ as they relate to the Rhode Island inshore lobster fishery.

2. Summary

For cases where data necessary for the application of dynamic pool models is not readily available, it is suggested that the delayed recruitment model provides a reasonable alternative for management decision making. In the surplus-yield models, CPUE is used as an index of population abundance. The limitations of this index have been discussed by Bennett (1976), Skud (1977), and Munro (1974). Ultimately the choice between model types reduces to a trade-off, in which the single function applications of the surplus-yield model are evaluated against the extensive data requirements of dynamic pool models. The criteria for this choice are established by the long- and short-term management objectives.

6. Population Dynamics of Clawed Lobsters

REFERENCES

Almon, S. (1965). The distributed lag between capital appropriations and expenditures. *Econometrica* **33**(1); 178-196.

Bayley, P. B. (1977). A method for finding the limits of application of the von Bertalanffy growth model and statistical estimates of the parameters. *J. Fish. Res. Board Can.* **34**, 1079-1084.

Beddington, J. R., and May, R. M. (1977). Harvesting natural populations in a randomly fluctuating environment. *Science* **197**, 463-465.

Bennett, D. B. (1976). The influence of parameter inputs on a possible crustacean yield model. *ICES Spec. Meet. Population Assessments Shellfish Stocks, Pap.* No. 3, 1-11 (mimeo.).

Boudréault, F.-R., Dupont, J.-N., and Sylvain, C. (1977). Modèles linéaires de prédiction des débarquements de homard aux Iles-de-la-Madeleine (Golfe du Saint-Laurent). *J. Fish. Res. Board Can.* **34**, 379-383.

Box, G. E. P., and Jenkins, G. M. (1970). "Time Series Analysis, Forecasting and Control." Holden-Day, San Francisco, California.

Caddy, J. F. (1976). The influence of variations in the seasonal temperature regime on the survival of larval stages of the American lobster (*Homarus americanus*) in the Southern Gulf of St. Lawrence. *ICES Spec. Meet. Population Assessments Shellfish Stocks, Pap.* No. 10, pp. 1-44.

Chapman, C. J., Johnstone, A. D. F., and Rice, A. L. (1975). The behavior and ecology of the Norway lobster, *Nephrops norvegicus* (L.). *Proc. Eur. Mar. Biol. Symp., 9th, 1974* pp. 59-74.

Conan, G. (1975). Périodicité des mues, croissance et cycle biologique de *Nephrops norvegicus* dans le Gulfe de Gascogne *C. R. Hebd. Seances Acad. Sci., Ser D* **281**, 1349-1352.

Cooper, R. A. (1970). Retention of marks and their effects on growth, behavior, and migrations of the American lobster, *Homarus americanus*. *Trans. Am. Fish. Soc.* **99**, 409-417.

Cooper, R. A., and Uzmann, J. R. (1971). Migrations and growth of deep-sea lobsters, *Homarus americanus Science* **171**, 288-290.

DeWolf, A. G. (1974). The lobster fishery of the maritime provinces: Economic effects of regulations. *Bull, Fish. Res. Board Can.,* **187**, 1-58.

Dhrymes, R. J. (1971). "Distributed Lags: Problems of Estimation and Formulation." Holden-Day, San Francisco, California.

Doubleday, W. G. (1976). Environmental fluctuations and fisheries management. *ICNAF Sel. Pap.* **1**, 141-150.

Dow, R. L. (1969). Cyclic and geographic trends in sea water temperature and abundance of American lobster, *Science* **164**, 1060-1063.

Dow, R. L. (1974). American lobsters tagged by Maine commercial fishermen. 1957-1959. *Fish. Bull.* **72**, 622-623.

Dow, R. L. (1977). Relationship of sea surface temperature to American and European lobster landings. *J. Cons., Cons. Int. Explor. Mer* **37**, 186-191.

Dow, R. L., Bell, F. W., and Harriman, D. M. (1975). Bioeconomic relationships for the Maine lobster fishery with consideration of alternative management schemes. *NOAA Tech. Rep., NMFS SSFR* **683**, 1-44.

DuPont, J.-N., and Boudréault, F.-R. (1976). Prédiction des débarquements de homard aux Iles-de-la-Madeleine. II. Régression multiple sur les composantes principales. *Rech., Div. Pech. Marit., Min. Ind. Commer., Cah. Inf.* No. 73, pp. 67-101.

Farmer, A. S. D. (1975). Synopsis of biological data on the Norway lobster, *Nephrops norvegicus*, (Linnaeus, 1758). *FAO Fish. Synopsis* **112** (FIRS/S112), 1-97.

Flowers, J. M., and Saila, S. B. (1972). An analysis of temperature effects on the inshore lobster fishery. *J. Fish. Res. Board Can.* **29**, 1221-1225.

Food and Agriculture Organization (1977). "Yearbook of Fishery Statistics," Vol. 40. FAO(UN), Rome.

Fox, W. W. (1970). An exponential yield model for optimizing exploited fish populations. *Trans. Am. Fish. Soc.* **99**, 80–88.

Gibson, F. A. (1967). Irish investigation on the lobster (*Homarus vulgaris* Edu). *Ir. Fish. Invest., Ser. B* **1**, 1–45.

Gulland, J. A. (1971). "The Fish Resources of the Oceans." Fishing News (Books) Ltd., West Byfleet, Surrey.

Hancock, D. A. (1973). The relationship between stock and recruitment in exploited invertebrates. *Rapp. P.-V. Reun., Cons. Int. Explor. Mer* **164**, 113–131.

Havinga, B. (1929). Kaisergranat. *Handb. seefisch. Nordeurop.* 3(2), 38–40.

Hewett, C. J. (1974). Growth and moulting in the common lobster (*Homarus vulgaris* Milne-Edwards). *J. Mar. Biol. Assoc. U.K.* **54**, 379–391.

Jones, R. (1974). Assessing the long-term effects of changes on fishing effort and mesh size from length composition data. *ICES CM 1974/F* **33**, 1–14 (mimeo).

Jones, R. (1976). The use of marking data in fish population analysis. *FAO Fish. Tech. Pap.* **153**, 1–42.

Lund, W. A., Stewart, L. L., and Rathbun, J. (1973). Investigation on the lobster. *Conn. Fish. Dev. Act. Final Rep.* **3-130-R**, 1–189. (mimeo).

McCaughran, D. A., and Powell, G. C. (1977). Growth model for Alaska king crab (*Paralithodes camtschatics*). *J. Fish. Res. Board Can.* **34**, 989–995.

McLeese, D., and Wilder, D. G. (1958). The activity and catchability of the lobster (*Homarus americanus*) in relation to temperature. *J. Fish. Res. Board Can.* **15**, 1345–1354.

Marchesseault, G. D., Saila, S. G., and Palm, W. J. (1976). Delayed recruitment models and their application to the American lobster (*Homarus americanus*) Fishery. *J. Fish. Res. Board Can.* **33**, 1779–1787.

Mauchline, J. (1977). Growth of shrimps, crabs, and lobsters—an assessment. *J. Cons., Cons. Int. Explor. Mer* **37**, 162–169.

Munro, J. L. (1974). The biology, ecology, exploitation, and management of Caribbean reef fishes. Part VI. The biology, ecology and bionomics of Caribbean reef fishes—Crustaceans (spring lobsters and crabs). *Univ. West Indies Zool. Dep., Res. Rep.* **3**, 1–57.

Orach-Meza, F. L., and Saila, S. B. (1978). Application of a polynomial distributed lag model to the Maine lobster fishery. *Trans. Am. Fish. Soc.* **107**, 402–411.

Paloheimo, J. E. (1963). Estimation of catchabilities and population sizes of lobsters. *J. Fish. Res. Board Can.* **20**, 59–88.

Pella, J. J., and Tomlinson, P. K. (1969). A Generalized stock production model. *Bull. Int.-Am. Trop. Tuna Comm.* **13**, 421–496.

Perkins, H. C. (1971). Egg loss during incubation from offshore northern lobsters (Decaopda: Homaridae). *Fish. Bull.* **69**, 451–453.

Poole, R. W. (1972). An autoregressive model of population density change in an experimental population of *Daphnia magna*. *Oecologia* **10**, 205–221.

Pope, J. G. (1972). An investigation of the accuracy of virtual population analysis using cohort analysis. *Res. Bull. Int. Comm. Northwest Atl. Fish.* **9**, 65–74.

Russell, H. J., Jr., and Borden, D. V. D. (1980). "Management Studies of Inshore Lobster Resources," Final Rep., State-Federal Lobster Contract #03-4-043-360 (in preparation).

Saila, S. B., and Flowers, J. M. (1965). A simulation study of sex ratios and regulation effects with the American lobster, *Homarus americanus*. *Gulf Caribb. Fish. Inst., Miami Univ., Proc.* **18**, 66–78.

Saila, S. B., and Flowers, J. M. (1968). Movements and behavior of berried female lobsters displaced from offshore areas to Narragansett Bay, Rhode Island, *J. Cons., Cons. Int. Explor. Mer* **31**, 342–351.

Saila, S. B., and Flowers, J. M. (1969). Geographic morphometric variation in the American lobster, *Syst. Zool.* **18**, 330-338.

Saila, S. B., Flowers, J. M., and Hughes, J. T. (1969). Fecundity of the American lobster, *Homarus americanus*. *Trans. Am. Fish. Soc.* **98**, 537-539.

Scarratt, D. J. (1973). Abundance, survival, and vertical and diurnal distribution of lobster larvae in Northumberland Strait, 1962-63, and their relationships with commercial stocks. *J. Fish. Res. Board Can.* **30**, 1819-1824.

Scarratt, D. J. (1964). Abundance and distribution of lobster larvae (*Homarus americanus*) in Northumberland Strait. *J. Fish. Res. Board Can.* **21**, 661-680.

Scarratt, D. J., and Elson, P. F. (1965). Preliminary trials of a tag for salmon and lobsters. *J. Fish. Res. Board Can.* **22**, 421-423.

Schaefer, M. B. (1954). Some aspects of the dynamics of populations important to the management of the commercial fisheries. *Bull. Int.-Am. Trop. Tuna Comm.* **1**, 27-56.

Schaefer, M. B. (1957). A study of the dynamics of the fishery for yellowfin tuna in the eastern tropical Pacific Ocean. *Int.-Am. Trop. Tuna Comm.* **2**, 245-285.

Schmidt, P., and Waud, R. N. (1973). The Almon lag technique and the monetary versus fiscal policy debate. *J. Am. Stat. Assoc.* **68**, 11-19.

Sissenwine, M. P. (1976). The effect of random fluctuations on a hypothetical fishery. *ICNAF Sel. Pap.* **2**, 137-144.

Skud, B. E. (1977). Soak-time and the catch per pot in an offshore fishery for lobsters (*Homarus americanus*). *ICES Spec. Meet. Population Assessments Shellfish Stocks* No. 8 (mimeo).

Sutcliffe, W. H. (1973). Correlations between seasonal river descharge and local landings of American lobster (*Homarus americanus*) and Atlantic halibut (*Hippoglossus hippoglossus*) in the Gulf of St. Lawrence. *J. Fish. Res. Board Can.* **30**, 856-859.

Sutcliffe, W. H., Jr., Drinkwater, K., and Muir, B. S. (1977). Correlations of fish catch and environmental factors in the Gulf of Maine. *J. Fish. Res. Board Can.* **34**, 19-30.

Symonds, D. J. (1972). The fishery for the Norway lobster, *Nephrops norvegicus* (L.) off the northeast coast of England. *Fish. Invest., (London), Ser. 2* **27**, 1-35.

Taylor, C. C., Bigelow, H. B., and Graham, H. W. (1957). Climatic trends and the distribution of marine animals in New England. *Fish. Bull.* **57**, 293-345.

Templeman, W. (1936a). Local differences in the life history of the lobster (*Homarus americanus*) on the coast of the maritime provinces of Canada. *J. Biol. Board Can.* **2**, 41-88.

Templeman, W. (1936b). The influence of temperature, salinity, light, and food conditions on the survival and growth of the larvae of the lobster (*Homarus americanus*). *J. Biol. Board Can.* **2**, 485-497.

Thomas, H. J. (1955). Observations on the sex ratio and mortality index in the lobster, (*Homarus vulgaris*). *J. Cons., Cons. Int. Explor. Mer* **20**, 192-305.

Thomas, J. C. (1973). An analysis of the commercial lobster (*Homarus americanus*) fishery along the coast of Maine, August 1966 through December 1970. *NOAA Tech. Rep., NMFS SSRF* **667**, 1-57.

Tracey, M. L., Nelson, K., Hedgcock, D., Schlesser, R. A., and Pressick, M. (1975). Biochemical genetics of lobsters: Genetic variation and the structure of American lobster (*Homarus americanus*) populations. *J. Fish. Res. Board Can.* **32**, 2091-2101.

Uzmann, J. R. (1970). Use of parasites in identifying lobster stocks. *Proc. Int. Conf. Parasitol., 2nd*, pp. 12-20.

Uzmann, J. R., Cooper, R. A., and Pecci, K. J. (1977). Migration and dispersion of tagged American lobsters, *Homarus americanus*, on the southern New England continental shelf. *NOAA Tech. Rep., NMFS SSRF* **705**, 1-92.

Wilder, D. G. (1963). Movements, growth, and survival of marked and tagged lobsters liberated in Egmont Bay, Prince Edward Island. *J. Fish. Res. Board Can.* **20**, 305-318.

Chapter 7

Spiny Lobster Fisheries Management

B. K. BOWEN

I.	Introduction	243
II.	Major Spiny Lobster Fisheries	244
III.	The Western Australian Fishery	246
	A. Description	246
	B. Potential for High Exploitation Rate	250
	C. Management Objectives	251
	D. Data Base for Management Decisions	252
	E. Process of Management Decision Making	255
IV.	The Limitation of Fishing Effort	256
	A. Data	256
	B. Results of the Introduction of Limited Entry	258
V.	Concluding Comments	261
	References	263

I. INTRODUCTION

In most of the major spiny lobster fisheries there has been a similar pattern of development; a rapid increase in catch and effort from about the late 1940s, followed by either a fall in catch (e.g., New Zealand) or a period of relatively stable catch with increasing effort (e.g., Western Australia). Both of these situations cause concern for the industry and those responsible for management. In addition, there is an increasing recreational interest in the capture of spiny lobsters following an explosive growth in high-speed sports craft and underwater diving.

The problems facing most fisheries administrators is well expressed by Beardsley *et al.* (1975) in a white paper on the Florida spiny lobster fishery:

> What data are available indicate that there have been serious declines in the catch per unit of fishing effort in recent years (Seaman and Aska, 1974). Divers find that areas which previously had good lobster fishing now have few lobsters. Commercial fishermen who in early years operated only 200 traps now use as many as 2000 traps to make the same harvest. In our opinion, based on communication with both fishermen and scientists, this intensified fishing pressure creates an exploitation rate for the Florida population that is high enough so that most lobsters at or near legal size in the Florida Keys are being caught each year.
>
> The rapidly intensifying conflict between recreational and commercial interests in the fishery must be resolved so the lobster resource may provide maximum benefit to society. If these problems are not addressed soon, not only may the ability of the fishermen to economically harvest this high quality protein source for society be eliminated, but the reproductive potential of the population may be reduced.

These extracts refer to the main interacting factors that fisheries managers must consider carefully in determining management measures for a spiny lobster fishery; i.e., (1) the benefits to society, (2) the economic return, (3) the size at first capture, (4) the level of fishing effort, and (5) the reproduction potential and its relation to recruitment levels.

The observations by Beardsley *et al.* (1975) on the state of the Florida fishery were made some 17 years after Smith (1958) had set out very clearly the kind of regulations available for the proper management of a fishery. Smith had also listed specific regulations pertaining to the Florida spiny lobster fishery, including closed seasons, size and weight limits, prohibition on the taking of animals carrying eggs, gear restrictions, and closed areas. Clearly, these restrictions were not sufficient in the face of increasing fishing pressure by both amateur and professional fishermen. The problem of increasing fishing pressure on spiny lobsters in the waters off Florida has been aggravated as a result of the Bahama bank being now mostly closed to United States fishermen. The displaced boats are intensifying efforts along South Florida, especially the Florida Keys.

In subsequent sections, a brief description is given of the major spiny lobster fisheries throughout the world. This is followed by a more detailed description of the fishery off the west coast of Australia, which has been regulated by license limitation since 1963, together with a discussion of the usual array of measures introduced in earlier years, such as a legal minimum length, closed seasons, and the protection of "berried" females.

II. MAJOR SPINY LOBSTER FISHERIES

The principal producers of spiny lobsters are South Africa, Australia, New Zealand, Cuba, Brazil, the United States, and Mexico.

The South African fishery is based mainly on *Jasus lalandii*. This species is

7. Spiny Lobster Fisheries Management

caught in a series of discrete fishing grounds along the coast of southern Africa, in waters less than 100 m in depth. The standard method of capture in deeper water involves traps (pots) operated from powered vessels of up to 20 m in length. In shallow waters, capture is made from dinghies using drop nets. The major portion of the catch is processed to a frozen tail pack for sale to the United States. Annual production of South African spiny lobsters is presently about 8000 tonnes, though during the 1950s and 1960s the yield was much greater (e.g., 24,000 tonnes in 1952 and 17,000 tonnes in 1962). The other commercial spiny lobster species taken in this area is *Jasus tristani,* caught in waters around Tristan, Inaccessible, Nightingale, and Gough Islands, and on Vema Seamount, and also *Palinurus gilchristi,* taken in deep waters off the south coast of Africa.

The fishery of the Australian west coast is based principally on *Panulirus cygnus,* and that of the southeast coast is based on *Jasus novaehollandiae.* Both are major fisheries, the west coast yielding approximately 8000 tonnes per year, and the southeast coast 4000 tonnes per year. The fishery for the west coast is described in greater detail in Section III. The catch from both western and southeastern Australia is processed mainly to a frozen tail pack for sale in United States. Both fisheries employ pots as the standard method of capture.

The New Zealand fishery is based principally on *Jasus edwardsii* taken from waters around the North and South Islands and the Chatham Islands to the east. Present annual production for the North and South Islands area is approximately 3500 tonnes, although it was much higher in earlier years (8500 tonnes in 1956 and 5000 tonnes in 1965). The waters around the Chatham Islands were first worked in 1965, and there was a rapid growth in production to about 6000 tonnes in 1968, but thereafter production fell. The present annual take is approximately 500 tonnes.

The principal catching method is identical to the Australian technique, i.e., pots are worked from vessels measuring to about 20 m in length. The product is processed to a frozen tail pack for sale in United States.

Major fisheries of Cuba, Brazil, and the United States (mainly Florida), are based on the spiny lobster *Panulirus argus.* This species is distributed throughout the area, from North Carolina (in the U.S.) to northeastern Brazil, including the Caribbean and the Gulf of Mexico. In 1974, the *Panulirus* fisheries of the United States produced approximately 6000 tonnes (including 1500 tonnes caught in international waters off foreign shores), while Cuba and Brazil produced 10,000 and 8000 tonnes, respectively. Lobster traps or pots are the principal catching method used by these countries, although other countries use a variety of methods, including hand nets, spears, and diving, to capture *P. argus* (Streetor and Weidner, 1976). United States and Brazilian production is primarily for the North American market, whereas the Cuban catch is mainly sold in Europe. Many other countries throughout the Caribbean also catch *P. argus,* but in far less quantities than the Unites States, Cuba, and Brazil.

Panulirus laevicauda is also taken in substantial numbers in Brazil, accounting for about one-fifth of that country's total catch.

Along the coast of California there is a relatively small fishery based on *Panulirus interruptus*, with an annual production of approximately 100 tonnes. This species is also the principal lobster product of Baja California, and Mexico. Annual production is roughly 1100 tonnes, the majority of which is exported to the United States.

Morgan (Chapter 5), in his review of the population dynamics of the Palinuridae, has listed those species which form the basis of a major fishery. In addition to those discussed above he has listed *Panulirus japonicus, Jasus verreauxi, Palinurus mauritianicus*, and *Jasus paulensis*.

III. THE WESTERN AUSTRALIAN FISHERY

A. Description

The western rock lobster, *Panulirus cygnus*, is found on the west coast of Australia from North West Cape (21° 45′S) to Cape Leewin (34° 22′S) (Fig. 1). Initially, spiny lobsters were caught in the immediate vicinity of two well established centers of population, Fremantle and Geraldton. During the late 1940s and through the 1950s the fishery expanded to the north and south. Today there are lobster boats operating along the entire coast, from Kalbarri in the north to Bunbury in the south. In addition, there is a substantial fishery at the Abrolhos Islands some 40–60 miles off the coast at Geraldton.

In the early years of the developing fishery the boats were small (up to 10 m) and the pots were lifted by hand. The areas of operation were thus determined by the suitability of the boats to cope with distances and prevailing weather conditions, and the capability of the skippers to locate reef areas that provide the habitat of the spiny lobster. Later, during the 1950s and 1960s, significant technological advances were made. Among these were the introduction of echosounders, the use of synthetic rope, pot winches and pot tippers, and the construction of boats designed specifically for the spiny lobster industry. Most of the newer vessels have high speed planing hulls that achieve speeds of up to 25 knots. They also have large deck space for carrying gear, allowing for pots to be shifted each day, thus increasing fishing pressure. The number and size of vessels now operating is shown in Table I.

Expansion of areas of operation in the early 1950s was achieved in two ways. Rough road tracks were made to anchorages from which fishing boats could be serviced so that the catch could be collected for return to processing establishments at Fremantle or Geraldton, and freezer boats were built, which were both catching and processing units. By this method important fishing centers have developed at Lancelin, Cervantes, Jurien Bay, and Leeman.

7. Spiny Lobster Fisheries Management

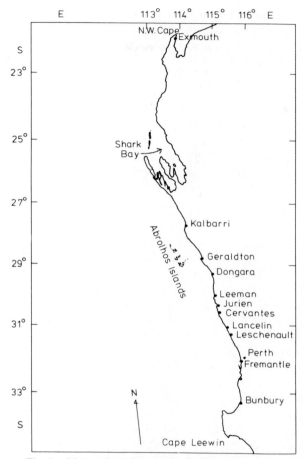

Fig. 1. Major spiny lobster centers in Western Australia.

The method of capture has altered little during the development of the fishery. Basically, there are three types of pots; wooden "batten," stick or cane "beehive," and wire "beehive" (Fig. 2). Research undertaken on the relative fishing power has shown that there is little difference between the three types, although wire "beehive" pots are not popular.

The "beehive" pots are used in waters of all depths, whereas "batten" pots are confined mainly to the shallow protected waters. Surface identification of a spiny lobster pot is made by means of a polystyrene float or floats, which are attached to the pot rope and appropriately marked with the fisherman's boat license number. Pots are lifted, baited, and reset each morning, weather permitting. The most popular bait is "Australian salmon" heads (*Arripis trutta*) and

TABLE I

Number of Vessels by Size Categories in the Western Australian Spiny Lobster Fishery

Length (m)	Number
Under 6	8
6 to under 9	376
9 to under 12	285
12 to under 15	99
15 and over	34
Total	802

bullock hocks or hides, but many other baits are also used (Morgan and Barker, 1975).

Generally the smaller spiny lobsters of up to 3–4 years of age (i.e., to about 70 mm carapace length) inhabit the shallow water reefs in depths to 7 m. The larger animals are found in deeper waters and are fished to 200 m. However, most of the catch is taken from depths ranging from 1 to 100 m. Each year, for a period of about 4 weeks after the season opening on November 15, substantial catches are taken of recently molted, immature animals (locally known as "whites") in the shallow waters. The vessels then move into deeper waters, following the "whites" as they move offshore, darkening in color and mingling with the "red" population (George, 1958a).

Spiny lobster processing establishments have been located at Geraldton, Dongara, Jurien Bay, Cervantes, Lancelin, Leschenault, Perth, and Fremantle (Fig. 1). Fishermen return to their anchorages between midday and 5 PM, depending on the area of operation, and the catch is taken immediately to the processing establishment, which operates until all the lobsters are processed. Quality control regulations require that all spiny lobsters must be alive at the commencement of processing. The major method of processing (over 95% of the catch) is by de-tailing and packing individually wrapped green (uncooked) tails for sale on the United States market. Smaller quantities are cooked whole for the local and European markets.

The method of handling catches from the Abrolhos Islands (Fig. 1) differs from the coast in that the catch is held for up to four days in floating holding crates awaiting bagging and transport to Geraldton by carrier boat. The majority of boats in this area have deck tanks with circulating water in which the spiny lobsters are placed immediately after capture.

The freezer boats were of importance in the early years of the developing spiny

7. Spiny Lobster Fisheries Management

Fig. 2. Three types of spiny lobster pots used in Western Australia. Top, batten pot; center, stick or cane beehive pots; bottom, wire beehive pots.

lobster fishery, when some 45 were in operation. But as fishing pressure increased and population density decreased there was evidence that some freezer boat skippers were processing lobsters of less than the legal minimum size. Since it was not possible to have inspectors on these vessels, the dayboat operators who are under rigorous inspection by fisheries inspectors at each processing center complained of illegal practices by the freezer boat skippers. As a result, the Government has discouraged their use, and so the freezer boats have greatly declined in number during the past decade.

B. Potential for High Exploitation Rate

The exploitation rate that can be generated by the total fishing effort on a particular unit stock of fish is dependent on a number of interacting factors. In the spiny lobster fishery of Western Australia, the development of a market, capital cost of entry to the fishery, distribution of the spiny lobster population, and the vulnerability of the lobsters to capture collectively interacted to produce a high exploitation rate. Spiny lobsters were always available out of Fremantle and Geraldton, but it was not until the late 1940s that the fishery expanded rapidly and a high exploitation rate developed.

1. Market

In the post-World War II period, the United States became a ready market for frozen green (uncooked) spiny lobster tails. This market has continued to accept all the available spiny lobster tails from Australia. Because it is a restaurant luxury item rather than a staple food source, the American price of spiny lobsters has tended to rise in response to product availability and not in relation to cost of production in Australia. The price paid to Western Australian fishermen has risen from 20 cents Australian per pound whole weight in 1952 to $1.00 in 1969 to $2.20 in 1976. The recovery rate of tail weight to whole Western Australian spiny lobsters is approximately 42%.

Australia, like all other spiny lobster producing countries except Cuba, is primarily dependent on the United States for its market success. There appears to be limited potential for the export of large quantities of spiny lobster to other countries and thus the strength of the market outlet and return to the fishermen depends heavily on the American economy. An example of this dependence is given by the Australian Department of Primary Industry (1977) in its Market Situation and Outlook Report:

> Generally tight supplies of cold-water tails were reflected in rising prices throughout the year. Demand was strong aided by a rise in economic activity especially in the first half of the year (1976); there was some slackening of demand in the second half of the year as the economy faltered and some consumer resistance developed towards high lobster prices.

2. Capital Cost of Entry to the Fishery

Spiny lobsters may be taken in shallow waters close to the coast so that, initially, the lower limit of finance required to begin a fishery is the cost of a small boat and pots. This allowed the spiny lobster fishery to grow by the acquisition of larger boats and more pots, without any great financial risk.

3. Distribution of Spiny Lobster Population

In Australia, not only was the inner boundary of the spiny lobster population close to the shoreline, but it was also close to centers of population at Geraldton and Fremantle. Fishermen were, therefore, able to center their activities on these two ports before expanding and investing money in the establishment of new fishing villages. A decision to spend risk capital in the early days of the fishery was not required.

4. Vulnerability of Spiny Lobsters to Capture

Over short periods, the catchability coefficient is influenced by the type of bait used, the sea swell, wind strength, tidal movement, and water turbidity. Over longer periods, it is influenced by lunar cycles, and during the year vulnerability changes are associated with changes in water temperature, salinity, and the percentage of spiny lobsters in premolt condition (Morgan, 1974). In addition, vulnerability decreases with an increase in size. Thus, large variations are observed in the catchability coefficient over a time span. Even so, the lobster pot is sufficiently effective as a catching method to generate a high exploitation rate at economically achievable levels of fishing effort. Studies by Bowen and Chittleborough (1966) and Morgan (1977) showed that the exploitation rate of legal-sized animals (76 mm CL) at the Abrolhos Islands had reached between 60 and 70%.

C. Management Objectives

In March 1897, a minimum legal whole weight of 8 oz was prescribed for the spiny lobster of Western Australia (Bowen, 1971), and in September of the same year this was increased to 12 oz. The legal minimum weight was based on the general understanding of size at first maturity and the market requirement. In subsequent years the method of measurement was changed to that of carapace length and the legal size to 3 inches (76 mm) CL. Around the turn of the century, administrators held the view that the act of fishing was unlikely to reduce a fish population substantially, providing that each animal was allowed to breed once. This view has now changed as a result of a better understanding of the dynamics of fish populations that have been subjected to varying degrees of fishing pressure.

As the Western Australian spiny lobster fishery developed in the 1950s and stock density began to fall, both administrators and fishermen began to discuss the effect of increasing exploitation on the stocks. At this time, publications on the dynamics of exploited fish populations and resource management became available (Schaefer, 1954; Beverton and Holt, 1957; Ricker, 1958; Crutchfield, 1965). From the discussions between Government and industry, three main objectives were developed and adopted, and they have remained as the principal management objectives of the Western Australian fishery (discussed below).

1. Optimum Utilization of the Resource

The spiny lobster resource is fished by both sportsmen and professional fishermen. Most of the sportsmen are aware of the rules, such as the legal minimum length and allowable number of pots, but they have very little understanding of the factors affecting population abundance and they will readily retain undersized animals. This was also the situation in the commercial fishery until the late 1960s. However, since then an expansion of research on the spiny lobster has led to greater contact with the professional fishermen, and an increased communication program has resulted in a very responsible attitude being adopted by most professional fishermen. The data must be presented in an understandable manner so that the fishermen can be shown what the basic field data collected actually represents. Possessed of this knowledge they will accept additional conservation measures when such measures are needed to maintain the fishing effort at a level that provides near maximum total production and an adequate catch per unit of effort for each individual operator.

2. Reasonable Economic Return

The fishermen have sought rules from the Government that would allow them to maintain a reasonable living standard. In Geraldton, approximately one-third of the work force is associated with the spiny lobster industry. The fishermen therefore represent a substantial group in the town, and they play an important role in the community life. They are keen to maintain their position, which includes a reasonable income consistent with the risks and capital involved in this industry. They also expect a fair price from the processing sector, and have achieved this by forming fishermen's cooperatives (one at Geraldton and two at Fremantle) that ensure a maximum return to the fishermen. In fact, the competition for product between spiny lobster processors and the fishermen's cooperatives has become a dominating influence in maximizing the prices that are paid to the fishermen.

3. Orderly Fishing

Operating in a common property resource creates difficulties for those participating. Fishermen have no specific area rights and are unable to make management decisions in isolation. There will always be the efficient operator who will achieve success, whatever the rules, and there will always be the inefficient operator who will not. However, fishermen do expect from the Government a set of rules that will reduce conflicts arising from a concentration of boats and pots on preferred grounds. The professional fishermen also expect measures from the Government that will provide for amateurs catching spiny lobsters for sport not in excess of their personal requirements.

D. Data Base for Management Decisions

Gulland (1971) has pointed out that the fisheries manager requires an array of data that describes both the developing fisheries and the heavily exploited stocks.

7. Spiny Lobster Fisheries Management

In relation to the development of a fishery, he states that "Policy-makers should then have before them three sets of figures: the best current estimates of the potential sustained yield from the stock; the present catch; and plans for increase in the catch." However, it was not until the last decade that biologists, economists, and adminstrators began talking to each other about their objectives, the data required, and the manner by which the data should be integrated to provide for a management strategy while the fishery being studied was still in a developing stage. Thus, for many fisheries detailed data were not collected until the fishery was shown to be in a state of heavy exploitation.

Royce (1975) described the use of yield models in fishery management. He states,

> With the development of ecological understanding, we began to develop the concept of ecological units or stocks from which we could take a sustainable yield. This concept which is the focus of attention here, emerged as a useful instrument during the 1920's, '30's and '40's. Baranov developed a yield-per-recruit model in 1918, but his work received little attention in fisheries until it was elaborated during the 1930's by Michael Graham, W. F. Thompson, J. Hjort, and later in the 1950's by W. E. Ricker, M. B. Schaefer, and the team of Beverton and Holt, to name only the leaders.

During the 1960s, these pioneers were followed by others who have developed the theory further.

Royce's first model (Fig. 3a) shows the relationship of yield to time in a managed and an unrestrained fishery. Figure 3b shows the first of the mathematical models depicting a long-run effort–yield relationship. This general production model is essentially a description of the outcome of the interrelationships of recruitment, growth, and mortality. His third model (Fig. 3c) introduces the basic population parameter of growth, fishing mortality, and natural mortality described by Beverton and Holt (1957). The yield per recruit curve is a simplification of the Beverton and Holt eumetric fishirg curve, which shows the concept of yield per recruit relationship with an optimal selection of effort and size at first capture. Figure 3d shows the relationship between stock size and recruitment. There is great variability in this relationship. Royce (1975) has stated "What seems to be emerging is a family of curves in which recruitment is related to fecundity, to the size of parent stock, to density of eggs and larvae, and to density of predators."

To further study the dynamics of the spiny lobster population of Western Australia, an attempt has been made to undertake research programs that lead to an understanding of the population dynamics of the fishery in terms of the general production model, the Beverton and Holt type yield per recruit model, and the stock/recruitment relationships (Fig. 3d). In the 1940s, the first phase of this approach was introduced, involving a system of collecting catch and effort data from the skipper of every spiny lobster vessel. Thus each month, every skipper is required by law to fill out a return setting out details of his catch, the

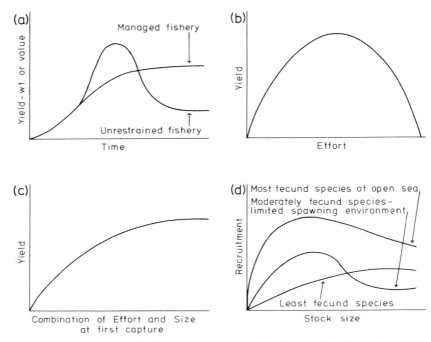

Fig. 3. Yield models used in fishery management: (a) The time trends in fish catches; (b) The effort-yield relationship; (c) The yield per recruit; (d) Some stock-recruitment relationships.

area of operation, the number of days worked, the number of pots used, and the number of crewmen on the vessel. The area of operation is recorded by grid number, with each statistical block being one degree of latitude by one degree of longitude. Thus, there developed a time series of catch and effort data that have been used in calculating a curve of catch against effort by the Schaefer method (see Fig. 2, Chapter 5 of this volume).

The work by George (1958a,b) and Sheard (1962) added substantially to an understanding of the biology of the western rock lobster (*P. cygnus*) and of the development of the fishery. Bowen and Chittleborough (1966) examined catch and effort data for the Abrolhos Islands and presented preliminary assessments of the stock levels. Following this work there was a rapid growth in research on the spiny lobster population off Western Australia by Chittleborough, Phillips, Morgan, and others in the attempt to provide an understanding of the movement of phyllosoma, the settlement of puerulus, the growth of juvenile animals, and the dynamics of the fishable stock, including an estimate of the relative abundance of the female breeding population. Research programs have recently been undertaken to study the octopus as a predator of pot-caught spiny lobsters, the

7. Spiny Lobster Fisheries Management

amateur fishery, and the effect of handling spiny lobsters that are returned to the sea (i.e., undersized animals and "berried" females).

Although economic research is not a continuing process, periodic surveys have been undertaken (e.g., Australian Department of Agriculture, 1974) to gain an understanding of the economic consequences of the regulations imposed on the fishery.

E. Process of Management Decision Making

Because Australia is a federation, and the spiny lobster fishery is within three miles of the coast and beyond, the responsible management authority is a combination of the State Department of Fisheries and Wildlife of the Government of Western Australia and the Fisheries Division of the Department of Primary Industry, Commonwealth of Australia. Biological research is undertaken by the Department of Fisheries and Wildlife and the Commonwealth Scientific and Industrial Research Organization (CSIRO). Economic research is the responsibility of the Fisheries Division of the Department of Primary Industry, with assistance from the Department of Fisheries and Wildlife. Although each research officer is responsible to the head of his organization, a coordinating committee has been established in Western Australia, called the Western Fisheries Research Committee, to foster research discussions and encourage cooperation in the execution of field programs. The Committee meets in full session once a year, so that each research officer can present a summary of research undertaken and a proposal for research for the year following. By this means, the programs of each officer are appraised by his colleague research officers and members of the committee in a spirit of cooperation. At the same time, results from the research programs that have implications for management are identified, discussed, and submitted to the managing authority.

The members of the Western Fisheries Research Committee are the Western Australian Director of Fisheries and Wildlife (Chairman); Chief of the Division of Fisheries and Oceanography, CSIRO; Chief of the Fisheries Division, Department of Primary Industry; Professor of Zoology, University of Western Australia; representative of the Fisheries Department of South Australia; and the Research Coordinator.

At the industry/government level, a Rock Lobster Industry Advisory Committee has been established to provide advice to the management authority on matters relating to the industry. This Committee includes the Director of Fisheries and Wildlife (Chairman); Chief of the Fisheries Division, Department of Primary Industry; an officer of the Department of Fisheries and Wildlife; three fishermen; two representatives of the Rock Lobster and Prawning Association of Australia (an industry association), and an amateur fisherman. The two Association members are, in fact, representatives of the processing sector of the industry.

The officer responsible for day-to-day administration, the Chief Research Officer of the Department of Fisheries and Wildlife, and the officer undertaking population dynamics research are always in attendance at the meetings of this Advisory Committee. The Committee is kept informed of the work undertaken under the auspices of the Western Fisheries Research Committee and receives detailed reports on management implications arising from the research results presented. The Committee also receives economic advice as it becomes available, and from planned discussion sessions with industry groups is able to understand the views of the fishermen on the general effect of the management regime. In addition, the Committee has received written submissions from fishermen giving their appraisal of the value of the present rules and suggestions for alterations.

The Committee may either provide advice on matters requested by the Minister and the Director of Fisheries and Wildlife or it may initiate advice from information it has before it. In practical terms, every proposed change to the management rules is submitted to the Committee for its advice. The composition of the Committee ensures that consideration of each proposal takes into account biological, economic, and social factors before recommendations are framed for the consideration of the Minister.

IV. THE LIMITATION OF FISHING EFFORT

A. Data

The Western Australian spiny lobster fishery experienced a rapid rise in catch during the 1950s as a result of the introduction of additional fishing units. By 1963, the number of vessels had risen to about 800 and the number of pots in the water to 97,000.

Although production had risen rapidly during the 1950s (Fig. 4), fishermen noted that during the four seasons commencing 1958/1959, additional units of effort had not resulted in proportional increases in annual production. Even with a general increase in the price per kilogram of spiny lobsters, fishermen were becoming concerned that the trend to reduced catches per boat and pot would eventually cause their operations to be unprofitable. There were overtures by many fishermen to the Government, through the then Fishermen's Advisory Committee, to introduce measures that prohibited the introduction of additional boats and that limited the number of pots per boat. At about the same time there was a move to limit the number of shrimp boat licenses in Shark Bay, north of Geraldton, to protect the capital investment in that developing fishery. Thus, although the reasons for the request were quite different, Government was being urged to introduce a limited entry philosophy for two important Western Austra-

7. Spiny Lobster Fisheries Management

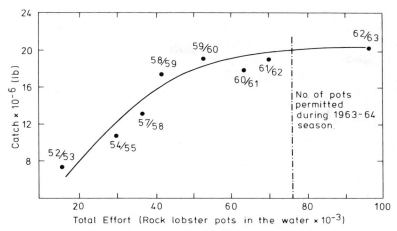

Fig. 4. Total effort against catches for years 1952-1953 to 1962-1963.

lian fisheries. At that time there were no limited entry fisheries in Australia, nor was the management measure generally in use outside Australia.

In 1962, a decision was made by the Government to introduce effort restrictions. Approximately four months notice was given that those vessels engaged in the spiny lobster industry as of March 1, 1963, would constitute the lobster fleet. To the present date further licenses have not been granted. A decision was also made to control the number of catching units, i.e., the lobster pots. Noting that the number of pots had increased by 66% from 1958/1959 to 1961/1962, while the production increase was only 8%, and noting that part of the increase in the number of pots was due to fishermen using more pots than they could efficiently work, it was decided to establish a table of pot quotas depending upon the size of the vessel. Following consideration of detailed submissions from industry, the number of pots was fixed at 3 per foot of boat length, with a maximum of 200. This decision had the effect of reducing the number of pots from approximately 97,000 in the 1962/1963 season to 76,000 during the 1963/1964 season. The Government was feeling its way in the administrative realities of a limited entry policy. It was not surprising, therefore, that further rules had to be introduced to complement the initial decision. For instance, there was no boat replacement policy, and a number of fishermen with small vessels made plans to build larger vessels and use the additional pots allowed under the three pots per foot of boat length rule.

On May 8, 1965, a policy decision was handed down that a fisherman replacing his boat would not be permitted to use additional pots even if his new vessel was larger. This policy had the effect of allowing fishermen to build larger, more comfortable vessels without the catching power (in terms of pots operating) to provide adequate return on capital. Some of these fishermen used more pots than their entitled number so that vessel repayments could be met. There is no doubt that this was one of the causes of considerable overpotting in the industry.

On December 14, 1965, an amended boat replacement policy was announced, essentially stating that new vessels must not be larger than those being replaced. However, for reasons of boat safety at sea, exceptions were made for boats less than 7.6 m in length. Such vessels could be replaced with vessels of up to 7.6 m.

Amateur spiny lobster fishermen have not been restricted in terms of the number of licenses issued. However, each license holder may use not more than two pots, and not more than two pots can be used or carried in any boat at any one time, irrespective of the number of licensed amateur fishermen in the boat. Amateurs may also take spiny lobsters by hand, but sharp instruments such as a speargun are prohibited. There is a daily bag limit of eight lobsters per license holder.

The amateur fishery accounts for a little less than two % of the annual spiny lobster catch. However, the potential effect of an increasing amateur fishery cannot be overlooked when developing a management strategy. Davis (1977) has shown that recreational divers reduced a population of *Panulirus argus* in Florida by 58% over a period of 8 months. He concluded that "The level of sport harvest examined in this study could not be sustained on an annual basis at Dry Tortugas without a significant reduction of the rock lobster population and a subsequent alteration of its role in the coral reef ecosystem."

An example of the growth in amateur activities on spiny lobsters is given by Odemar *et al.* (1975), who noted that the number of charter boat diver-days in southern Californian waters increased from 275 to 15,871 during the period from 1958 to 1973.

B. Results of the Introduction of Limited Entry

Of the array of management measures available, an entry limitation policy intrudes most on the freedom of choice of the individual fishermen. Those who hold authorizations to operate in Western Australia are restricted in terms of size and replacement of vessel, number of pots, and area of operation. Those who do not hold authorizations are restricted to the point of not being permitted to fish the resource, and if the economics of that fishery are attractive, they seek ways and means of obtaining an authorization.

Administratively, this policy has required the establishment of a separate record system that contains the details of every vessel holding an authorization and an administrative group within the Department to oversee the collection of data and to consider requests on matters such as vessel replacements, fishing zone changes, and ownership transfers. Administration is therefore generally aware of

7. Spiny Lobster Fisheries Management

the vessel price being paid to a person leaving the industry. The purchaser is permitted to retain the authorization attached to a spiny lobster vessel when a purchase is made, provided that both parties submit written evidence to the administration of the sale and the purchase. In the first years of the limited entry policy, vessels changed hands at about vessel price. However, toward the end of the 1960s it was evident that the vessel price also reflected the value ascribed to a vessel endowed with a spiny lobster authorization. During the last ten years the value of an authorization has become more precisely known because of a policy of allowing certain vessels (the freezer boats) to be withdrawn from the industry. The pots of those vessels were made available for purchase by existing boat owners, and the price per pot approximates the value of the catch that one lobster pot would yield in a year (i.e., the value of $\simeq 100$ kg of spiny lobster). This has added substantially to the capitalization of the industry.

Even with these quite rigorous limitation measures, the total fishing effort in terms of the number of pot lifts has increased since 1963. This has partly resulted from a general upgrading of the fleet, so that fishermen have operated on more days during the scheduled season (November 15 to the following August 14) than they had been able to in past years.

Morgan (Chapter 5) has shown that effective fishing effort has increased considerably and that the level is now similar to that of the 1962/1963 season, before the pot numbers were reduced (Fig. 2, Chapter 5 of this volume). Morgan has also shown that there has been a marked reduction in the index of abundance of the spawning stock. Noting a 70% exploitation rate that exists for legal-sized lobsters, further increases in fishing effort could cause the recruitment of the puerulus larvae to decline (see Chapter 5). Consequently, concern was expressed by fisheries scientists and by the Rock Lobster Industry Advisory Committee regarding the increasing fishing effort. Thus, subsequent to a recommendation by the Committee, the managing authority announced a six week reduction in the season as a first step toward controlling the growth in fishing effort.

The economic effects of boat and pot limitations in the Western Australian lobster fishery is discussed in a Fisheries Report by the Australian Department of Agriculture (1974). That report, which covers the period between 1966–1967 and 1968–1969, states

> There is, however, little doubt that in the absence of license limitation and pot restrictions, there would have been a larger number of boats operating in the fishery. Total fishing effort would have been substantially higher, and with total catch relatively stable the average catch per boat would have been lower and as a consequence the average profitability of the fleet would have been below that of the fleet operating at the time of the survey. Thus, the management measures have not only prevented the unrestrained exploitation of the resources, they have resulted in considerable economic benefit accruing to those fishermen licensed to operate in the fishery.
>
> The analysis in the foregoing sections indicates that, whatever measure of capital investment is chosen, rock lobster fishermen earned a very satisfactory return on their capital

TABLE II

Comparison of Income and Expenditure for Vessels in the Three Major Western Australian Spiny Lobster Fishing Zones[a]

	1968/1969			1973/1974		
	Zone A	Zone B	Zone C	Zone A	Zone B	Zone C
Gross Income	23,096	22,365	23,650	27,407	19,860	22,253
Trip costs	2,623	2,253	2,818	3,654	2,508	3,115
Boat costs	2,763	2,912	2,678	4,673	2,962	3,969
Administrative costs	967	1,087	1,060	1,885	1,307	1,595
Total operating costs	6,353	6,252	6,556	10,212	6,777	8,679
Depreciation	1,244	1,108	1,410	3,015	2,608	2,548
crew payment	2,327	3,153	4,648	3,880	2,451	5,420
Total costs	9,924	10,513	12,614	17,107	11,836	16,647
Return to owner/ skipper and capital	13,172	11,752	11,036	10,300	8,024	5,606

[a] From the Fisheries Division, Australian Department of Primary Industry.
[b] *Trip costs* comprise mainly fuel and bait; *Boat costs* comprise boat repairs, maintenance, and purchases of and repairs to fishing gear; *Administrative costs* comprise insurance, license fees, vehicle running costs, and other land-based costs associated with running the business side of the operation; and *Crew payment* includes payment to all crew members other than the skipper.

invested during the years covered by the survey. This is true of all regions and the various groups of boats within sectors with the possible exception of freezer boats. Complete freedom of entry would have resulted in lower returns to boats on average, and most probably inefficient use of labour and capital. The profitability of the fishery is reflected in the value added to boats by the possession of a rock lobster license. For non-freezer boats this amounted to approximately $17,000 and for freezer boats about $37,000.

A further economic study for the period from 1972/1973 to 1974/1975 has been undertaken, and the report is in press. The latest report shows that the financial position of the fishermen was not as sound in 1973/1974 as it was in 1968/1969 (Table II) because costs had risen faster than prices. However, since 1973/1974, prices have risen from $1.20/lb to $2.20/lb, and fishermen are still prepared to pay high prices to acquire a "spiny lobster" vessel, and thus gain entry to the lobster fishery.

V. CONCLUDING COMMENTS

The need for management of a fishery resource has been well established by Gulland (1971). He has presented the options available to the managers of living marine resources, as well as the biological and economical consequences of each action. Copes (1978) has explored management options for the South Australian spiny lobster fishery, taking biological and economic factors into account. Studies of this type will become increasingly important to the fisheries administrator who is responsible for introducing sound management measures to gain optimum benefit from lobster resources.

Since spiny lobster populations are subject to heavy exploitation, the introduction of a sound management strategy is most important if the fishery is to provide the maximum benefit to the community. In the spiny lobster fishery off the west coast of Africa (between 20 and 30°S latitude) an experimental series of reductions and finally suspension of the legal minimum length resulted in a consistent fall in annual production (Beyers, 1979). Other spiny lobster fisheries have declined, even with the maintenance of the traditional management measures of legal minimum size and protection of berried female animals. The New Zealand lobster fishery at the Chatham Islands is perhaps the most striking example (Annala et al., 1977). Odemar et al. (1975) reported that the commercial landings of the California spiny lobster "decreased from 933,449 pounds in 1950 to 233,179 pounds in 1973. Log books reported 160,025 pounds taken during the 1974-75 season, the lowest season on record." It is evident, for the Western Australian lobster fishery at least, that the management strategy must include a consideration of the two important elements of legal minimum size and total amount of fishing effort.

As the exploitation rate increases, so too does the importance of obtaining information about the appropriateness of the legal minimum length for the new exploitation rate. Beardsley *et al.* (1975) state that for the Florida spiny lobster fishery "... intensified fishing pressure creates an exploitation rate for the Florida population that is high enough so that most lobsters at or near legal size in the Florida Keys are being caught each year." Although the data may be available on which to establish a legal minimum size, its value in terms of gaining maximum benefit from the resource may be reduced if administration does not take appropriate action to ensure its effectiveness. Spiny lobsters are a luxury item and much sought after by members of the public. Legal animals are expensive and there is considerable temptation for amateur fishermen to retain illegal or "short" lobsters if their catch of legal-sized animals is small. In addition, professional fishermen may find a ready market for undersized lobsters that should have been returned to the sea. Odemar *et al.* (1975) concluded that "the take of undersize lobsters is the largest single factor contributing to the continued decline in lobster stocks" of *Panulirus interruptus* in southern California. It is important, therefore, for administration to ensure that the inspection system and legal processes are adequate to deter the massive retention of undersized animals.

Of course, there are other causes of mortality. Animals returned by fishermen to grounds that are either not suitable for survival or that contain major predators will obviously suffer a higher mortality rate. Also, animals that have lost appendages through handling are less likely to survive. At the very least, they will grow slower than those that have remained on the sea bed. Escape gaps in lobster pots have been shown to be successful in reducing the quantity of undersize spiny lobsters that are captured (Bowen, 1963). Escape gap regulations have been introduced in a number of countries.

The data available from most of the major lobster fisheries show that the annual total production is likely to fall quite dramatically unless controls are placed on the total fishing effort. However, a strategy such as boat license limitation introduces a new set of management problems, e.g., the capitalization of the value attached to a fishing license and the administrative machinery required to record the sale, transfers, and replacement of licensed vessels. Inevitably, there will still be an increase in effective effort by each fishing unit that will increase the total fishing effort above the previously designated limit. In Western Australia, the effective fishing effort has increased by about 25% of the 1963 figure, when license limitation was first imposed. Fisheries administration is a dynamic process, so that judgements are frequently made about the importance of such increases. After examining the problem from either a biological or economic point of view, it is decided whether or not further action is warranted. Copes (1978) has discussed two methods of reducing fishing effort; one is to not allow the transfer of licenses, thus reducing the number of boats as operators leave the industry, and the other is to introduce a "buy-back" scheme that will encourage withdrawal by providing financial incentives.

If the effort increase is likely to produce serious biological consequences as a result of declining population abundance, a combination of effort limitation and limited total allowable take has its attractions. However, in practical terms, this can probably be achieved only if the number of operators controlling the catching sector is relatively small. Other measures available to the administrator are restrictions on gear, season and area, but use of these measures do not produce the same long-term beneficial effects as that produced by reducing the number of vessels.

From the experience gained by managing spiny lobster fisheries since the 1950s, some guiding principles appear to be emerging (1) Precise catch and effort, and length–frequency data are required for a first understanding of the effect of increasing fishing pressure on the population; (2) The setting of a legal minimum size with adequate inspection and legislative backing is needed; (3) Fishing effort restrictions, which may be combined with catch quotas, are needed to assist in obtaining maximum benefit from a spiny lobster resource; (4) The effect of high exploitation rates on the stock/recruitment relationship is not well understood, and long run data on indices of recruitment success should be collected and analyzed; and (5) An effective and continuing communication system with both the professional and amateur spiny lobster fishermen needs to be established to ensure that the objectives of management and methods employed are well understood.

REFERENCES

Annala, J., Booth, J., and McKoy, J. (1977). Catch-effort trends in the New Zealand rock lobster fishery. *Catch '77* **4**(2), 3–5.

Australian Department of Agriculture (1974). "Economic Survey of the W.A. Rock Lobster Fishery," Fish. Rep. No. 10. Fish. Div., Aust. Dep. Agric.

Australian Department of Primary Industry (1977). "Market Situation and Outlook Report—Rock Lobster." Aust. Dep. Primary Ind.

Beardsley, G. L., Costello, T. J., Davis, G. E., Jones, A. C., and Simmons, D. C. (1975). The Florida spiny lobster fishery. *Florida Sci.* **39**(3), 144–149.

Beverton, R. J. H., and Holt, S. J. (1957). On the dynamics of exploited fish populations. *Fish. Invest. (London), Ser. 2* **19**, 1–532.

Beyers, C. J. de B. (1979). Stock assessment and some morphological and biological characteristics of the rock lobster *Jasus lalandii* on Marshall Rocks, its main commercial fishing area off South West Africa, 1971–1974. *Investl. Rep. Sea Fish. Brch. S. Afr.* **17**, 1–26.

Bowen, B. K. (1963). Preliminary report on the effectiveness of escape-gaps in crayfish-pots. *West Aust. Dep. Fish. Wild. Rep.* **2**, 1–13.

Bowen, B. K. (1971). Management of the western rock lobster (*Panulirus longipes cygnus*, George). *Proc., Indo-Pacific Fish. Coun.*, **14**(2) 139–154.

Bowen, B. K., and Chittleborough, R. G. (1966). Preliminary assessments of stocks of the western Australian crayfish, *Panulirus cygnus* George. *Aust. J. Mar. Freshwater Res.* **17**, 93–121.

Copes, P. (1978). "Resource Management for the Rock Lobster Fisheries of South Australia." Steering Committee for the review of Fisheries of the South Australian Government.

Crutchfield, J. A., ed. (1965). "The Fisheries - Problems in Resource Management." Univ. of Washington Press, Seattle.

Davis, G. E. (1977). Effects of recreational harvest on a spiny lobster, *Panulirus argus*, Population. *Bull. Mar. Sci.* **27**(2), 223-236.

George, R. W. (1958a). The status of the "white" crayfish in Western Australia. *Aust. J. Mar. Freshwater Res.* **8,** 476-90.

George, R. W. (1958b). The biology of the western Australian commercial crayfish, *Panulirus longipes*. Ph.D Thesis, University of Western Australia.

Gulland, J. A. (1971). "Management, " FAO/UNDP Indian Ocean Programme, IOFC/DEV/71/4. FAO, Rome.

Morgan, G. R. (1974). Aspects of the population dynamics of the western rock lobster *Panulirus longipes cygnus* George. II. Seasonal changes in the catchability coefficient. *Aust. J. Mar. Freshwater Res.* **25,** 249-259.

Morgan, G. R. (1977). Aspects of the population dynamics of the western rock lobster and their role in management. Ph.D. Thesis, University of Western Australia.

Morgan, G. R., and Barker, E. H. (1975). The western rock lobster fishery 1973-1974. *West. Aust. Dep. Fish. Wild. Rep.* **19,** 1-22.

Odemar, M. W., *et al.,* Bell, R., Haughan, C. W., and Hardy, R. A. (1975). Report on California spiny lobster, *Panulirus interruptus* (Randall) research with recommendations for management. State of California, The Resources Agency, Dept. Fish. Game. 1-53.

Ricker, W. A. (1958). Handbook of computations for biological statistics of fish populations. *Bull., Fish. Res. Board Can.* **119,** 1-300.

Royce, W. F. (1975). Use of yield models in fishery management. *in* optimum sustainable yield as a concept in fisheries management. *Am. Fish Soc., Spec. Publ.* No. 9, 9-12.

Schaefer, M. B. (1954). Some aspects of the dynamics of populations important to the management of the commercial marine fisheries. *Inter-Amer. Trop. Tuna Comm. Bull.,* **1**(2), 26-56.

Seaman, W., and Aska, D. Y. eds. (1974). Research and information needs of the Florida spiny lobster fishery; Proceedings of a conference held March 12, 1974, in Miami, Fla. State Univ. Syst. Fla., Sea Grant Program Rep. 74-201. 1-64.

Sheard, K. (1962). "The Western Australian Crayfishery, 1944-1961. Paterson Brokensha, Perth, Australia.

Smith, F. G. W. (1958). The spiny lobster industry in Florida. *Fla. Bd Conserv., Ed. Ser.* **11,** 1-36.

Streeter, D. H., and Weidner, D. M. (1976). Caribbean spiny lobster fisheries surveyed. *Mar. Fish. Rev.* **36**(7), 31-33.

Chapter 8

The Clawed Lobster Fisheries

R. L. DOW

I.	Introduction	265
II.	Methods of Capture	267
	A. Traps and Boats	267
	B. Lobster Bait	269
	C. Methods of Storage and Shipment of European Lobsters	270
III.	*Nephrops* Fisheries	271
	A. Introduction	271
	B. History of the Fishery	272
	C. Marketing	273
	D. Management	274
IV.	*Metanephrops* and *Nephrosis*	276
V.	*Homarus* Fisheries	277
	A. *Homarus gammarus*	278
	B. *Homarus americanus*	283
VI.	Effects of Sea Surface Temperature Cycles on Landings of *H. americanus*, *H. gammarus*, and *Nephrops*	300
	Comparison of European, Canadian, and American Trends	302
VII.	Conclusions	304
	Appendix	306
	References	313

I. INTRODUCTION

Fisheries for *Homarus americanus*, *Homarus gammarus*, and *Nephrops norvegicus* are distributed around the perimeter of the mid-North Atlantic Basin, almost everywhere that the animals are found (see distribution maps in Chapters 3 and 4). Other clawed lobster species are potential food resources on a worldwide basis. Although the three major species are intensively exploited, fisheries for *H. gammarus* and *Nephrops* have not been officially reported from all geographical areas that they are known to occupy. The annual catch of clawed lobsters is approximately 70×10^3 tonnes, with the *Nephrops* fishery making up about 50%, *H. americanus,* 46%, and *H. gammarus,* 4% of the total. Trends of landings in the three fisheries are given in Fig. 1. The retail value is very great in the North American fishery alone, where it is estimated to be at least U.S. $250 million. Income to the fishermen throughout the United States and Canada amounted to more than U.S. $100 million annually in 1976 and 1977. Although comparable data for *Nephrops* and *H. gammarus* landings are not available, it is expected that the total landed value of both species may be equal to that of *H. americanus*.

All clawed lobsters are now considered to be luxury foods, but this was not true in the past. Early accounts from the United States indicate that in the eighteenth century, *H. americanus* was used as agricultural fertilizer and in the nineteenth century as codfish bait, subsistence, and for semicommercial purposes. France now prohibits the use of some sizes of *Nephrops* for fertilizer or animal feeds, suggesting that in former times such procedures may have been commonplace.

This chapter reviews the history and present status of the clawed lobster fisheries, including fishing and management practices, the influence of environmental and economic factors, and some of the attitudes of fishery participants. European lobster management schemes are presented in Chapter 9 of this volume.

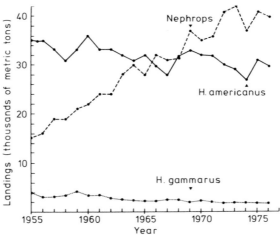

Fig. 1. Landings of clawed lobsters in Europe, the United States, and Canada from 1955 to 1976.

II. METHODS OF CAPTURE

A. Traps and Boats

Fishing devices used in the *Homarus* lobster fisheries include traps, pots, or creels (both experimental and traditional), which may be made of wood, metal, or plastic (Fig. 2). The design may be rectangular, bowed, or circular.

In Canadian, United States, and European inshore fisheries, traps are set on the ocean bottom singly or in strings of ten or more. The trap is not an efficient fishing device, yet saturation of the bottom in many areas effectively reduces the legal population by 90% or more each year (Dow *et al.*, 1962; Thomas, 1973). In the United States trap fishery, escape vents for sublegal size lobsters have been used by a small number of commercial fishermen for many years, and in some areas they are required by law (See Table X). Studies in progress, as well as

Fig. 2. Traps used in the clawed lobster fisheries.

scuba observations of entrapped lobsters in lost ("ghost") traps indicate the need for developing and utilizing biodegradable portions in wood, metal, and plastic traps in order to reduce the noncatch losses of lobsters associated with the trap fishery (Sheldon and Dow, 1975).

Vessels for the *Homarus* inshore fishery in Canada, the United States, and Europe range widely in size and capability, but are generally less than 15 m in length. During the depression years between World Wars I and II, much of the fishing was done by oared or wind-powered boats.

Incidental catches of *H. americanus* from offshore stocks date from 1891, when the beam trawl was first used (Schroeder, 1959). When it became economically feasible to fish directly for lobsters offshore, primarily in the canyon and slope areas, the otter trawl was employed. In more recent years, this fishery has been extensively replaced by one using large metal traps (Anonymous, 1969). Large, high-speed lobster boats similar to the inshore lobster boats (Fig. 3) once supplemented and now largely replace the trawlers for offshore fishing. The high-speed boats were designed for one- or two-day trips, in contrast to the longer trips by otter trawlers. Minimum overall boat length for offshore lobstering by otter trawler is about 25 m.

Otter trawlers are employed extensively in the *Nephrops* fishery, since this species occurs on deeper and less rocky bottoms. *Nephrops* constitutes a by-

Fig. 3. Lobster boats tied up at a dock in Maine.

catch in the whitefish trawl fishery and an incidental catch in the shrimp (*Pandalus borealis*) fishery. Where a directed fishery for *Nephrops* has developed, a trawl designed especially for the purpose is used, i.e., cod ends of 60–70 mm mesh to as small as 30 mm (Thomas, 1970). Experimental results with an electrified trawl show promise for increased catches during periods when lobsters are active (Stewart, 1974). Creels have been used to take *Nephrops* in some areas since 1887, where ledge outcrops in mud bottom, make the use of conventional otter trawls impossible.

B. Lobster Bait

Lobsters are both omniverous and opportunistic in their feeding habits. Their response to various baits, natural and synthetic, appears to be partly conditioned by food preferences of the moment, with different animals responding differentially to the same bait. Scuba divers have observed and photographed lobsters raiding traps to consume the bait on the premises or to depart to their burrows with the pilfered food. Use of Redfish (*Sebastes marinus*) bait increases trap catch from 25 to 30% over that gained by use of the traditional bait, herring (*Clupea harengus*) (Table I). Experiments using artificial and other natural baits indicate that lobsters respond most actively to a natural bait treated with urea. Among the natural, untreated baits, crushed green crabs (*Carcinus maenas*) also produced very strong reactions on the part of lobsters (Dow and Trott, 1956). The most complete report on the early artificial bait experiments is contained in Prudden's book "About Lobsters" (1962).

Unbaited traps will frequently catch lobsters more consistently than some baits, since these readily attract shelter-seeking or displaced animals. Captive lobsters themselves may also attract untrapped lobsters. A 2-year study of ghost trap effects on the lobster resource demonstrated a more than 100% attraction of

TABLE I

Comparisons of the Attractiveness of Several Bait Types for *H. americanus*[a]

	Number of lobsters	% of total catch	Fishing efficiency rating
Herring	846	16.97	100.0
Green Crabs (*Carcinus maenas*)	932	18.70	110.2
Alewives (*Alosa pseudoharengus*)	1052	21.10	124.3
Redfish (salted)	1062	21.30	125.5
Redfish (frozen fresh)	1093	21.93	129.2
Total	4985	100.00	

[a] Data from Dow and Trott, 1956.

lobsters in the second year. Apparently, new lobsters were drawn by the lobsters already in the traps, rather than by the shelter potential of the traps themselves (Sheldon and Dow, 1975).

Studies reported by Hancock (1974) indicate the possible role that attraction and avoidance would have in the development of specific artificial lobster baits. In support of Hancock's findings, experimental evaluation of an early herring-based bait by the Maine Department of Marine Resources scientists in 1948 and 1949 demonstrated that unbaited control traps were more attractive to lobsters than the artificial bait being tested, which actually seemed to keep lobsters away from the traps.

C. Methods of Storage and Shipment of European Lobsters

Wood boxes or crates are used for short term, in-water storage. For long-term storage, saltwater ponds or relatively small impoundment areas have been developed, constructed of concrete and rock dams with grates and drainage sluices. Normally, storage of recently molted lobsters is not attempted until August. Large claws are tied or banded to prevent cannibalism. In Europe, unlike North America, plugging the large claws is discouraged because of bacterial infection, loss of claws, or reduction in food quality. In some European countries, banding and tying are the only permitted methods of immobilizing the large claws.

For shipment to market, lobsters are placed in boxes with adequate packing material to reduce movement and prevent damage. Shavings, ground peat, sawdust, and straw are among the materials used to accomplish this objective. Ice or artificial refrigerants reduce shipment mortalities and result in a better quality product. For air shipment, containers of waxed cardboard or plasticized material without ice are employed. Since adequate packing material will immobilize lobsters and eliminate activity, refrigerants are not required for good quality lobsters.

Grading by weight improves marketing. Higher prices are paid for lobsters between 338 and 900 g, while above and below this range, the value decreases. Depending on air temperature and humidity, lobsters will lose a varied amount of weight during shipment. This loss can be as little as 3% when ice is used and as much as 8–10% with dry packaging (Anonymous, 1952).

There are marked differences between the live lobster trade practices of *H. gammarus* and *H. americanus*. In North America, the emphasis has been on volume rather than on quality, as in Europe. This difference can be partly accounted for by the differences in attitude toward the product as well as the smaller inventories and greater average value of the European lobster. European dealers cannot accept the losses tolerated in North America as a matter of course.

Relatively few lobster claws are immobilized in North America by using bands or string. Instead, plastic or wood plugs are inserted into the joint of large claws, a practice which invariably results in bacterial infection and meat discoloration. Mortalities occurring in the *H. gammarus* trade appear to be much less than the average 25% loss sustained in a normal year by North American fisheries or the 40% or more loss of an exceptional year, primarily from *Aerococcus viridans homari* during storage in tidal impoundments, but also in flow-through or recirculated tank systems. In addition to storage losses, extensive mortalities from cannibalism in traps and predation on sublegal releases have been observed (Sheldon and Dow, 1975).

III. *Nephrops* FISHERIES

A. Introduction

The principal fishing grounds and presumed centers of major *Nephrops* abundance reported by Farmer (1975) are both coasts of Scotland, the south and west coasts of Iceland, the North Sea, the west coast of France, the Irish Sea, the Kattegat, the Skagerrak, the Bay of Biscay, the Adriatic, and the Tyrrhenian Sea. All the nephropid lobsters are relatively deepwater animals that occupy burrows in soft sediments. They spend much of their time there, making brief forays for feeding or other purposes under acceptable light conditions, i.e., that generally associated with the subdued intensity of sunrise and sunset, turbid water, or great depth (see Chapter 4). Trawl fishermen depend on these periods to make their catches (Chapman *et al.*, 1975).

Since most of the *Nephrops* lobster landings are made by trawls, it is impossible to estimate the extent of the resource as accurately as it has been done for the more selective *H. gammarus* and *H. americanus* trap fisheries. Gulland (1971) estimated a potential annual *Nephrops* lobster production of 60,000 tonnes from the Northeast Atlantic. Landings from the Mediterranean have remained very stable since 1960 (Farmer, 1975). The unit price of *Nephrops* varies by individual sizes (Figueiredo and Thomas, 1967), so that larger lobsters generally bring a higher price than their smaller counterparts. The landed value is reported to be highest in Germany, with lesser prices in Spain, Portugal, and Italy.

The Council of ICES General Assembly has been concerned for many years with the possible overfishing of *Nephrops* lobster stocks, and has urged more adequate research to be undertaken in each country having a commercially important fishery. The *Nephrops* resource had been given little attention until the development of the otter trawl, which permitted fishing in deeper water where much of the population lives.

B. History of the Fishery

From a total reported catch of about 5000 tonnes (t) in the 1930s and 1940s, the fishery rapidly increased to 15,000 t in 1955, peaking at 44,000 t in 1976 (Appendix AIII). According to Wood (1957), the volume of *Nephrops* landed in Scotland increased sevenfold between 1950 and 1956, while landed value increased eighteenfold. In 1938, Algeria, Belgium, Denmark, France, Germany, Portugal, Sweden, Yugoslavia, England, and Wales all reported landings ranging from less than 500 t in Algeria to 3400 t in France. It was not until 1941 that Spanish landings were reported, and by 1948 Spain was the principal producer, with a catch of 2100 t for the year. The eightfold increase in total catch from 1949 to 1973 was related to improved economic conditions for the consumer, the discovery of new fishing grounds, an improvement in the catch effectiveness of gear, as well as more favorable sea temperature after 1953. In recent years, the countries with the highest catches of *Nephrops* have been France, Scotland, Spain, Ireland, Iceland, and Italy (see Appendix AIII).

Landings of *Nephrops* follow a seasonal northward drift pattern in the peaking of annual catch where fishing is unrestricted (Table II). Some differences in the seasonal concentration of *Nephrops* landings can probably be attributed to differences in sea temperature from south to north, with more southerly stocks becoming available earlier in the year. Ex-vessel prices and landings are significantly inversely correlated in Sweden from 1951 to 1976 ($r = -0.724$, from data supplied by B. I. Dybern and B. Lindfors, personal communications) and in Norway from 1955 to 1976 ($r = -0.547$). In Scotland, the correlation coefficient is positive for the years from 1954 to 1976 ($r = 0.847$), in which the price increased from £91/t to £574/t while landings rose from 576 to 11,046 t (Table III). Additionally, Swedish fishing effort, as indicated by number of annual vessel trips, is highly significantly correlated with total annual landings in the period 1959–1976 ($r = 0.948$, data supplied by F. G. Howard, personal communication. The ex-vessel unit value of *Nephrops* in the England and Wales

TABLE II

Seasonal Concentration of Norway Lobster Landings

Country	Season	% of annual catch	Reference
Denmark	June–November		Jensen (1962)
Scotland	July–October	49.4	Thomas (1970)
Ireland	June–October		Hillis (1972)
Portugal	April–August	69.4	Figueiredo and Barraca (1963)
England	September–December	68.8	Symonds (1971)
France	April–June		Fontaine and Warluzel (1969)
Germany	June–October	83.6	Aker and Tiews (1965)

TABLE III

The Catch, Price, and Trips of Scotland *Nephrops*

Year	Catch (10^3 tonnes)	£/tonne ex-vessel	Number of trips (10^3)
1954	.6	91	—
1955	1.1	95	—
1956	1.1	94	—
1957	1.4	94	—
1958	1.1	109	—
1959	2.2	123	4.2
1960	2.0	133	4.0
1961	2.9	130	4.8
1962	3.5	139	10.9
1963	3.7	145	12.0
1964	4.9	148	15.0
1965	5.2	168	15.6
1966	6.3	176	20.9
1967	6.7	193	23.5
1968	7.2	225	25.6
1969	8.2	248	25.9
1970	8.2	244	24.0
1971	9.0	234	23.5
1972	10.8	358	29.1
1973	9.8	510	35.6
1974	8.3	462	30.6
1975	8.2	476	28.3
1976	11.0	574	29.8

fishery between 1950 and 1976 rose two and one-half times, although the total landings increased more than 100-fold. In the Federal Republic of Germany fishery in the Kattegat and Skagerrak, annual landings of *Nephrops* ranged from 2 to nearly 150 t ($\bar{x} = 60$ t) over the last quarter-century (statistisches Bundesamt, Wiesbaden). Here too, landings and ex-vessel prices were significantly inversely correlated ($r = -0.506$).

C. Marketing

The principal concentration of *Nephrops* in relatively deepwater areas, in contrast to the more shallow depths occupied by *Homarus,* was not discovered until the development of the trawl fishery in the early 1900s. This discovery led to moderate commercial exploitation, which was markedly intensified after World War II. In this period, considerable quantities of packaged, frozen *Nephrops* tails arrived in the United States market, imported primarily from Ice-

TABLE IV

Recent *Nephrops* Exports by Denmark and Norway[a]

| | Denmark | | Norway[b] | |
Year	Tonnes (10^3)	Value ($)	Tonnes (10^3)	Value ($)
1968	0.09	2,486	0.6	1,613
1969	0.8	1,952	0.6	1,882
1970	0.9	2,156	0.5	1,544
1971	1.0	2,347	0.4	1,589
1972	1.6	4,449	0.4	1,778
1973	1.4	4,635	0.4	1,928
1974	1.4	4,619	0.3	1,639
1975	2.4	8,874	0.3	2,028

[a] From F. A. O. Yearbooks, with permission.
[b] Includes *H. gammarus*.

land. Since they competed favorably in price and quality with *H. americanus*, some U.S. coastal states attempted to control imports by legislation.

Nephrops lobsters have been marketed both whole and as tails (Figueiredo and Thomas, 1967). With increasing production, onshore processing plants were built. The recommended methods of handling the catch before processing have been described in detail by Early (1965). Table IV shows that exports from Norway have declined somewhat to about 300 t in recent years, although exports from Denmark rose to 2400 t in 1975. Approximately 3% of the 1976 total catch was exported to the United States, primarily from Iceland, the United Kingdom, and France. During international food shows in (Cologne and Munich), discussions with dealers from representative European countries indicated a comparatively high level of integration from top to bottom in the *H. gammarus* and *Nephrops* fisheries, with most of the production being channeled through major fishing ports (Dow, 1966a). High density recirculated seawater holding tanks were employed extensively for *H. gammarus* and correspond in date of development to research in Canada and Maine in the late 1940s (Bramsnaes and Boëtius, 1953; Wilder, 1953; Anonymous, DMR 1953).

D. Management

Management regulations imposed on the *Nephrops* fishery differ among countries. Much of the following has been summarized from Farmer (1975). At present, with few exceptions, berried females are not protected. Under French law, the landings, sale, and transportation of *Nephrops* carrying eggs has been prohibited since 1967, although, the Directors of Marine Affairs may authorize the landing, transportation, and sale of berried lobsters for the purpose of restocking reserves. Similarly, fishing for, possession on board, buying, selling, transporting, and the use for any purpose (particularly as animal fodder and fertilizer) of *Nephrops* of less than 115 mm (fished north of 48° N) and 100 mm (fished south of 48° N) in total length (tip of rostrum to the end of the telson) have been prohibited under French law since 1964. In some areas, protection of berried lobsters is accorded at times either by regulation or voluntarily by fishermen. Since most *Nephrops* are taken in trawl nets, the principal management efforts have been directed toward mesh sizes. In the French fishery, mimimum mesh size is 50 mm. In Icelandic waters, the *Nephrops* fishery is permitted only in water deeper than approximately 90 m between May 15 and August 31. The minimum stretched mesh size permitted is 80 mm. Total allowable catch is estimated each year. Log books showing trawling time, area fished, and catch are required. Berried females are protected, and infractions may result in fine or loss of license (P. Guðmundsson, personal communication; U. Skuladottir, 1965).

From January 1936, the minimum size of *Nephrops* permitted to be landed in Denmark was 160 mm total length. A convention in 1952 among Denmark, Norway, and Sweden introduced a common total minimum length of 150 mm (Jensen, 1959, 1967). In 1959, this was amended to 130 mm, but Denmark retained the minimum total length of 150 mm. In 1965, the Danish minimum legal size for *Nephrops* was altered to 147 mm, and a minimum tail length (from front edge of the first segment to the rear edge of telson) of 80 mm was introduced. At present, the total length minimum is 130 mm, and the minimum tail length is 72 mm (B. I. Dybern, personal communication).

In Scottish waters, the minimum permitted stretched mesh size is 70 mm, although seine nets with meshes of 55 mm have been previously permitted (Thomas, 1960, 1969, 1970; Pope and Thomas, 1965; DAFS, 1964). The use of trawl nets for the capture of *Nephrops* was permitted in 1962 in certain areas of the Firth of Clyde, the Moray Firth, and the Firth of Forth, subject to restrictions on vessel size, the number, material of construction, and dimensions of the otter boards, as well as the relative percentage by weight of fish landed. The maximum size of vessels permitted to fish for *Nephrops* was 17 m in the Firth of Forth and 21 m in the Firth of Clyde and Moray Firth. The permitted fishing season was from May 1 to September 30 in the Firth of Clyde, from April 1 to October 31 in the Moray Firth, and for a period of 3 years in the Firth of Forth. In 1968, the requirement that otter boards should be made of wood was relaxed, and fishing in the Moray Firth and Firth of Clyde became legal throughout the year (Thomas, 1960; Pope and Thomas, 1965).

In the United Kingdom, the principal fishery of *Nephrops* uses trawls with a minimum stretched mesh size of 70 mm (F. G. Howard, personal communication), except in the Irish Sea, where the legal minimum mesh size is 50 mm when the fishing is exclusively directed toward *Nephrops*. In a mixed fishery for whiting, the minimum is 60 mm.

There are many inconsistencies in the *Nephrops* regulations among various countries. In Spain, the minimum overall length is 120 mm (from eye socket to base of telson). In Sweden, Norway, and Denmark, the total minimum legal length is 130 mm, but only 100 mm in France. Iceland has a minimum abdominal length of 70 mm and Denmark of 72 mm, and Iceland also has a minimum abdominal weight of 10 g, comparable to the Greek regulation prohibiting the catching and sale of *Nephrops* having a total weight of less than 100 g, with a tolerance of not more than 10 g (Farmer, 1975). Berried lobsters are not protected legally, but at times they have been discarded by Swedish fishermen (Dybern, 1977). In Norway, a mixed fishery for *Nephrops* and the northern shrimp, *Pandalus borealis*, uses a 35 mm mesh size. In Sweden, the cod end of trawls is generally 70 mm mesh size. In some areas, 60–65 mm mesh is permitted. In a mixed fishery with *Pandalus borealis*, the same minimum landing size as in Norway is in effect (Dybern, 1977). In Ireland, cod end mesh sizes range from 40 to 45 mm. In Portugal, mesh sizes range from 60 to 70 mm. In the Netherlands, the fishery is incidental to that for plaice and sole and there appear to be no regulations specifically pertaining to lobsters. In Spain, *Nephrops* is the by-catch in the trawl fishery for other species. In Faeroese waters, the maximum horsepower permitted for vessels fishing for *Nephrops* is 40 hp. Also, up to six small cutters are allowed to fish for *Nephrops*. Fishing is permitted only from June 15 to August 14 (Thomas, 1970), and there is a 150 mm minimum size.

IV. *Metanephrops* AND *Nephropsis*

Holthuis (1974) described 18 Atlantic species of Nephropidea. *Nephropsis aculeatus*, from the east coast of Florida and in the Gulf of Mexico, is the only species to have been fished commercially (Roe, 1966). Since most of the Nephropidea are deepwater animals, adequate gear for taking them in commercial quantities has been a major limiting factor. Common names frequently confuse identity with the name "langostino," or its orthographic synonyms applied to various species of lobsters and crabs in different countries or regions. With such diversity as *Metanephrops rubellus* and *Cervimunida johnstoni* (the Chilean crab, trademark "Langostinos"), it is presumed that some unaccounted-for landings of Nephropsida other than *N. norvegicus* and *Nephropsis aculeatus* are occurring from the Caribbean to the southwest Atlantic.

Twelve species of *Metanephrops* in the Pacific and Indian Oceans have been reported by Jenkins (1972). In a personal communication, J. C. Yaldwyn states that *Metanephrops challengeri* is a minor commercial species in New Zealand,

and he anticipates that its commercial importance will increase when more deepwater trawling is undertaken. Supporting his assumption of increasing industrial interest are the comments of R. G. Wear (personal communication), who stated that *M. challengeri* "has been the subject of relatively extensive exploratory fishing effort as at times they are caught in considerable abundance." He further stated that Russian trawlers have had "spectacular success, from time to time returning with many tonnes from southern New Zealand."

Japan, Ecuador, Hong Kong, and Pakistan have all reported occasional landings of lobster (homard) which have been recorded by FAO. It is presumed that these reports refer to catches of one or more of the several species of *Metanephrops* found in the Indo-Pacific Oceans.

V. *Homarus* FISHERIES

The fisheries for *Homarus americanus* in the northwest Atlantic and *H. gammarus* in the northeast Atlantic are very similar. The gear, seasonal movements, temperature requirement, and habitats are virtually identical. In some areas there is a year-round fishery, but in others, fishing is limited to the summer months. The two species show peak catches in the same months where there is an unrestricted seasonal fishery (Fig. 4). The lobster year appears to coincide quite closely with molting of the new crop to legal size, which occurs in July, plus or minus about 30 days. Nineteenth century reports of the *H. gammarus* fisheries of Norway, England, and Wales could well be applied today to the American lobster fisheries of Canada and the United States.

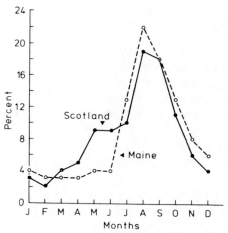

Fig. 4. Average percent of annual *Homarus* catch by months, 1950–1954; Scotland, *H. gammarus;* Maine, *H. americanus.*

A. *Homarus gammarus*

1. Early History

Landings of European lobsters (*H. gammarus*) in Norway for export to England through British companies have been reported in "Om det norske Hummerfiske og dets Historie" (A. Boeck, 1869) from 1821 to 1865. From 1821 to 1824, more than one million lobsters were exported annually. The catch increased to 1,429,703 lobsters in 1827 and 1.5 million in 1828, with an annual average of 1,268,000 for the years 1825-1830. During the next 5-year period the catch declined to an annual average of 640,000 lobsters, and attempts were made by public officials and others to regulate the fishery. Various recommendations, including a minimum total length of approximately 203 mm, closed seasons, and closed areas were turned down by the Norwegian parliament largely on the basis of the objections of fishermen and the English companies responsible for the export market. The government requested Prof. Rathke, Prof. C. Boeck, and Prof. G. O. Sars (at that time a clergyman) to make a study and report the situation, but their recommendations did not agree. Boeck recommended a minimum size, which the government accepted. However, the Storthing refused to enact it into law on the grounds that fishermen would believe that such a regulation limited their liberty. Despite a decree to the contrary in 1728, there was an underlying belief that the lobsters belonged to the landowners as far as 10 fathoms from shore at low tide. While lobster landings continued to decline and prices increased, more studies were requested. Management regulations, especially closed periods during the spring-fall seasons, were strenuously opposed by the English companies that purchased the lobsters, and by the local fishermen as well. Finally on January 1, 1849, after 20 years of controversy, provisions of the new law were adopted.

1. No fishery from July 15 to September 30.
2. Local exceptions to the closure shall not include the month of August.
3. Fines for fishing or selling were provided.
4. Identified police courts and procedures.
5. Time limitations on sale and export of lobsters in relation to the closed period.
6. Effective date.

The regulations reduced exports from 600,000 during the years 1846-1848 to 408,310 in 1849 and 427,600 in 1850. Contributing to the decline in landings in 1849 and 1850 was a reduction in landed price paid by the English companies. In one lobster district, annual exports of 26,000 in 1846-1848 declined to 7960 in 1849, 1664 in 1850, and none at all from 1851 to 1854. A recovery to 14,470 export lobsters in 1855 was attributed largely to purchases by Danish importers.

Bingley (1800) described the *H. gammarus* fishery of Wales, and F. Buckland and S. Walpole *et al.* (1877) reported on the lobster fisheries of England, Scotland, and Ireland, without including any information on the Wales fishery (suggesting that Wales was far less important in lobster production than were the other areas). The fishery appeared to be primarily one of subsistence, and sales were limited to the immediate vicinity of the fishing ports. In general, conditions in the vicinity were comparable to the primitive conditions existing in the *H. americanus* fisheries of Canada and the United States at that time.

In "Reports on the Crab and Lobster Fisheries of England and Wales," Buckland and Walpole (1877) concluded that the increase in price paid lobster fishermen for their catches had more than offset the decline in supply. Better transportation capability to expanded markets had increased the area being fished and also the number of fishermen in comparison with the fishery of a half-century earlier. Overfishing occurred in easily accessible areas and those protected from adverse weather conditions. Higher prices induced improvements in gear efficiency, and as overfishing developed, fishermen expanded and intensified their operations in order to continue earning a living. Hatcheries to rear juvenile lobsters, a minimum carapace size, protection of berried lobsters, and seasonal closures during the molting as well as the egg-hatching period were among the remedies suggested by fishermen, dealers, and observers. It was concluded that the sale of berried lobsters should not be prohibited, because it would then not be worthwhile for fishermen to continue the fishery. Closed seasons were unacceptable on the basis of geographical differences in growth and other conditions which also excluded a uniform minimum size.

2. Current Trends

The report resulting from the Bergen, Norway meeting of the Working Group on *Homarus* Stocks of the ICES Shellfish and Benthos Committee (May 1977) contained the following statement:

> The catches of the European lobster *Homarus gammarus* have continued to decline in the traditional fisheries of Sweden, W Norway, E Scotland, Wales and SW Ireland. In England and France catches have been maintained at recent levels, which are below average. Catch-per-unit-effort (cpue) is low and falling in many areas. Fishermen are attempting to compensate for falls in cpue by increasing the number of traps fished. Catches and cpue have increased in E Norway and W Scotland, the latter mainly as a result of French, English and Channel Island vessels fishing previously unexploited stocks. Part-time fishermen continue to increase in many areas. In Sweden it is estimated that only 40% of the total catch is landed by full-time fishermen. The value per kg of lobsters has increased considerably in all fisheries.

Denmark, Norway, Sweden, France, Spain, Scotland, Ireland, England, and Wales have consistently reported *H. gammarus* landings since 1950. Only Ireland, England, and Wales have reported landings for all years from 1930. Total landings reported from 1950 through 1975 were the following:

Location	Quantity (tonnes)	Percent
Scotland	17,720	26.2%
Norway	12,496	18.4%
England and Wales	11,397	16.8%
France	11,032	16.3%
Ireland	6,337	9.4%
Sweden	3,223	4.8%
Denmark	1,842	2.7%
Spain	945	1.4%
All others	2,725	4.0%
Total	67,717	100%

Homarus gammarus catches have declined in all countries from peak landings during the period from 1930 to 1975. The most consistent landings have been those of England and Wales, where catch has ranged from 297 tonnes in 1941 to 611 t in 1948 (mean: 459 t). Landings in Ireland have ranged from 388 t in 1931 to 102 t in 1948 (mean: 228 t). French landings increased from 193 t in 1930 to a high of 625 t in 1958, an increase of 224%, followed by a variable decline of approximately 36% to the present. Between 1946 and 1973, Scottish catch declined 50% from a high of 1011 t in 1946 to 500 t in recent years. The greatest reported declines have occurred in Denmark (99.1%), from 1234 t in 1946 to 11 t in 1974, in Norway (92.4%), from 1313 t in 1932 to 100 t in 1973, and in Sweden (89.6%), from 299 t in 1933 to 31 t in 1975. Spain has shown the greatest percentage fluctuation in landings, from 5 t in 1963 to 161 t in 1967 to 12 t in 1974.

The ex-vessel prices of *H. gammarus* in Norway and Sweden since the 1950s show highly significant inverse correlations with landings. In Sweden, correlation coefficient between landings and prices for the period of 1951–1976 is -0.900. Responses in Sweden to the decline of nearly 85% in landings have been an increase of more than 450% in ex-vessel price and a decrease of 54 and 55% in the number of fishermen and in the units of gear, respectively (Table V). The relationship between the landed value and the catch of *H. gammarus* has been about the same in Norway as it has been in Sweden, i.e., the inverse correlation coefficient between the two series is -0.934. Dannevig (1951) related the impact of sea temperature on molting and its sequential effect on the magnitude and timing of landings. He recommended an increase in the legal minimum size because "as the female will not grow fertile till a length of approximately 22 cm is reached, it is evident that a very high percentage will not have a chance of reproducing themselves." At that time, the taking of berried lobsters was permitted in the Norwegian *H. gammarus* fishery. Various methods

8. The Clawed Lobster Fisheries

TABLE V

Swedish *Homarus gammarus* Fishery

Year	Number of fisherman	Number of creels (10³)	Catch (tonnes)	Price (kr/kg)
1951	1821	61	252	6.99
1952	1837	62	210	7.36
1953	1823	62	216	7.82
1954	1779	61	188	7.98
1955	1702	60	167	8.65
1956	1674	57	178	8.72
1957	1536	53	148	9.37
1958	1498	52	164	9.52
1959	1411	50	137	11.25
1960	1353	48	168	9.75
1961	1273	45	147	10.91
1962	1204	43	119	13.30
1963	1258	42	105	14.70
1964	1239	38	92	16.02
1965	1305	39	86	19.41
1966	1358	41	78	22.44
1967	1015	30	64	21.84
1968	891	27	66	24.65
1969	835	27	66	25.18
1970	—	—	73	23.59
1971	—	—	51	24.41
1972	—	—	54	23.04
1973	—	—	45	28.31
1974	—	—	38	34.16
1975	—	—	31	39.26
1976	—	—	41	38.78

of improving the fishery were under investigation, including the purchase of berried lobsters from fishermen to be held until the beginning of the closed season when they would be liberated. Other schemes included hatching of eggs and release of fry as well as rearing of larvae through metamorphosis to bottom stage before liberation.

Data of the England–Wales *H. gammarus* fishery are limited to landings and ex-vessel values (Table VI). The two series are inversely correlated at a highly significant level for the period of 1954–1976, and except for 1970, 1971, and 1973, the landings have declined since 1964. The most recent 8 years of landings average nearly 100 tons less than the first 8 years. Oddly, during 6 of the 23 years, annual landings were reported to be exactly 408 tons.

The *H. gammarus* fishery of Scotland peaked at 1011 t in 1946 then declined slowly to 528 t in 1976. In that time, the landed value increased almost without

TABLE VI

England and Wales *H. gammarus* Catch and Values

Year	Landings (tonnes)	Value (£)	£/kg
1954	451	156,966	0.348
1955	508	193,489	0.380
1956	509	203,373	0.400
1957	531	236,921	0.446
1958	497	238,708	0.480
1959	491	232,066	0.472
1960	466	238,335	0.511
1961	567	293,392	0.517
1962	470	279,489	0.594
1963	481	287,232	0.597
1964	479	353,851	0.739
1965	408	332,000	0.814
1966	408	379,000	0.929
1967	408	356,000	0.873
1968	408	383,000	0.939
1969	357	447,000	1.252
1970	510	609,000	1.194
1971	459	600,000	1.307
1972	408	639,000	1.566
1973	459	854,000	1.860
1974	408	832,000	2.039
1975	342	943,000	2.757
1976	348	1,254,776	3.606

interruption from £305/t in 1950 to £4089/t in 1976 (F. G. Howard, personal communication). For the 1950–1961 period, there is a significant coefficient of correlation between ex-vessel price and landings ($r = 0.690$), and from 1967 to 1976 there is a significant correlation coefficient ($r = 0.716$) between catch per unit effort in the inshore fishery and total landings.

Reported landings of *H. gammarus* in Italy have declined from a peak of nearly 1000 t in 1962 to no reported production since 1965. The recorded Yugoslavian catches have been of minor importance, ranging from less than 100 t to none. In Algeria, maximum reported landings of 100 t in 1964 have declined to no landings at the present, which compares with the situation in Belgium. Of the other minor producers of *H. gammarus* Turkey experienced an increase in production from 500 t in 1957 to 2200 t in 1964, and a subsequent decline that is presently listed as "no landings." There is no Soviet fishery for *H. gammarus* in the Black Sea, although the taking of occasional specimens has been reported (S. A. Mileikovsky and B. G. Ivanov, personal communication).

8. The Clawed Lobster Fisheries

B. *Homarus americanus*

On the assumption that supply and market problems are relatively the same in all the clawed lobster fisheries, an explanation of the Maine lobster industry is presented in greater detail as being representative of all three fisheries.

1. Early History

A specific reference to the American lobster fishery is contained in "Narrative of Weymouth's Voyage to the Coast of Maine" (Rosier, 1605); "and towards night we drew with a small net of twenty fathoms very nigh the shore; we got about thirty very good and great lobsters... which I omit not to report, because it showeth how great a profit the fishing would be...."

Just when lobsters were first marketed in North America is not clear, but it was probably by the early 1700s. However, the fishery did not become commercially important until after 1840. During the 1840–1850 decade, Massachusetts dealers looked to Maine and Canada for a continuing supply, since readily available local stocks for the Boston market had become critical by 1812, when the first North American lobster conservation law was enacted in Massachusetts. Demand for live lobsters in the New York and Boston markets induced both resident and nonresident fishermen to specialize in lobster fishing. Gradually, by the late 1840s the fishery had extended eastward to Penobscot Bay, Maine and to Eastport, Maine by 1855.

Canneries provided the first major commercial outlet for *H. americanus* in Canada and New England. Employing canning methods obtained from France by way of Scotland and Nova Scotia, the first successful United States cannery was built at Eastport in 1842 and was supplied with lobsters transported by smack vessels from the western coast of Maine. Portland and Boston investors furnished most of the needed capital, and by 1890, 23 plants had been built along the coast as far west as Portland. European markets absorbed about one-half of the canned lobster pack and the southern and western United States used the remainder. One plant specialized in canning whole lobsters in the shell to be used for garnishing dishes in Europe. The development of canneries in Canada was stimulated by expanding European markets and a decline in Maine lobster production. By 1880, 200 plants were in operation, 40% of which (by value) were owned or controlled by U.S. firms. In the next 15 years, the number of Canadian canneries expanded to nearly 650, and by 1900 to 900 plants. Lobster canning restrictions in the United States dated from 1872. In 1895, a law limiting the overall length to 267 mm put an end to canning and permitted the live lobster industry to take over commercialization of the resource, except for limited canned and frozen specialty products. In Canada, lobster canning has continued at a declining rate to the present, largely for export to Europe and the United States.

Competition between the live market and the canneries for the supply of lobsters increased the price paid to fishermen. The nineteenth century price figures that now exist show substantial increases after 1880 (Cobb, 1901).

Year	Cents per kilogram
1880	4.2
1892	8.4
1898	19.6

2. Live Lobster Industry

The canning industry induced the first exploitation on a broad commercial scale of Canadian and United States lobster resources. It encouraged diversification of fishing activities and broadened the economy of coastal areas. It also stimulated competition from the live lobster industry and forced the latter to improve handling, transportation, and distribution methods and facilities.

The Canadian Lobster Commission of 1898 was concerned that as the lobster fishery became more deeply involved in the live lobster trade, it would tend to market lobsters from all parts of the coast at the same time, and thus glut the market. This condition frequently occurred in the United States, where seasonal restrictions on lobster fishing no longer existed with the demise of canning. To avoid catch peaking simultaneously with the United States, five fishing areas were established in Canada covering successive periods of time along different portions of the coast. Tidal impoundments ("pounds") were also constructed for storage of lobsters to supply markets during the winter, when landings are appreciably reduced (Fig. 5).

In 1875, the first North American pound was built on Vinalhaven Island in Penobscot Bay. By 1904, nearly 30 such facilities with a total capacity of 1.5 million lobsters had been built in Maine. Currently there are approximately 75 tidal pounds in Canada and the United States, with an estimated capacity of some 7 million lobsters. In addition, floating storage cars and flow-through tank systems will hold several million more lobsters on a short-term basis. Such holding devices provide much the same marketing service to the fresh market area today as did the canneries to the canned foods market years ago, within the limitations imposed by transportation and distribution facilities. Live-storage pounds have made possible the development and maintenance of more stable marketing conditions (Dow et al., 1959).

One of the major problems of successful tidal pound storage is the high incidence of the disease gaffkemia, a bacterial blood disease that was first identified in Maine during the 1946 epidemic from collections made by Taylor and Dow (Snieszko and Taylor, 1947). Breaks in the shell or membrane of lobsters caused by fighting, plugging claws, or punching a hole or cutting a notch in the

Fig. 5. Tidal impoundment for holding live lobsters.

tail fan greatly damage the stored product, consequently affecting the industry. Gaffkemia and other lobster diseases are discussed in Chapter 6, Vol. I.

3. The Lobster Dealer

The lobster dealer evolved from the commercial fisherman toward the middle of the nineteenth century. His primary functions of purchase, storage, transportation, and distribution grew from the economic and geographic isolation of Canadian and Maine lobstermen and the need for adequate supplies of lobsters in the concentrated retail market served by Boston and New York. The lobster dealer purchases lobsters at a favorable price, stores them, and eventually sells them under the best terms obtainable. One of the important speculative functions of the dealer involves the attempt to keep the stored lobsters alive.

Wholesale dealers sell their lobsters in Boston and New York wholesale market or directly to hotels, restaurants, and market customers. In addition, a continuous trade between and among Canadian and U.S. dealers accounts for a large sales volume at the wholesale level. New York City's Fulton Fish Market now buys and sells about 5000 metric tons of lobsters annually; nearly one-fourth of the combined Maine landings and Canadian imports. The bulk of the Massachusetts catch is sold locally or in the Boston market. Washington, D.C., is a major market area for lobsters from Virginia to New Jersey, which augments imports from Maine and Nova Scotia.

Approximately 50 Fulton Fish Market seafood dealers and uptown brokers constitute the New York wholesale lobster market. Less than ten of these firms account for 90% of the trade, and three of the larger brokers conduct about 50% of the business. The term "broker" is not used in a technical sense. The "broker" may take title to the lobsters, but may be forced to sell them before delivery because of inadequate storage facilities. Fulton Market dealers were placed at a competitive disadvantage by the brokers. Continued growth of the New York market is doubtful. In fact, a future decline is likely to be caused by the continuing development of closer ties between the Maine or Canadian dealers and their direct customers, which has been the case through the 1970s. Most brokers deal only in lobsters. Direct deliveries are made from the trucker's freight station to the broker's customers, a relatively more efficient operation than that conducted in the Fulton Market.

The New York market determines lobster price. In periods of short supply, prices tend to be higher there than elsewhere because of the demand inelasticity of specialty restaurants and other customers who must have their normal supplies regardless of price. Dealers supplying these markets bid up prices to obtain required stocks, thereby influencing all lobster prices. In periods of seasonal oversupply, surplus production is absorbed by the New York market. Almost any quantity of lobsters can be sold there at relatively low prices, since demand becomes very elastic at these lower prices. During these market adjustments, New York prices are lower than those in more isolated and more stable markets.

4. Price Determination

Most economists consider the wholesale market for lobsters an "imperfect" one. Buyers are widely scattered and badly informed regarding marketing conditions. Many of the primary dealers have very limited financial resources and, consequently, are sometimes forced to dump their holdings in order to obtain cash for their next purchase. The scarcity of supplies in the period immediately preceding the sale of the Canadian early spring catch enables large wholesalers to make or break the market. An artificially low price can be established a few days prior to the arrival of the first Canadian landings. Although a loss must be taken on the short supplies on hand, the net result is that the large dealer is able to stock his pounds at relatively low prices. The long-term profit potential in such an operation may be great.

Product differentiation enables many small dealers to conduct profitable enterprises. Good appearance, high meat content, and hardiness are among the characteristics used to judge premium lobsters. Most dealers buy from fishermen on an ungraded basis, paying the same price for all lobsters. Small dealers are forced to sell whatever stock they have on hand to make room for new supplies and to raise the cash to pay for the next purchase. They are able to do so because the New York wholesale market will absorb large quantities of ungraded lobsters at relatively low prices. It has been estimated by persons associated with the industry that the elimination of 10–30% of the poorer grades from the live market would benefit the entire fishery.

8. The Clawed Lobster Fisheries

The difference between New York price quotations and the prices received by fishermen is accounted for by the dealer's or wholesaler's costs of grading and culling, transportation, lobster mortality, packaging, and profit. Profit rates are lower during the summer when the peak sales volume is marketed, but they tend to be higher for the smaller volume handled in the winter and early spring.

Price breaks are indicative of the market's inability to absorb abrupt changes in the supply of lobsters. Nevertheless, the New York market serves as an ultimate source of liquidity for Maine and, by inference, Canadian dealers. Despite the imperfections in the wholesale market, it does respond actively to the prevailing supply and demand situation. Abundant supply and lighter demand may force the seller to dispose of his stock quickly and at a lower price. Generally, a seller's market exists during the winter and early spring. Light supplies during this period are not augmented until Canadian shipments arrive at the end of April. Peak production from Maine during the summer and fall, plus Canadian imports, ensure relatively heavy supplies at that time of year. Wholesale prices are equalized among the major markets, except for differentials based on transportation costs. However, the prices received by fishermen may vary widely. Isolation of the fishermen from alternative marketing outlets and the imperfections in the marketing system make the fishermen vulnerable to relatively rapid price changes (Dow et al., 1961).

5. Trends in the Maine Fishery after 1939

The most noteworthy occurrence in the Maine lobster fishery was the rapid increase in total annual landings from 3000 t in 1939 to 10,000 t in 1953 (Table VII). In the same period, the landed value (ex-vessel price) rose from $300 to $900/t. These lobster-year (July 1–June 30) data were used to identify molt classes, rather than year classes, based on size frequency distribution resulting from stratified sampling of the lobster catch from 1939 to 1957. Since lobsters of the same age do not molt at the same time or increase the same amount at molt, it is preferable to use the term "molt class" rather than "year class" and to group lobsters by carapace length in relation to some common base, i.e., legal minimum or postlarval size.

There was a highly significant correlation ($r = 0.902$) between the ex-vessel price of one year and the total annual landings of the following year, during the 14-year period 1939–1952. When ex-vessel prices increased or decreased, landings followed suit (Table VIII). Price and landings moved toward an equilibrium point at the intersection of the price-landing axes, but when price declined in 1946, 1947, and 1949, fishermen responded by reducing landings in 1947, 1948, and 1950. These declines in landings stimulated higher ex-vessel prices, which in turn led to greater annual catches. This relationship can be likened to the rabbit-fox cycle in the wild, the corn-hog cycle in commercial agriculture, or the cobweb theorem (Fig. 6), where the oscillations of supply and demand are associated with dynamically changing economic situations (Samuelson, 1964).

TABLE VII

Catch of *H. americanus* (in Maine) per Year and per Molt Class

Lobster year	Total catch by lobster year	Recruits	First molt	Second molt	Third molt	Total catch molt class
1939–1940	6,157,919	5,383,792 86.95%	614,621	142,262	17,244	X
1940–1941	7,250,857	6,421,359 89.14%	655,478 10.59%	162,419	11,601	X
1941–1942	8,198,890	7,447,051 87.02%	607,538 8.43%	134,462 2.17%	9,839	X
1942–1943	7,322,423	6,306,071 71.72%	842,811 9.85%	155,235 2.15%	18,306 0.30%	6,192,038 100.01%
1943–1944	10,320,341	8,193,532 75.45%	1,886,369 21.46%	220,833 2.58%	19,607 0.27%	7,203,739 99.99%
1944–1945	13,572,315	10,810,349 79.06%	2,159,355 19.88%	555,108 6.31%	47,503 0.55%	8,558,198 100.00%
1945–1946	15,821,538	13,025,872 83.30%	2,254,569 16.49%	496,797 4.57%	44,300 0.50%	8,791,848 99.99%

1946–1947	14,386,624	11,671,896	2,124,131	580,489	10,108	10,859,792
		79.55%	13.58%	4.25%	0.09%	99.99%
1947–1948	15,720,848	12,483,601	2,735,701	473,245	28,301	13,673,708
		84.15%	18.64%	3.03%	0.21%	100.01%
1948–1949	14,337,737	12,318,782	1,796,698	207,918	14,339	15,637,587
		81.11%	12.11%	1.42%	0.09%	100.00%
1949–1950	16,366,018	13,529,787	2,278,150	500,800	57,281	14,672,796
		81.66%	15.00%	3.38%	0.39%	100.00%
1950–1951	16,540,417	13,528,708	2,452,699	504,432	54,578	14,835,677
		84.03%	14.80%	3.32%	0.37%	100.01%
1951–1952	17,532,735	14,780,096	2,128,474	538,255	85,910	15,187,274
		85.82%	13.22%	3.25%	0.57%	100.00%
1952–1953	17,563,959	15,101,492	2,026,881	388,163	47,423	16,568,164
		83.58%	11.77%	2.41%	0.29%	100.00%
1953–1954	19,227,049	16,431,436	2,384,154	357,623	53,836	16,099,181
		84.23%	13.19%	2.08%	0.33%	99.99%
1954–1955	18,957,626	16,095,024	2,303,352	502,377	56,873	17,221,473
			11.81%	2.78%	0.33%	100.00%
1955–1956	18,838,009	15,419,253	2,646,475	691,286	80,995	18,069,018
				3.54%	0.45%	100.00%
1956–1957	18,171,189	14,255,689	3,248,684	585,054	81,762	19,507,836
					0.42%	100.00%

TABLE VIII

Maine Lobster Industry 1939–1953[a]

Year	Landed value (10^2 \$/tonne)[b]	Landings (10^3 tonnes)[c]	Year of landings
1939	3	3	1940
1940	4	4	1941
1941	4	4	1942
1942	5	5	1943
1943	6	6	1944
1944	6	9	1945
1945	9	9	1946
1946	8	8	1947
1947	8	7	1948
1948	9	9	1949
1949	8	9	1950
1950	8	9	1951
1951	8	9	1952
1952	9	10	1953

[a] Catch size is closely correlated with value of the catch in the preceding year.
[b] Rounded to nearest hundred.
[c] Rounded to nearest thousand.

Contemporary reports by federal biologists in 1941, covering the critical condition of the Maine fishery completely ignored economics and ascribed depressed landings wholly to biological factors. This conclusion was made even though many fishermen had been forced by low prices to use oar-powered boats rather than the combustion-engine vessels that had been in service prior to World War I. The more nearly optimum sea temperatures after 1939 may have improved environmental conditions for the lobster supply, but it is likely that price alone stimulated the extraordinary increase in lobster landings.

6. Temperature-Effort

The major environmental factor influencing changes in *H. americanus, H. gammarus,* and *Nephrops* abundance and availability throughout their range has been the cyclic fluctuations in sea temperature (Dow, 1978). It is not completely understood how this influence functions, but it appears to be dependent on the rate at which sublegal lobsters are recruited by molting to legal or commercial size, as well as the survival of a year class to that size (Dow, 1969b, 1977b; Flowers and Saila, 1972).

Interacting factors of fluctuating sea temperature influencing supply and variable fishing effort affecting yield have been evident throughout the history of the inshore *H. americanus* trap fishery. Less than 10% of unaccounted variance in

8. The Clawed Lobster Fisheries

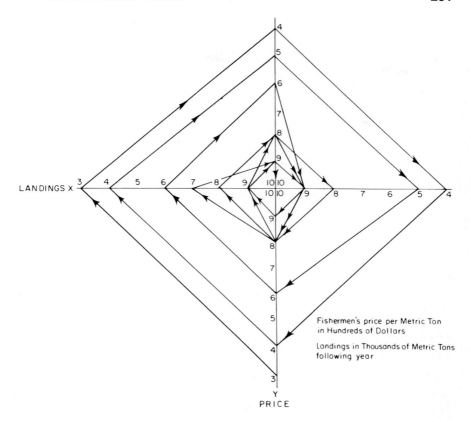

Fig. 6. Cobweb theorem for the Maine lobster fishery 1939–1952.

supply and yield can be attributed to other factors. Since 1946, predictions of landings in the Maine fishery have been made annually or for a longer time period, with a standard error of ± 8.16% and a range of errors from 0 to 24%. These predictions have been based on (1) the landed value during the preceding year (1946–1956), (2) the mean annual sea temperature measured at Boothbay Harbor (1957–1968), and (3) fishing effort and sea temperature (1969–1977) (Dow and Trott, 1956; Dow, 1964, 1969b, 1976, 1977b).

From 1954 through 1956, landings and ex-vessel prices were generally stable; a period of transition before entering the subsequent 12-year cycle during which landings were significantly dominated by sea temperature ($r = 0.958$) (Fig. 7). As temperature declined, landings decreased. In the years from 1954 to 1964 and

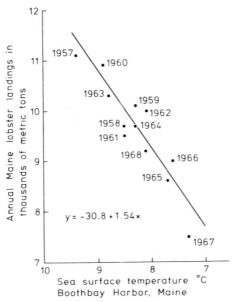

Fig. 7. The relationship between sea surface temperature and annual Maine lobster landings in the years 1957–1968.

1969 to 1977, the sea temperature averaged 8.9°C, but landings in the earlier period averaged 10,100 t or 2000 t more per year than during the more recent period. Annual fishing effort between 1954 and 1964 averaged 654,000 traps, whereas 1,433,000 traps were used annually during the recent 9-year period, indicating that overfishing has reduced expected landings by approximately 257 t for each 100,000 traps above 650,000, or ⅓ t/1000 traps. Between 1969 and 1977, the annual sea temperature at Boothbay Harbor averaged 8.9°C, which is more nearly optimal for lobsters, yet landings during this period averaged only 8000 t compared to an annual average 9600 t from 1957 to 1968, when the temperature averaged 8.3°C. In addition, a delayed time series regression analysis of the 1969–1977 data indicated a highly significant inverse correlation between fishing effort (number of traps) and landings ($r = -0.910$). Within several differential levels of effort and supply, fluctuations in yield are measurably related to corresponding fluctuations in temperature or in fishing effort. When fishing effort increases, yield expands to the capacity of the legal supply. Subsequent fluctuations in supply are correlated with variations in sea temperature irrespective of further increases in effort (Fig. 8).

Higher than optimum temperature results in early peaking of molt and greater concentration of the crop in the first 2–3 months of the lobster year. When

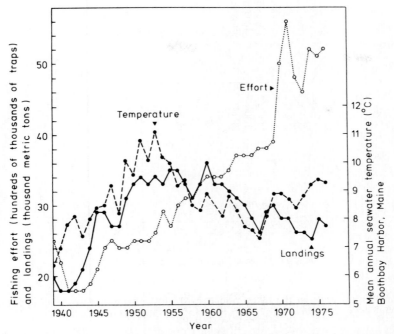

Fig. 8. The effort, sea surface temperature, and landings of inshore American lobsters, Newfoundland to New York, 1939–1977.

temperature is below optimum, the frequency of molting is less and the lobsters require more time to reach minimum legal size. The number of lobsters molting annually is reduced, and the magnitude of the crop diminishes. Historic temperature trends appear to be sufficiently consistent to permit generalized forecasts to be made. Seasonal temperature deviations result in corresponding deviations in legal lobster supply. If later compensating temperature fluctuations occur, lobster supplies will adjust to these changes. For example, low spring temperatures result in late recruitment of legal stock. If this recruitment enters the fishery before effort declines seasonally, it will show up in the October to December catches. If recruitment enters the fishery later, it will show up in the following spring fishery. Lower temperatures postpone the time and decrease the magnitude of recruitment (Dow, 1966b).

Meterological studies (H. C. Willett, personal communication; United Press International, 1976) indicate that following a warmer cycle between the middle and late 1970s (less evident than the warm cycle of the 1950s), summer sea temperatures probably will decline to the lowest levels yet recorded and "probably be as low as they are likely to get for most of the next century." These

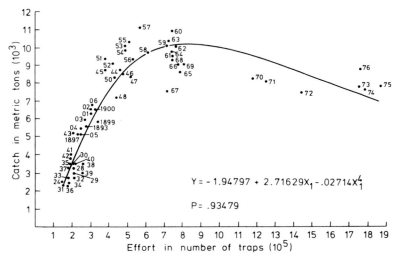

Fig. 9. Maine lobster catch-effort data, 1897–1976.

forecasts indicate that the catch of the major lobster-producing areas of Canada and the United States will probably continue to decline through the remainder of the century, with a strong possibility of some recovery in 1980–1983, resulting from near optimum spawning year temperatures between 1974 and 1976. This expected recovery may be neutralized by suboptimum sea temperatures during the year of recruitment to legal size. Markedly greater fishing effort expressed by an increase of 100% in the number of traps being fished between 1968 and 1972 continued to expand at a more modest rate toward the end of the decade (Fig. 9).

7. Management

In the United States, early regulations were designed to hedge the inefficiency of participants in the lobster industry, rather than to protect the resource. The North American lobster industry has never supported regulations that were designed to provide for maximum abundance of the resource, nor has encouragement been given to biological research that would develop information as a basis for such regulations.

Canadian areal and seasonal regulations have been built around the seasonally unrestricted U.S. fishery, as an economic reaction to the normal peaking of catch during the summer and fall. In doing so, the Canadians avoid recently molted, poor quality lobsters and are able to supply hard shelled, high meat yield lobsters that are more capable of withstanding the stress of shipment to major market areas.

The increasingly intensive fishery in the United States has relied heavily on seasonally recruited lobsters from premolt, sublegal stocks, to the extent that fluctuations in the population have become more evident. Catch has become less flexibly responsive to changes in fishing effort induced by higher landed value. The lack of flexibility in catch has been associated with the decline in abundance, and the decline in abundance appears to have been the result of suboptimum temperature followed by overfishing (see Figs. 7 and 8).

For years, economists and biologists in Canada and the United States have recommended a reduction in *H. americanus* fishing mortality. Economists argue that spreading employment is not economically desirable, since the policy does not provide the necessary incentive for good management or the efficient use of the resource. Unrestricted entry increases competition, dilutes the profitability for those who are competent fishermen, and generally results in the misuse of the resource and a decrease in operational efficiency. Biologists have reported the repeated capture and release of sublegal lobsters, with consequent physical damage and frequent mortality from predation and mishandling (Krouse, 1976).

In 1960, the most recent peak year, the inshore trap catch of lobsters from those areas which normally produce more than 90% of the total North American yield was 36,000 t, only a modest fraction of the potential world market. Using published data on offshore stocks and historical information on population trends of the inshore resource, it has been estimated that the total annual *H. americanus* catch in Canada and the United States during periods of optimum climatic conditions would probably not exceed 45,000 t (Dow, 1971). Conversely suboptimum conditions would likely reduce yield to approximately 25,000 t annually. The most biologically sound management might increase yield by 5000–9000 t.

Resource management has rarely been exercised effectively in regulating the clawed lobster fisheries. Adequate efforts to slow production declines in the *H.*

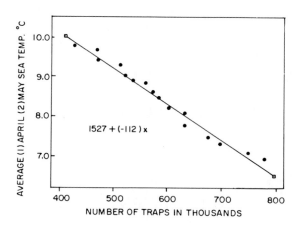

Fig. 10. Plan to sustain annual yield of 10,000 t in Maine lobster fishery by adjusting annual fishing effort (number of traps) to sea temperature trends.

TABLE IX
Seasons, Size Limits, and Maximum and Minimum Trap Limits by Lobster Districts of the Canadian Atlantic Provinces, 1975[a]

District	Season	Carapace size limit (mm)	Maximum trap limit	Minimum trap (class A)
1	November 15–June 24	81	375	75
3	October 15–December 31	81	300	75
	March 1–July 20			
4 West	Last Tuesday of November–May 31	81	375	75
4 East	Last Tuesday of November–May 31	81	250	50
5	April 10–June 30	81	250	50
6A	May 20–July 20	81	300	75
6B	May 16–July 15	70	375	75
7A	May 1–June 30	81	375	75
7B	May 1–June 30	63.5	400	100
7C	May 1–June 30	63.5	375	100
8	August 10–October 10	63.5	250	100
9	May 10–July 10	76	300	300
10A	May 20–July 31	76	300	n.a.
10B	June 15–August 15	76	300	n.a.
10C	May 10–July 27	76	300	n.a.
10D	May 1–July 17	76	300	n.a.
11	April 20–July 5	81	n.a.	n.a.
12	May 5–July 10	81	n.a.	n.a.
13	April 20–July 15	81	n.a.	n.a.
14	April 20–June 30	81	n.a.	n.a.
Offshore	January 1–December 31	81	n.a.	n.a.

[a] District number refers to Fig. 11.

8. The Clawed Lobster Fisheries

gammarus and *H. americanus* fisheries by reducing fishing effort or increasing minimum legal size have not been attempted until recently. A management program for the State of Maine has been described, which was derived from the actual historic fishing practices of the industry (Dow, 1964). The plan was intended to stabilize catch at approximately 10,000 t by adjusting annual fishing effort to sea temperature trends (Fig. 10). During the period of 1941–1963, fishermen adjusted their fishing effort to maintain constant production or possibly, annual income. The volume of catch was lower than it would have been with a larger minimum size, but fishing effort had not yet reached the catastrophic level that it did during the 1970s.

Declines in landings of *H. gammarus* and *H. americanus*, despite the recent more favorable sea temperature, indicate the need for better management. Based on predicted climatological conditions during the remainder of the century, landings of both *H. gammarus* and *H. americanus* will not otherwise improve or maintain their current catch level. Efforts to develop a unified management plan in the United States resulted in the establishment of a sub-board, composed of state fishery administrators and the Regional Director of The National Marine Fisheries Service, to provide overall guidance to facilitate implementation of a management program through existing legal and institutional channels. It has proposed, on biological grounds, to increase the present carapace size limits of

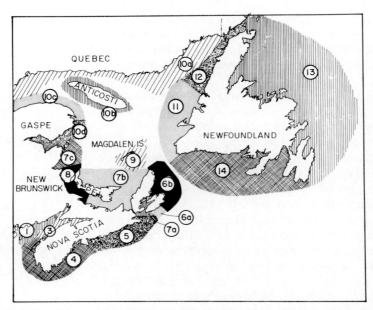

Fig. 11. Canadian lobster fishing districts. Numbers refer to districts in Table IX.

TABLE X

U.S. Lobster Regulations, by State (1977)

Regulation	ME	NH	MA	RI	CT	NY	NJ	DE	MD	VA	NC
1. License requirements											
No license required									X		X
Required to fish lobster	X	X	X	X	X	X		X		X	
Required to land lobster	X	X	X	X	X	X				X	
Required to deal in lobster	X	X	X	X	X						
2. Legal provisions for aquaculture enterprises	X				X	X	X				
3. Fishermen classification											
None											
Commercial	X	X	X	X	X	X	X	X	X	X	X
Noncommercial		X			X	X		X			
4. Catch/effort reporting											
Not required	X		X								
Required annually		X			X	X	X	X	X	X	X
Requires daily record				X	X						
5. Gear regulations											
None							X		X		
By license class		X	X	X	X	X				X	

6. Fishing activity regulations

	C1	C2	C3	C4	C5	C6	C7	C8	C9
Quantity allowed	X							X	
Type allowed	X	X	X				X	X	
Owner identification required	X	X	X	X			X	X	
Escapement opening in catching device specified	X	X					X		X
None									
By license class or method		X		X		X		X	
Number of licenses		X	X						
Catch quotas	X	X				X		X	
Area	X	X				X			
Season	X	X		X	X		X	X	
Day or time of day	X	X	X	X	X		X	X	X
Landing of lobster meat regulated	X	X	X	X	X		X	X	X
Landing of lobster parts regulated	X	X	X	X		X	X	X	X
Landing of gravid female lobsters prohibited	X			X	X	X	X	X	
Landing of v-notched female lobsters prohibited					X				
Landing of lobsters regulated by size (carapace length)	X	X	X	X		X	X	X	
127 mm Maximum allowed	X				X				
79 mm Minimum allowed		X	X				X	X	
81 mm Minimum allowed	X			X		X	X	X	

78.6, 79, and 81 mm to a uniform 89 mm or larger in the United States. Escape vents for sublegal lobsters have been introduced in some states.

Canadian scientists working cooperatively with the U.S. State-Federal Scientific Committee hope to increase the size limit from 63.5 to 76 mm in some Canadian areas. Further changes in size limits are expected in the next few years. In an attempt to reduce the fishing effort, a license buy-back scheme has been introduced in one province (D. E. Aiken, personal communication). The offshore fishery is subject to quota, seasonal, and gear restrictions (A. B. Stasko, personal communication). Canadian management favors full-time fishermen (Table IX and Fig. 11). In the United States, where regulations are the responsibility of the several states and not, as in Canada, the federal government, there exists a wide range of legal restrictions (Table X).

Numerous biological, economic, and bioeconomic models designed for lobster fisheries management have been developed since World War II. Among them are those published by Saila and Marchesseault (see Chapter 6), Dow *et al.* (1975), Hewett (1974), DeWolf (1974), Wilson (1974), Rittgers (1973), Thomas (1973), Acheson (1972), Hall (1972), Bell (1972), Flowers and Saila (1972), Huq and Hasey (1972), Gibson (1971), Dow (1964), and H. J. Thomas (1951). Generally these models have been designed by either economists or biologists, without the benefit of mixing the two disciplines. Among the major biological problems of the three lobster species is a lack of adequate information about migration, geographical location of larval and juvenile sources, methods of accurate aging, identification of stocks, growth rates, and nonfishing mortalities. For these and possibly other reasons, it does not seem wise to review the models listed above.

In 1972, a committee composed of marine scientists from the 11 coastal states (Maine to North Carolina) where *H. americanus* occurs and two federal representatives (a biologist and an economist) was appointed for the purpose of preparing a lobster management plan for both inshore and offshore populations. Since only Maine was prepared for such a contingency with the necessary research information, the next several years were spent developing management recommendations for the other lobster areas. Data on fishing and natural mortality, molt increment, frequency of molt, and size at maturity have been used to estimate total yield from the resource and to formulate a management plan within the limitations imposed by climatic cycles.

VI. EFFECTS OF SEA SURFACE TEMPERATURE CYCLES ON LANDINGS OF *H. Americanus, H. Gammarus,* AND *Nephrops*

Results of previous studies demonstrated highly significant correlation coefficients between the sea surface temperature at Boothbay Harbor, Maine, U.S.A., and that at Torungen, Aust Agder, Norway, from 1905 to 1974. Highly signifi-

8. The Clawed Lobster Fisheries

cant correlations also exist between Maine landings of *H. americanus* and the total *H. gammarus* landings from 1950 to 1972, between Boothbay Harbor sea temperatures (Welch, 1977) and Maine landings, and between Boothbay Harbor temperatures and total *H. gammarus* landings during the same period (Dow, 1977b).

Subsequent research has developed evidence of highly significant correlation coefficients between Boothbay Harbor sea surface temperatures 6 and 7 years preceding and the total inshore trap landings of *H. americanus* between 1955 and 1975 ($r = 0.852$). In addition, there are highly significant correlations between Boothbay Harbor temperatures 6 and 7 years preceding and the total *H. gammarus* landings during the same period ($r = 0.900$), as well as a highly significant inverse correlation between Boothbay Harbor temperatures 6 and 7 years preceding and the total *Nephrops* landings during the same period ($r = -0.854$) (Fig. 12). These data indicate, in addition to the probable age of the average lobster in the catch, that the optimum temperature range as measured at Boothbay Harbor, Maine, for the *H. americanus* catch is between 9.2° and 10.7°C, with a catch range during those years from 31.0 to 36.0 × 10^3 t. The optimum temperature range for *H. gammarus* as measured at Boothbay Harbor appears to be identical to that of *H. americanus,* with a catch range of 2.5–4.2 × 10^3 t during

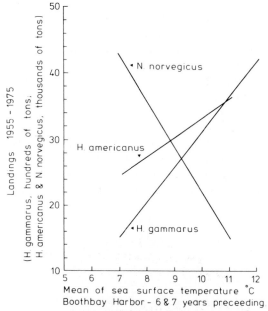

Fig. 12. Mean of sea surface temperature (Boothbay Harbor, Maine) and lagged landings of *H. gammarus* (hundreds of tonnes), *H. americanus,* and *N. norvegicus* (thousands of tonnes), 1955–1975.

optimum temperature years. *Nephrops* apparently has an optimum temperature range as measured at Boothbay Harbor of 7.5°–8.5°C, corresponding to a Torungen range of 8.0°–8.5°C. The range of landings from 35.0 to 44.0 × 10^3 t is associated with optimum temperature years.

In view of the predicted lower sea temperatures during the 1980s and 1990s (H. C. Willett and J. T. Prohaska, 1963; H. C. Willett, personal communications), it is probable that both *H. gammarus* and *H. americanus* abundance and availability will continue to decline. Although the same environmental conditions will likely be more nearly optimal for *Nephrops*, present knowledge of this resource is inadequate to make forecasts.

Comparison of European, Canadian, and United States Trends

The landings of *H. gammarus* in Norway increased with minor interruptions to a peak of 1,956,276 in 1865. In the northwest Atlantic during this period, fragmentary sea temperature records suggest that these years were warmer than they had been during the early decades of the nineteenth century. The cyclic trends of seawater temperature are similar on the two sides of the North Atlantic, and higher temperatures may have had a favorable influence on the *H. gammarus* fishery in Norway.

H. gammarus landings in Norway from 1876 to 1946, with the exception of the depression of the 1930s, followed the trends that were experienced in Maine, i.e., production peaks occurring in the 1880s, 1910, and the middle 1940s, and low production occurring in the 1890s, before and after 1920, and in the late 1930s and early 1940s (Dannevig, 1951). Only during the early and middle 1930s were there major differences in production trends between Norway and Maine. Production in Maine was suppressed by economic conditions, but a comparable lack of markets may not have existed in Norway at that time.

Dannevig (1951) also reported "at least 50 percent of the marketable lobster is taken every year" (based on marking/tagging experiments). Observations made between 1937 and 1945 by fishermen and scientists noted a scarcity of sublegal lobsters, which Dannevig explained as perhaps being the result of higher sea temperatures that had increased the growth rate to bring more and younger lobsters to legal size in a shorter period of time. The increased landings of the 1950s and 1960s in the northwest Atlantic have also been in part accounted for by an increased growth rate during those years of higher sea temperature (Taylor *et al.*, 1957; Dow and Trott, 1956).

Thomas (1951) reported marked fluctuations in the lobster populations of the Scottish coast between 1928 and 1948. The rapid increase in Scottish landings on a percentage basis was even greater than the increase in Maine landings of *H. americanus* during and following World War II (Dow, 1955, 1977a). Minimum

Maine landings occurred in 1936 (2323t), 4 years prior to the Scottish minimum reported by Thomas. This time interval corresponds to the 2–4 year lag in sea temperature trends between the northwest and the northeast Atlantic (Dow, 1977b). Even the period from 1928 to minimum landings years in Maine and Scotland corresponds in uniformity of catch with minor fluctuations characterizing the two fisheries. In Maine, the fishery was economically depressed so that the prices paid to fishermen were well below the pre-World War I level (Dow and Trott, 1956). It is therefore suggested that similar economic conditions prevailed in Scotland during the same period, especially in areas where some alternative employment opportunities may have existed.

Simpson (1958) studied the fishery of Wales in 1954 and 1955 and noted that the results for one area suggested a total mortality of 58% during the fishing season. In those years, fishing mortality alone in the Maine fishery was estimated to be 86% of the survivors, following a 25 to 35% loss of the immediate sublegal prerecruit class associated with trap and release mortalities (Dow, 1964; Sheldon and Dow, 1975).

Gibson (1971) concluded that the greater catch per boat in the Irish fishery between 1955 and 1959 was related to lobster abundance, an observation that is supported by corresponding increases in *H. americanus* landings in Canada and northern New England and the total *H. gammarus* landings during the same period (Dow, 1977b).

Seasonal fluctuations of the English *H. gammarus* fishery reported by Hepper (1971) are typical of the New England *H. americanus* fishery, where fluctuations in catch and effort are consistently associated with corresponding seasonal fluctuations in sea temperature. Lobsters are most catchable in August and September, but having recently molted, and with sea temperatures at their highest levels, quality is poor and prices are depressed (Anonymous, 1952). Many lobsters do not survive to market during this period. In the fall and winter, prices are higher when landings are down. Survival to market is much better during the cold months. With the onset of declining sea temperature in the late fall and early winter, both inshore *H. americanus* and *H. gammarus* move into deeper and warmer water, thus influencing the geographical location of the winter fishery (McLeese and Wilder, 1958; Dow, 1964; Hepper, 1971).

Conflicting reports on inshore–offshore migration of *H. americanus* appear to result from inadequate identification of water depth and, by inference, sea temperature. Cooper *et al.* (1975) reported relatively constant seasonal abundance of *H. americanus* in waters of 6–12 m, 18–24 m, and 30–60 m depths. They did not report on lobsters at depths of less than 6 m, which may be critical in determining winter migrations to deeper water and the return summer migration to shallow water. Both Harriman (1952) and Cooper (1970), who examined lobster migration in the vicinity of Monhegan Island, Maine, found little evidence of anything more than random movements in the relatively deep waters (up to 100 m) surrounding this offshore island. Sterl (1966) and Dow (1969a) found that all lobsters entering and occupying the shallow depths (<4 m) of Cousins Island Cove at Yarmouth, Maine, during the summer would move into deeper water in November.

Increases in landings resulting from previously unexploited stocks indicate a marked difference in available supplies between the *H. gammarus* and *H. americanus* fisheries. Offshore populations of the North American east coast appear to be fully utilized insofar as economic limitations permit, whereas an increase in *H. americanus* abundance at the southern end of its range associated with more favorable sea temperature was anticipated as early as 1967 (Dow, 1969b), well before information on subsequent catches became available.

H. americanus is a classic example of overexploitation of a renewable marine resource. Declining yield associated with less frequent periods of optimum sea temperature conditions has apparently been accelerated by continued increases in fishing effort. In many countries, *H. gammarus* seem to share the same distinction. Since 1950, only Sweden shows a concurrent reduction in the number of fishermen and creels with declining landings.

VII. CONCLUSIONS

Fisheries for the three lobster species (*H. gammarus, H. americanus,* and *Nephrops*) appear to be in biological and economic trouble—*Nephrops* for lack of meaningful management research to identify problem areas and the *Homarus* species from overfishing and inadequate biological knowledge for effective management. All three species are obviously subject to natural fluctuations in abundance, which appear to be associated with sea temperature fluctuations during climatic cycles. The fact that these fluctuations in abundance occur need not be a major obstacle to effective fishery management, unless migration into and out of fishing areas is much more extensive than is presently recognized.

Except on the extremities of their north–south range, as contrasted with the southern Gulf of St. Lawrence and the Gulf of Maine, it is unlikely that fluctuations in natural abundance from maximum to minimum are greater than two to one. Therefore, stable landings can be controlled by reducing or increasing fishing mortality, presumably by increasing fishing mortality during periods of reduced abundance and by reducing fishing mortality on a percentage basis during periods of greater abundance.

Since all three lobster species appear to be long-lived, and temperature subcycles are approximately one decade in length, natural abundance ranges can be artificially manipulated by fishing regulations. An example of this type of management is illustrated in Fig. 10, which was constructed from actual effort, catch, and temperature data during the period of coverage.

8. The Clawed Lobster Fisheries

Traditionalism and certain attitudes on the part of the industry have played dominant roles in determining management policies. Fishermen in the *H. americanus* fishery appear reluctant to alter their fishing practices and gear to reduce incidental mortalities associated with their fishing operations, including escape vents and degradable panels to offset in part the deleterious effects of ghost traps. No records have been found indicating any effort to improve lobster fishing devices in terms of the lobster resource.

The attitudes held by many fishermen regarding management changes are essentially identical in both Canada (Anonymous, 1975) and the United States (Dow, 1969a) *H. americanus* fisheries. Fishermen are suspicious of change and are unwilling or unable to understand why the resource, in terms of scientific management, should be given first consideration rather than the industry or the public.

APPENDIX AI

Annual *Homarus americanus* Landings from 1928 to 1976 by Province (Canada) or State (U.S.A.)[a,b,c]

Year	Newfoundland	New Brunswick	Prince Edward Island	Nova Scotia	Quebec	Maine	New Hampshire	Massachusetts
1928	2087	2631	2994	7802	1179	3221	59	926
1929	1270	3719	3357	8618	1225	3003	57	740
1930	1089	4128	3674	9435	1270	3509	82	1061
1931	1089	4309	4264	10161	1043	3240	65	1019
1932	907	4491	5216	10796	1452	2743	100	974
1933	816	3402	4173	8029	1452	2675	98	874
1934	726	2948	3493	8392	1633	2439	—	—
1935	1089	2494	2903	8029	1089	3487	88	819
1936	2041	2540	2676	6577	998	2323	—	—
1937	1860	3311	2631	7212	907	3333	71	966
1938	1769	3084	3221	6849	862	3474	77	1050
1939	1270	3674	3175	6577	862	3005	81	1057
1940	1216	2777	2503	5996	701	3467	107	1106
1941	1079	2691	2674	6358	699	4054	—	—
1942	969	3195	2902	5772	809	3812	132	1022
1943	862	2924	2708	6979	749	5202	518	1143
1944	1040	3182	3000	7807	999	6376	311	1243
1945	1057	3415	3673	8563	1101	8677	374	1301
1946	1266	4213	3894	8491	1074	8517	—	1489
1947	1545	3438	2666	7522	1046	8290	236	1665
1948	1686	3804	2959	8321	1086	7223	182	1457
1949	2286	4263	3104	9022	940	8742	189	1615
1950	2295	5140	4127	9969	1033	8324	278	1412

1951	1960	4792	3784	11012	1083	9415	232	1686
1952	1689	4708	3799	10462	1050	9087	186	1586
1953	2031	3914	3174	10726	1200	10112	277	1712
1954	2378	3639	3338	10545	1272	9818	352	1569
1955	2499	4100	3778	10408	1246	10302	386	1589
1956	2188	5231	4400	10093	1657	9316	347	1475
1957	1904	4740	3871	8241	1401	11068	413	1500
1958	2131	4383	3615	8134	1220	9665	372	1415
1959	1699	4233	3808	9553	1442	10126	388	1569
1960	2045	5404	4599	9043	1611	10889	424	1497
1961	1779	4932	4314	8994	1549	9485	373	1294
1962	1884	4243	3951	9074	1918	10013	329	1421
1963	2038	3833	3346	9174	1737	10344	339	1382
1964	2046	3293	3549	8670	1437	9713	386	1703
1965	1695	2773	4009	8410	1494	8556	347	2869
1966	1579	2777	3563	7255	1697	9034	366	2180
1967	1415	2344	4065	6579	1501	7479	326	2121
1968	1818	2699	3936	7166	1274	9300	342	2160
1969	1730	3051	3718	8577	1079	8997	332	2234
1970	1463	2789	3946	7161	1194	8243	312	2560
1971	921	2661	3717	8343	1108	7964	307	2773
1972	1237	2523	3336	6946	1009	7374	306	3629
1973	1263	2415	3505	7983	980	7731	225	2541
1974	1326	2257	2832	6832	1005	7465	226	2696
1975	1696	3045	3897	7647	1203	7719	218	2520
1976	2229	2744	3859	5702	1247	8619	216	3018

(continued)

APPENDIX AI (*continued*)

Connecticut	Rhode Island	New York	New Jersey	Delaware	Maryland	Virginia	North Carolina	Total
315	743	—	—	6	—	1	—	21,958
269	614	285	337	5	—	2	—	23,502
333	614	243	457	5	—	—	—	25,900
226	571	215	295	5	—	0	—	26,502
272	570	180	213	5	—	5	—	27,924
153	321	157	163	6	6	0	—	22,325
—	—	—	—	—	5	0	—	19,636
248	281	191	99	2	10	0	—	20,829
—	—	—	—	2	7	0	—	17,164
237	351	166	122	2	—	—	—	21,169
251	322	111	148	1	0	1	—	21,220
267	288	106	196	3	0	0	—	20,561
216	169	87	182	1	—	1	—	18,529
—	—	—	—	—	0	0	—	17,555
186	196	59	150	1	—	0	—	19,205
104	132	82	238	2	—	—	—	21,643
127	106	73	222	2	0	1	—	24,489
132	121	54	131	2	0	—	—	28,601
181	163	126	143	—	0	0	—	29,557
204	178	178	50	—	0	0	—	27,018
133	174	88	136	—	1	0	—	27,250
171	161	125	126	0	0	0	—	30,744
100	113	52	115	—	—	2	—	32,960
141	96	58	43	—	0	1	—	34,303
127	41	117	51	—	0	1	—	32,904

105	55	166	50	—	0	5	—	33,527
90	86	114	47	—	0	6	—	33,254
102	101	94	76	—	—	7	—	34,688
83	91	93	111	—	0	4	—	35,089
83	118	78	55	—	0	1	—	33,473
83	118	85	91	—	0	3	—	31,315
107	185	122	89	—	1	11	—	33,333
108	206	106	80	—	0	14	—	36,026
94	176	115	84	—	1	6	—	33,196
113	118	79	49	—	1	10	—	33,203
114	76	65	20	—	1	10	—	32,479
132	95	112	34	—	2	13	—	31,185
337	816	294	462	—	1	16	—	32,079
355	759	331	347	—	2	11	—	30,256
409	885	398	399	—	8	89	2	28,020
403	1394	530	549	—	2	51	14	31,638
424	1926	641	651	—	10	82	9	33,461
305	2357	747	833	—	6	104	2	32,022
237	2444	812	600	13	11	106	0	32,017
245	1524	519	593	10	7	401	—	29,659
247	1258	405	618	13	9	91	1	29,285
293	1420	332	502	12	17	124	—	27,339
295	1635	303	370	12	27	41	—	30,628
269	1440	269	292	12	53	32	—	30,001

[a] Anonymous, 1928–1976.
[b] 0̸, Magnitude known to be more than zero but less than half the unit or final digit used.
[c] All landings from Newfoundland to New Hampshire are inshore, trap-caught lobsters only; Massachusetts to New York from 1928 to 1964, inshore trap-caught only; Massachusetts to New York, 1965 to 1976, include both inshore and offshore catches; New Jersey catches include both trap and trawl; Delaware to North Carolina, offshore.

APPENDIX AII

Annual *Homarus gammarus* Landings by Country from 1930 to 1976[a,b]

Year	Denmark	Norway	Sweden	France	Spain	England and Wales	Scotland	Ireland	Belgium	Netherlands	All other countries
1930	122	636	229	193		564	673	374	—	23	
1931	150	745	250	383		499	673	388	—	17	
1932	167	1313	298	374		527	721	361	—	19	
1933	183	1103	299	393		498	753	325	—	32	
1934	150	1093	287	389		546	688	253		32	
1935	163	926	223	334		493	667	253	4	22	
1936	162	1037	259	352		517	653	185	5	23	
1937	170	1098	278	322		515	529	191	—	20	
1938	160	1046	239			533	546	157	5	15	
1939	134	998	197			456	—	159		13	
1940	150	892	194		40	323	376	135	0		
1941	118	856	167		⋮	297	379	126	0		
1942	191	597	148		⋮	368	502	157	0		
1943	394	463	171			430	655	173	⋮	⋮	
1944	511	384	213		18	434	691	164	⋮	⋮	
1945	883	615	245		13	513	890	179			
1946	1234	941	224		12	560	1011	183			
1947	1186	850	211		24	494	903	177			
1948	800	720	203		20	611	997	102			
1949					30	540	946	114			
1950	216	969	215	304	19	352	569	170			45
1951	157	862	252	368	29	346	466	139			37
1952	186	712	210	449	32	331	460	164			32

310

1953	145	848	216	485	37	403	461	200	37
1954	124	648	188	499	34	451	433	189	36
1955	108	632	167	497	34	508	481	253	30
1956	101	708	178	537	32	509	499	308	30
1957	74	655	148	568	53	531	527	270	35
1958	75	714	164	625	68	497	704	300	29
1959	72	684	137	401	57	491	819	347	1130
1960	85	787	168	497	37	466	889	267	30
1961	76	692	147	509	26	567	991	180	36
1962	67	555	119	437	24	470	899	167	34
1963	71	502	105	318	5	481	804	153	35
1964	50	380	92	388	23	479	793	217	50
1965	35	410	86	426	20	408	643	205	31
1966	30	312	78	446	20	408	586	278	219
1967	30	240	64	422	161	408	567	279	262
1968	24	313	66	361	99	408	616	287	258
1969	25	228	66	340	17	357	568	298	39
1970	22	210	73	324	47	510	602	277	72
1971	15	166	51	310	20	459	678	285	10
1972	16	167	54	373	16	408	585	221	49
1973	13	141	45	400	13	459	545	300	—
1974	11	139	38	351	12	408	600	253	82
1975	14	128	31	397	14	342	503	330	77
1976	12	121	40	336	21	348	528	369	⋯	77

[a] Data for some years were supplied by individual countries; for other years, Report of the working Group on *Homarus* Stocks, Shellfish and Benthos Committee, ICES, 1975, and FAO Yearbook.

[b] — none, magnitude known to be nil or zero; ∅, magnitude known to be more than zero but less than half the unit or final digit used. · · · data not available or unobtainable;

APPENDIX AIII

Total Annual Catches (10^3 tonnes) of *Nephrops* by Fishing Area and by Country from 1955 to 1976[a,b]

Country	1955	1956	1957	1958	1959	1960	1961	1962	1963	1964	1965	1966	1967	1968	1969	1970	1971	1972	1973	1974	1975	1976
Algeria	X	X	X	0.1	0.1	0.1	0.1	...	...	0.1	0.1	0	0.1	0.1	0	0.1	0.1	0	0	0	0	0.1
Belgium	0.4	0.4	0.6	0.8	0.9	0.8	0.9	0.7	0.8	0.8	0.5	0.7	0.5	0.5	0.5	0.5	0.4	0.3	0.4	0.4	0.5	0.4
Denmark	1.0	1.5	1.7	1.7	1.5	2.2	1.5	1.7	1.8	2.2	1.7	1.2	1.5	1.7	1.2	1.2	1.2	2.1	1.3	1.7	2.6	1.6
Faeroe Islands	...	...	...	...	0	0.1	0	0	0.1	0.1	...	...	...	...	...	0	0.1	0	—	—	0	0
France	5.1*	5.2*	7.4*	6.8	7.2	8.2	8.9	8.3	8.7	9.7	7.8	9.0	7.7	8.3	11.3	10.0	9.0	9.6	12.1	12.5	12.8	12.2
Germany (Federal Republic)	0.1	0.1	0.1	0.1	0.1	0.1	0.1	0.1	0.1	0.1	0.1	0	0.1	0.1	0	0	0	0	0	0	0	0
Iceland	...	...	...	0.7	1.4	2.1	1.5	2.5	5.2	3.5	3.7	3.5	2.7	2.5	3.5	4.0	4.7	4.3	2.8	2.0	2.4	2.8
Ireland	0.2	0.2	0.3	0.6	0.8	0.4	0.7	0.8	1.5	1.0	0.8	1.3	0.9	1.5	1.4	2.0	1.8	1.8	2.2	1.4	1.1	1.9
Italy	...	...	...	1.5	1.3	1.5	1.6	...	1.3	1.3	2.0	2.1	2.1	2.0	1.8	1.6	1.7	1.9	1.8	2.1	2.3	2.9
Morocco	...	...	...	0	...	...	0	0	0	0	0	0	0	0	0	—	—	0	0	0	0	0
Netherlands	...	...	...	...	...	...	...	...	...	—	—	—	—	—	—	—	—	—	—	—	—	—
Norway	0.1	0.1	0.2	0.1	0.1	0.1	0.1	0.1	0	0.1	0.2	0	0	0.1	0.1	0.2	0	0	0	0	0	0
Portugal	0.3	0.4	0	0.1	0.1	0.1	0.1	0.1	0.1	0.2	0.2	0.2	0.3	0.2	0.3	0.2	0.1	0.1[F]	0.1[F]	0	0	0
Spain	...	...	...	2.5	2.6	2.3	2.9	2.7	2.7	3.2	3.8	4.2	4.7	4.5	4.6	3.6	3.6[F]	4.3	5.6	4.4	5.7	5.3
Sweden	0.7	0.7	0.8	0.7	0.7	0.7	0.7	0.5	0.6	0.8	0.5	0.4	0.5	0.6	0.4	0.3	0.4	0.5	0.5	0.6	0.4	0.3
U.K. (England and Wales)	0.3	0.2	0.3	0.4	0.3	0.4	0.8	0.3	0.3	0.4	0.4	1.1	0.8	1.0	0.9	0.6	1.0	0.9	0.8	0.7	1.2	1.7
U.K. (Scotland)	1.1	1.1	1.4	1.1	2.2	2.0	2.9	3.5	3.7	4.9	5.2	6.3	6.7	7.2	8.2	8.2	9.0	10.8	9.8	8.3	8.2	11.0
U.K. (N. Ireland)	...	...	...	0.8	0.7	0.5	0.9	0.8	1.4	1.3	0.9	1.4	2.0	1.9	2.7	2.8	2.9	4.0	4.0[F]	2.5	3.4	3.2
Yugoslavia	0.2	0.2	0.3	0.4	0.2	0.3	0.2	0.1	0.1	0.2	0.2	0.2	0.2	0.2	0.2	0.2	0.2	0.2	0.2	0.3	0.2	0.3
Greece	0.6	0.7	...	...	...	...	...	...	...	...	...	...	...	...	...	...	...	...	...	...	...	...
Total (nearest 10^3 tonnes)	15.0	16.0	19.0	19.0	21.0	22.0	24.0	24.0	28.0	30.0	28.0	32.0	31.0	32.0	37.0	35.0	36.0	41.0	42.0	37.0	41.0	44.0

[a] From FAO (1947) and M. J. Figueiredo and H. J. Thomas (1967).

[b] ..., Data not available, unobtainable; —, none, magnitude known to be nil or zero; [F] data estimated or calculated by FAO; X, category or entry not applicable; $\emptyset$, magnitude known to be more than zero but less than half the unit or final digit used; * including Algeria.

8. The Clawed Lobster Fisheries

REFERENCES

Acheson, J. M. (1972). Territories of the lobstermen. *Nat. Hist., N.Y.* **81**(4), 60–69.
Aker, E., and Tiews, K. (1965). The German *Nephrops* fishery. *Rapp. P.-V. Reun., Cons. Int. Explor. Mer* **156**(27), 147–149.
Anonymous (1928–1976). "Fishery Statistics." U.S. Dept. of Commerce, NOAA, NMFS, Washington, D.C.
Anonymous (1952). "Practical Hints for Lobster Fishermen." Scottish Home Department, Edinburgh.
Anonymous (1969). Steel lobster pots go to sea. *New Englander* May, p. 13.
Anonymous (1975). Environment Canada, Fisheries & Marine Service "Lobster Fishery Task Force," Final Rep., pp. 51–60.
Anonymous (1976). Scientist predicts world cold spell for 20 years, UPI clipping, Nov. 22. Portland Maine Press Herald.
Bell, F. W. (1972). Technological externalities and common property resources: An empirical study of the U.S. northern lobster fisher. *J. Polit. Econ.* **80**, 148–158.
Boeck, A. (1869). Om det norske hummerfiske og dets historie. *Tidsskr.* Fisk. **3**, die Aargangs, Kjobenhavn, pp. 28–43 and 145–189.
Buckland, F., and Walpole, S. *et al.* (1877). "Reports on the Crab and Lobster Fisheries of England and Wales, of Scotland and of Ireland." H.M. Stationery Office, c. 1695, London.
Chapman, C. J., Johnstone, A. D. F., and Rice, A. L. (1975). The behavior and ecology of the Norway lobster, *Nephrops norvegicus* (L.). *Proc. Eur. Mar. Biol. Symp., 9th*, pp. 59–74.
Cobb, J. N. (1901). The lobster fishery of Maine. *Bull. U.S. Fish Comm.* **19**, 257.
Cooper, R. A. (1970). Retention of marks and their effects on growth, behavior, and migration of the American lobster, *Homarus americanus*. *Trans. Am. Fish. Soc.* **99**, 409–417.
Cooper, R. A., Clifford, R. A., and Newell, C. D. (1975). Seasonal abundance of the American lobster, *Homarus americanus*, in the Boothbay Region of Maine. *Trans. Am. Fish. Soc.* **104**(4). 669–674.
DAFS (1964). Norway lobsters. *Rep. Fish Scotl.* p. 83.
DAFS (1969). *Nephrops* (Norway lobsters). *Rep. Fish. Scotl.* pp. 26–27.
Dannevig, A. (1951). Lobster and oyster in Norway. *Rapp. P.-V. Reun., Cons. Int. Explor. Mer* **128**, 92–96.
DeWolf, G. A. (1974). The lobster fishery of the Maritime Provinces: Economic effects of regulations. *Bull., Fish. Res. Board Can.* **187**, 1–59.
Dow, R. L. (1955). "Lobster Maximum Size Restrictions." Dept. of Sea and Shore Fisheries, Augusta, Maine. unnumbered report.
Dow, R. L. (1964). Supply, sustained yield, and management of the Maine lobster resource. *Commer. Fish. Rev.* **26**(11a), 19–26 (Sep. No. 716).
Dow, R. L. (1966a). "Report on Live Lobsters, IKOFA - Munich." Maine Dept. of Marine Resources, Augusta.
Dow, R. L. (1966b). Limitations on measurement of effort-yield in the Maine lobster fishery. Vol. 5 No. 8 pages 32–36, *Fish. News Int.*
Dow, R. L. (1969a). "Lobster Culture," Fish. Circ. No. 23. Maine Dept. of Marine Resources, Augusta.
Dow, R. L. (1969b). Cyclic and geographic trends in seawater temperature and abundance of American lobster. *Science* **164**, 1060–1063.
Dow, R. L. (1971). Changes in the abundance of the Maine lobster resource with sea temperature fluctuations and increases in fishing effort. *Mar. Technol. Soc. Meet., Fish Expo '71* 1–8.
Dow, R. L. (1976). Yield trends of the American lobster resource with increased fishing effort. *Mar. Technol. Soc. J.* **10**, 17–25.

Dow, R. L. (1977a). Maine's maximum size lobster law. *Natl. Fisherman*, July.
Dow, R. L. (1977b). Relationship of sea surface temperature to American and European lobster landings. *J. Cons., Cons. Int. Explor. Mer* **37**, 186–190.
Dow, R. L. (1978). Effects of sea surface temperature cycles on landings of American, European, and Norway lobsters. *J. Cons., Cons. Int. Explor. Mer.* **38**(2) 271–272.
Dow, R. L. and Trott, T. T. (1956). "A Study of Major Factors of Maine Lobster Production Fluctuations." Dept of Marine Resources, Augusta, Maine.
Dow, R. L., Harriman, D. M., and Scattergood, L. W. (1959). The role of holding pounds in the Maine lobster industry. *Commer. Fish. Rev.* **21**(5), 1–14.
Dow, R. L., Harriman, D. M., Pontecorvo, G., and Storer, J. (1961). "The Maine Lobster Fishery." United States Fish and Wildlife Service, Washington, D.C. (unpublished).
Dow, R. L., Goggins, P. L., Harriman, D. M., and Hurst, J. W., Jr. (1962). "The Lobster Resource of Maine." Maine Dept. of Sea and Shore Fisheries, Augusta (unpublished).
Dow, R. L., Bell, F. W., and Harriman, D. M. (1975). Bioeconomic relationships for the Maine lobster fishery with consideration of alternative management schemes. *NOAA Tech. Rep., NMFS SSRF* **683**.
Dybern, B. I. (1973). Lobster burrows in Swedish waters. *Helgol. Wiss. Meeresunters.* **24**, 401–414.
Dybern, B. I. (1977). Cooperative Research Report. ICES, Charlottenlund, Denmark (personal communication).
Early, J. C. (1965). "Processing Norway Lobsters," Torry Advis. Note No. 29. Torry Research Station, Aberdeen, Scotland.
Farmer, A. S. D. (1975). Synopsis of biological data on the Norway lobster, *Nephrops norvegicus* (Linnaeus, 1758). *FAO Fish. Synopsis* No. 112.
Figueiredo, M. J., and Barraca, I. F. (1963). Contribuição para o conhecimento da pesca e da biologia do lagostim (*Nephrops norvegicus* L.) na costa portuguesa. *Notas Estud. Inst. Biol. Mar., Lisb.* No. 28, pp. 1–44.
Figueiredo, M. J., and Thomas, H. J. (1967). *Nephrops norvegicus* (Linnaeus, 1758) Leash—a review. *Oceanogr. Mar. Biol.* **5**, 371–407.
Flowers, J. M., and Saila, S. B. (1972). An analysis of temperature effects on the inshore lobster fishery. *J. Fish. Res. Board Can.* **29**, 1221–1225.
Fontaine, B., and Warluzel, N. (1969). Biologie de la langoustine du Golfe de Gacogne *Nephrops norvegicus* (L.). *Rev. Trav. Inst. Peches Marit.* **33**(2), 223–246.
Gibson, F. A. (1971). Catch and effort in the Irish lobster/crawfish fisheries 1951–1969. *Shellfish Benthos Comm. ICES C.M.* **K:4**.
Gulland, J. A. (1971). (Comp.) "The Fish Resources of the Oceans." Fishing News (Books) Ltd., West Byfleet, Surrey; ref. ed. of *FAO Fish. Tech. Pap.* **97**, 1–425 (1970).
Hall, D. C. (1972). Common property resources: A comment on an empirical application to the U.S. northeastern lobster fishery. Graduate Thesis (on file: Dept. of Marine Resources, Augusta, Maine).
Hancock, D. A. (1974). Attraction and avoidance in marine invertebrates—their possible role in developing an artificial bait. *J. Cons., Cons. Int. Explor. Mer.* **35**(3), 328–331.
Harriman, D. M. (1952). "Progress Report on Lobster Tagging—1951–1952." Maine Dept. of Marine Resources, Augusta.
Hepper, B. T. (1971). An apparent relationship between catch per unit of effort and temperature in the English lobster fishery. *Shellfish Benthos Comm., ICES C.M.* **K:8**.
Hewett, C. J. (1974). Growth and moulting in the common lobster (*Homarus vulgaris* Milne-Edwards). *J. Mar. Biol. Assoc. U.K.* **54**, 379–391.

Hillis, J. P. (1972). Studies on the biology and ecology of the Norway lobster *Nephrops norvegicus* (L.) in Irish waters. Ph.D. Thesis, National University of Ireland.

Holthuis, L. B. (1974). The lobsters of the superfamily Nephropidea of the Atlantic Ocean (Crustacea:Decapoda). *Bull. Mar. Sci.* **24**(4).

Huq, A. M., and Hasey, H. I. (1972). "Draft of Final Report, Socio-economic Impact of Changes in the Harvesting Labor Force in the Maine Lobster Fishery," Manpower Res. Proj. University of Maine, Orono.

Jenkins, R. J. F. (1972). Metanephrops, a new genus of late pliocene to recent lobsters (Decapoda, Nephropidae). *Crustaceana* **22**, 161–117.

Jensen, A. J. C. (1959). Norway lobster in the Skagerrak and northern Kattegat. *Ann. Biol.* **14**, 212–4.

Jensen, A. J. C. (1962). The *Nephrops* in the Skagerrak and Kattegat (length, growth, tagging experiments and changes in stock and fishery yield). *ICES C.M.* **36**, 1–7 (unpublished).

Jensen, A. J. C. (1967). The Norway lobster, *Nephrops norvegicus*, in the North Sea, Skagerrak and Kattegat. *Mar. Biol. Assoc. India, Symp. Ser.* **2**(4), 320–327.

Krouse, J. S. (1976). Incidence of cull lobsters, *Homarus americanus*, in commercial and research catches off the Maine coast. *Fish. Bull.*, **74**, 719–724.

McLeese, D. W., and Wilder, D. G. (1958). The activity and catchability of the lobster (*Homarus americanus*) in relation to temperature. *J. Fish. Res. Board Can.* **15**(6), 1345–1354.

Pope, J. A., and Thomas, H. J. (1965). A summary of Scottish comparative fishing experiments on Nephrops norvegicus (L.). *Rapp. P.-V. Reun., Cons. Int. Explor. Mer* **156**(36), 190–201.

Prudden, T. M. (1962). "About Lobsters." Bond Wheelwright Co., Freeport, Maine.

Rittgers, J. C. (1973). "Allocation of Fishing Rights in Fisheries Management." Natl. Mar. Fish. Serv., Gloucester, Massachusetts (ms.).

Roe, R. B. (1966). Potentially commercial Nephropsids from the Western Atlantic. *Trans. Am. Fish. Soc.* **95**(1), 92–98.

Rosier (1605). "Narrative of Weymouths' Voyage to the Coast of Maine in 1605." Eastern Times Press, Bath, Maine (1860).

Samuelson, P. A. (1964). "Economics—An Introductory Analysis," 6th ed., pp. 396–398. McGraw-Hill, New York.

Schroeder, W. D. (1959). The lobster, *Homarus americanus*, and the red crab, *Geryon quinquedens*, in the offshore waters of the western North Atlantic. *Deep-Sea Res.* **5**, 266–282.

Sheldon, W. W., and Dow, R. L. (1975). Trap contributions to losses in the American lobster fishery. *Fish. Bull.* **73**(2), 449–451.

Simpson, A. C. (1958). The lobster fishery of Wales. *Fish. Invest., Ser. 2* **22**, No. 3.

Skuladottir, U. (1965). The *Nephrops* fisheries of Iceland. *Shellfish and Benthos Comm., ICES C.M.* No. 63.

Snieszko, S. F., and Taylor, C. C. (1947). A bacterial disease of the lobster (*Homarus americanus*). *Science* **105**, 500.

Sterl, B. S. (1966). "Preliminary Lobster Study in Cousins Island Cove." Maine Dept. of Marine Resources, Augusta, Maine.

Stewart, P. A. M. (1974). Norway lobster fishing with an electrified trawl. *Scott. Fish. Bull.* No. 41.

Symonds, D. J. (1971). The *Nephrops* fisheries of England and Wales. *ICES C.M.* **K:18**, 1–10 (unpublished).

Taylor, C. C., Bigelow, H. B., and Graham, H. W. (1957). Climatic trends and the distribution of marine animals in New England. *Fish. Bull.* **57**, 293–345.

Thomas, H. J. (1951). Fluctuations in the lobster (*Homarus vulgaris*) population of the Scottish coast. *Rapp. P.-V. Reun., Cons. Int. Explor. Mer* **128**, 84–91.

Thomas, H. J. (1960). *Nephrops*. 1. The commercial fishery for Norway lobsters in Scotland. *ICES C.M.* **177**, 1-8 (unpublished).

Thomas, H. J. (1969). Recent fluctuations in the Norway lobster fishery of the Firth of Forth. *ICES C.M.* **K:31**, 1-10 (unpublished).

Thomas, H. J. (1970). The *Nephrops* fisheries of the ICES area. *Shellfish Benthos Comm., ICES C.M.* **K:17**.

Thomas, J. C. (1973). An analysis of the commercial lobster (*Homarus americanus*) fishery along the coast of Maine. August 1966 through December 1970. *NOAA Tech. Rep., NMFS SSRF* **667**, 1-57.

Welch, W. R. (1977). "Monthly and Annual Means of Surface Sea Water Temperature, Boothbay Harbor, Maine, 1905 through 1977." Dept. of Marine Resources, Boothbay Harbor, Maine.

Wilder, D. G. (1953). Holding live lobsters in aerated artificial sea water. *Fish. Res. Board Can., Biol. Stn., St. Andrews, N.B., Gen. Ser. Circ.* **21**.

Willett, H. C., and Prohaska, J. T. (1963). "Long-term Solar-climatic Relationships," Final Sci. Rep., NSF Grant 14077. M.I.T. Cambridge, Massachusetts.

Wilson, J. A. (1974). "Economic Aspects of Fisheries Management—The Northern Inshore Lobster Fishery," Report on Contract No. N-043-30-72. University of Maine, Orono.

Wood, H. (1957). The Norway lobster or "prawn." *Scott. Fish. Bull.* No. 8, p. 14.

Chapter 9

Perspectives on European Lobster Management

DAVID B. BENNETT

I.	Introduction	317
II.	The European Lobster	318
	A. State of the European Lobster Stocks	318
	B. Regulations	320
	C. Management Options	323
	D. Future Management Strategy	326
III.	The Norway Lobster	327
	A. The *Nephrops* Fishery	328
	B. Regulations	328
	C. Future Management Strategy	328
	References	331

I. INTRODUCTION

For many centuries, the lobster has graced the table and delighted the palate of the European aristocracy. It cannot be denied that in Europe the lobster is a luxury food item, but it also plays an important socioeconomic role in many coastal fishing communities that depend on the European clawed lobster, *Homarus gammarus* (L.), as a major source of income.

The smaller clawed lobster, *Nephrops norvegicus* (L.), has a less glamorous image, but has developed in recent years into a major fishery with landings nearly twenty times greater than those of *H. gammarus*. The method of capture for the two species differs; *H. gammarus* is caught mainly by baited traps inshore and *N. norvegicus* is caught by trawls in the deep, offshore water.

The management of the *H. gammarus* fishery is steeped in history and provides an interesting example of the considerable efforts of fishery scientists and administrators to provide rational management. The Norway lobster, *N. norvegicus,* is sometimes known as the Dublin Bay prawn or as scampi when listed on a menu, but it is generally called *Nephrops*. This lobster seems to have survived until recently with the minimum of regulations. The present management problems of this fishery lie in its effect on the recruitment of finfish, as the small-mesh *Nephrops* trawls catch considerable quantities of juvenile gadoids. Because the two species of lobsters have such different management problems they will be dealt with separately in this chapter.

II. THE EUROPEAN LOBSTER

Fisheries for the European lobster, *Homarus gammarus,* have existed for several hundred years. In the early years, stock abundance was high and exploitation rates were low, but eventually the need for management regulations was recognized. Thus, some of the existing regulations have their origins in the nineteenth century. Despite this long historical experience, the present-day regulations are few in number and fairly simple in their application. This section summarizes the state of the European stocks, considers the existing regulations, and discusses their rationale, and looks at possible future management strategy.

A. State of The European Lobster Stocks

In this chapter, a brief introductory overview is provided. For a full discussion of the topic, see Chapter 8 of this volume.

The total recorded catch of the European lobster is around 1900 t, which is about 6% of the catch of the American lobster (ICES, 1978). In Europe, the catches from the well-established fisheries in Norway, Sweden and Denmark have fallen alarmingly from a combined annual average of 1059 t in 1950-1959 to 201 t for 1970-1977 (Table I). The catches from the east coasts of Scotland and England, Orkney and Shetland, have also decreased substantially. However, the overall landings in some countries have been maintained or increased by the discovery of previously unexploited stocks (e.g., English Channel, west coast of Scotland) or by an increase in fishing effort (e.g., Ireland).

The overall catch is being maintained in some countries by the development of offshore fisheries, which fish previously unexploited stocks up to 50 miles off the mainland. These stocks when first fished, give a high catch per unit effort and

TABLE I

European Lobster (*H. gammarus*) Landings[a,b]

Year	Denmark	England and Wales	France	Ireland	Norway	Scotland	Spain	Sweden	All European countries
1970	22	491	324	277	202	602	47	71	2108
1971	15	451	310	285	133	678	20	50	1952
1972	16	429	373	221	161	585	16	43	1893
1973	13	457	420	258	150	545	13	42	1898
1974	11	377	336	253	140	600	12	38	1825
1975	14	382	385	330	127	503	14	36	1826
1976	12	383	328	369	121	528	29	41	1852
1977[c]	14	444	353	339	99	541	69	33	1911
Averages									
1950–1959	126	439	573	234	743	690	40	190	3078
1960–1969	49	442	414	233	420	736	43	99	2536
1970–1977	15	427	354	292	142	573	22	44	1899

[a] From Bulletin Statistique, ICES.
[b] Data in tonnes (t).
[c] Provisional

initially support a high fishing intensity, but they are quite quickly fished down to a catch rate that is uneconomical for the larger boats. Within 4–5 years of the discovery of the offshore English Channel grounds, many of the boats switched to the west coast of Scotland.

B. Regulations

1. Minimum Landing Size

The main lobster regulation used in Europe is the minimum landing size. This is set at 74–83 mm carapace length (CL), although in some countries total length (TL) is used as the method of measurement (Table II). None of the size limits have been chosen following a yield assessment. Some are believed to have been chosen in relation to size at maturity. However, size at first maturity is about 80 mm CL for many European stocks (ICES, 1978). It is quite clear that the management of the majority of the stocks depends on a minimum size that has little relevance to protection of the breeding stock and has not been (knowingly) set to optimize the yield per recruit.

Within Europe, there have been some changes in the minimum landing sizes in recent years. Norway's minimum size of 210 mm TL, suggested in 1737 and introduced in 1879, was increased to 220 mm in 1964. Britain increased its size limit from 8 inches (203 mm) to 9 inches (228 mm) total length in 1951, and in 1976 changed the method of measurement to carapace length at 80 mm. This latter change, already carried out in Ireland in 1963, was introduced to aid enforcement, since total length measurement was subjected to abuse by fishermen stretching the abodmen during measurement.

2. Protection of Ovigerous Females

The landing of ovigerous (berried) females is banned only in Spain and in one area of Denmark (Limfjord) and in Portugal for a short period (Table II). The use of this particular regulation has been contentious, particularly in Britain. For many years the majority of local Sea Fisheries Committees in England and Wales had bylaws prohibiting the landing of berried lobsters. These bylaws were adopted largely as a result of investigations made at the turn of the century (Meek, 1925). Eventually these bylaws were consolidated into national legislation in 1951. The northeast coast of England was convinced by Meek of the value of this measure, but other parts of the country protested and evaded the law by the practice of scrubbing the eggs off. Many fishermen have an emotional belief in the need to protect the egg-carrying female. Perhaps this psychological effect would not occur if the eggs were carried internally.

The value of this law was debated for several years until, in 1966, it was repealed. Two main reasons were given for this change of policy: (a) it was

TABLE II
European Lobster (*H. gammarus*) Regulations

Country	Minimum landing size (mm)[a] CL	Minimum landing size (mm)[a] TL	Berried protection	Closed seasons	Other regulations
England and Wales	80	—	None	None	None
Scotland	80	—	None	None	None
Northern Ireland	80	—	None	None	None
Ireland	83	—	None	None	Diving banned
France	~80	230	None	None	15 Sanctuaries, hatchery-reared juveniles released, boats licensed, lath spacing
Portugal	~80	200[b]	January 1–March 1	October 1–January 1	None
Spain	~80	200[b]	Year-round	September 1–May 31	None
Denmark	~74	210	In one area only	None	None
Germany	~84	240	None	July 15–September 1	None (?)
Norway	~77	220	None	June 1–October 1	None
Sweden	~77	220	None	June 15–September 20 (N) July 15–September 15 (S)	Diving banned, trap limitations, lath spacing

[a] CL is the carapace length, TL is the total length.
[b] Eye socket to telson.

difficult to enforce, because of scrubbing, and (b) it could not be justified as being of benefit to recruitment, since nothing was known about the stock-recruitment relationship (Thomas, 1965). The opinion expressed was that a more effective protection of the breeding stock would result by increasing the minimum landing size above the size at first maturity. As far back as 1913, the French authorities accepted the view that the administrative difficulties of enforcing legislation protecting berried lobsters were too great, and repealed their existing decree. Three areas of England have retained local bylaws prohibiting the taking of berried lobsters from a fishery. However, these bylaws only apply up to 3 miles from the shore and are difficult to enforce. In one area the main fishing season is in July when most lobsters have hatched their eggs, and, in any case, the majority of the catch is immature. Most people involved in the British lobster fishery today consider that raising the minimum size above the size at first maturity is the practical way to protect the breeding stock. It is interesting to reflect that if only the lobster were as "well adapted" as the European edible crab, this problem would never have arisen. Berried edible crabs, *Cancer pagurus* L., are rarely caught in traps by fishermen.

3. Closed Seasons

Closed seasons exist in Portugal, Spain, Germany, Norway, and Sweden (Table II). Those in Portugal and Spain are during the winter period, when fishing activity would be low anyway. The closed season in Norway is aimed at the protection of the breeding stock and recently molted lobsters. In Sweden, the closed season during the summer reduces fishing effort, particularly by hobby fishermen. A natural closed season in the winter exists in all areas due to bad weather and reduced catchability.

4. Other Regulations

The only attempt to limit fishing effort, other than by closed seasons, is in Sweden, where limits of 40 traps per one fisherman, 75 traps per two fishermen, and 35 traps per fisherman for three or more men per boat have been set.

Only in France is there a licensing scheme administered by the "Comité Interprofessionnel des Crustacés." This covers all full-time fishermen, but the number of licenses is unlimited. The main purpose of the license is the use of its suspension or withdrawal as a deterrent in the enforcement of minimum size and closed area regulations. France has pioneered the establishment of 15 sanctuaries on the Atlantic and Channel coasts (Audouin *et al.*, 1971). The purpose of these is the establishment of reservoir breeding stocks. Berried lobsters have been purchased and released into the sanctuaries, and hatchery-reared postlarvae and 1-year-old juveniles have also been released in certain areas.

Diving as a method of capture is banned in Ireland and Sweden, mainly to prevent hobby fishing and to placate full-time trap fishermen. Specially designed

9. Perspectives on European Lobster Management

escape gaps are not in use in Europe, although France and Sweden do have lath spacing regulations. (Much of the information on regulations came from a questionaire sent to colleagues on the ICES *Homarus* working group. The author wishes to acknowledge their help in providing this information).

C. Management Options

Before considering the directions in which European lobster management is at present moving, let us consider the options available to both fishery scientists and administrators.

1. Minimum Landing Size

The main European management control is the minimum landing size. The question must be asked, "Can a lobster fishery be managed by minimum size alone?" The experience of countries on both sides of the Atlantic and the present poor state of the lobster stocks clearly indicate that effective management by minimum size regulations alone is not possible. However, the control of the minimum size does have an important role in lobster management. The appropriate size should be selected in relation to size at first maturity and optimal yield per recruit.

Selection of the optimum size at first capture should ideally be made using some yield assessment model. Most such models require certain basic input parameters, in particular recruitment, growth, and mortality rates. Information on recruitment is very limited and likely to remain so for many years, a problem that is usually overcome by using yield per recruit theory. This still leaves the problems of quantifying growth and mortality rates. A considerable amount of molt-increment data are available, but such is not the case for data on molt frequency under natural conditions, which are necessary to determine annual growth rates (ICES, 1978). Persistent tagging techniques now in use are providing estimates of both growth and mortality rates. In the last few years, several European countries have been collecting the data necessary for yield assessments and optimisation of minimum landing size can be expected.

The European lobster market is an international one. Live lobsters are exported and imported between countries on a considerable scale. The majority of the Irish, Scottish, English, and Welsh catches are exported to France, Belgium, Holland, and Spain. This international traffic in lobsters could create enforcement problems if different minimum landing sizes were enforced in each country. As far as is known growth and natural mortality rates do not vary appreciably from area to area within most countries, but may vary from country to country. Fishing mortality rates vary considerably both within and between countries, and therefore, to achieve the optimum yield, different minimum sizes may be required. However, the international enforcement problems and market size requirements must be carefully considered. With such a dependence in Europe on the minimum size, effective enforcement is essential. The temptation for fishermen to land undersized lobsters could be reduced by the use of escape gaps. In addition, the release of undersized lobsters via escape gaps on the sea bed would reduce the likelihood of damage while sorting on the fishing boat, as well as possible predation losses when undersized lobsters are returned to the sea (Brown, 1978).

2. Regulation of Fishing Effort

Most European lobster fisheries are free access public fisheries. The most appropriate way of controlling fishing effort would be to make them limited entry fisheries. If some form of licensing were introduced, and fishing were controlled by restricting the number of licenses available, it could discriminate against a particular group of fishermen, e.g., part-timers or divers. In addition, a condition of the license could be the provision of catch and effort statistics, and the suspension or withdrawal of a license would be an excellent deterrent to enforce other regulations, such as minimum size.

Although the number of fishermen or boats may be controlled by restricted entry and licensing, unless trap limits are also imposed, the licensed fishermen can still increase their fishing effort by increasing the number of pots fished.

Closed seasons have very little value in directly controlling fishing effort, unless coupled with other effort controls. Where a mixed fishery for lobsters and crabs exists, as in most of Europe, closed seasons for lobsters alone are very difficult to enforce. Closed seasons during certain times of the year can be used to restrict the activities of part-time fishermen. Sweden has a closed summer season, primarily for biological reasons, but also considered valuable for stopping leisure fishing for lobsters (H. Hallbäck, personal communication). In many areas, natural closed seasons occur as a result of bad weather, and cold water temperatures induce inactivity, thus reducing catchability.

Quotas have been used extensively in European finfish fisheries, where it has been necessary to apportion a total allowable catch (TAC) between nations. Most European lobster grounds are close to the shore and within traditional exclusive fishery limits. If a quota were imposed in a national fishery, it would tend to create a short period of intense fishing at the beginning of a season when the quota would quickly be caught. It would also be difficult to know when the quota had been reached. In most countries, the lobster catch is inadequately recorded (ICES, 1978).

Selection of the appropriate level of effort to achieve the optimum yield should ideally (as for the minimum size) be made using some yield assessment model. This need not be a complex mathematical model. There are many European lobster fisheries where it is believed that a reduction in fishing effort would result in increased catch rates, therefore increasing the economic efficiency of the industry. Social considerations must play a part in decision making when controlling fishing effort, particularly where local communities are dependent on fishing for employment.

If the data on growth and mortality are not available for yield assessment models, it may be necessary to make decisions on controlling effort (and/or minimum size), which may only be judged in light of future catches. The collection of adequate catch and effort information is therefore essential, so that the effects of management changes on the fishery can be observed. Even where sufficient data exists for a population model, it is necessary to prove the model by observing catch and effort statistics.

3. Improvement of Recruitment

It would appear quite logical to set the minimum size above the size at maturity, particularly where the rate of exploitation is high. This would ensure that breeding females and mature males are available in the stock. Since male lobsters are able to mate with more than one female, the stock of mature males need not be as large as that of the females (ICES, 1978). However, no information is available on the stock-recruitment relationship, and it is thus impossible to decide how large a stock of breeding females is required for adequate annual recruitment. The question of whether ovigerous (berried) females should be landed depends, again, on the unknown stock-recruitment relationship. France and Great Britain have repealed their legislation banning the landing of berried lobsters. The necessity and wisdom of having such a regulation depends on the relationship between the minimum size and size at maturity, the rate of exploitation, the origin of larval recruitment to the stock, and the feasibility of adequately enforcing the regulation.

An alternative approach is to buy ovigerous females from fishermen and release them back into the fishery, preferably into closed sanctuary areas as in France (Audouin *et al.*, 1971). To determine the value of this action, the future catches are being monitored, and it will be interesting but difficult to evaluate this approach. A simpler method involves closing off areas of the fishery to allow a large unfished breeding stock to build up. If larval drift occurs to other areas or juveniles disperse, this would increase recruitment to the fishable stock.

The releasing of hatchery reared juveniles into the fishery or sanctuaries has been tried in several European countries, in an attempt to overcome the high mortality occurring between the hatching of eggs and the settling of juveniles onto the sea bed (Thomas, 1964). Only France still continues with hatchery-reared releases. There are many problems in trying to evaluate the usefulness of this approach. It has not been possible to demonstrate an increase in commercial landings as a result of releasing juveniles.

4. Habitat Improvement

Improvement of the habitat by the construction of artificial reefs has been suggested many times (Unger, 1966). The main problem with artificial reefs is cost effectiveness. There may be some value in establishing artificial reefs in areas where the habitat is not suitable for lobsters. In most areas where fishing intensity is high and stock abundance is low, the availability of suitable habitat would not appear to be a limiting factor.

The ICES *Homarus* Working Group (ICES, 1978) recognized that there was relatively little information available on the effects of various pollutants on adult, juvenile, and larval lobsters. It was felt that the larvae are probably more sensitive to pollutants than the adults. Although it has not been possible to demonstrate any significant mortality due to pollution, the Group felt that more toxicological work, particularly with the larvae, would enable a closer assessment of the possible effects of pollutants on lobster stocks.

D. Future Management Strategy

The present-day lobster regulations in Europe have evolved slowly, and in many cases without a clear conservation or management rationale. Recent increases in the level of fishing effort and efficiency have outstripped the usefulness of regulations brought into operation many years ago. In many European countries, administrators have waited for fishery scientists to take the initiative and offer effective management advice based on sound scientific studies. The application of the principles of mathematical modeling to crustacean fisheries is many years behind that for finfish. Crustacean biologists have faced particular difficulties in attempting yield assessments. The use of population dynamics has been hindered, not by the problem of formulating suitable yield models, but by the difficulty of obtaining reliable estimates of the main parameter inputs to a simple yield model. This is due to the fact that crustaceans grow by ecdysis, which results in discontinuous growth, an apparent lack of aging structures, difficulties in distinguishing year classes or molt classes from polymodal size frequency distributions, and the necessity of developing persistent tags.

In recent years, more growth and mortality data have become available, and crustacean fishery scientists are beginning to carry out stock assessments that enable them to offer advice on the management of lobster stocks (ICES, 1978). Within certain constraints regarding the probable range of some parameter inputs to yield per recruit models, it is possible to conclude that for the majority of European lobster stocks, the size at first capture is too low and the fishing mortality too high. Although the stock-recruitment relationship is not known, common sense suggests that in highly exploited fisheries the minimum size should be set above the size of maturity.

The immediate short-term losses in catches resulting from increases in minimum landing size, and particularly any reduction in fishing effort, will be difficult to accept in socioeconomic terms. This chapter began by drawing attention to the fact that although lobsters are a luxury food, many fishing communities within Europe depend on the income generated from lobster fishing. Most fisherman, merchants, consumers, scientists, and administrators recognize that many European lobster stocks are overfished. The long-term future of lobster fishing in Europe is bleak unless all involved accept the fact that conservation and effective management hurts in the initial stages. Some short-term sacrifices will have to be made to ensure the future viability of lobster fishing.

Future management stragegy for the European lobster can be expected to include a general increase in size at first capture. In addition to optimizing yield per recruit, it will be necessary to consider the need to protect the breeding stock by choosing a minimum size above the size at first maturity. The effectiveness of the minimum size regulation could be enhanced by the use of escape gaps.

The managers of lobster stocks in Europe will, in the near future, have to face the problem of excessive fishing effort. Most European lobster fisheries have unrestricted free access. The lobster fishery attracts large numbers of part-time and holiday fishermen, including scuba divers, as well as the full-time fishermen. They have been encouraged by the relatively low capital investment needed to start inshore lobster fishing. There is a common call from full-time fishermen to ban or at least severely restrict the activities of those who do not depend on the sea for their livelihood. If effort control is going to be a feature of our future lobster management strategy, then there is a need for someone, i.e., the politician, to make a choice between commercial and recreational fishing. In many areas, the catch per unit effort is so low that only part-time fishermen can now fish economically. Excessive fishing capacity is common to nearly all European fisheries. The uncertainties following the extension of fishing limits to 200 miles and the failure of the European Economic Community (EEC) to formulate an agreed common fisheries policy have only made the problem worse. The efficiency of the modern fisherman, with his numerous technological inovations, is too much for the lobster. An effective management strategy for lobsters in Europe must include fishing effort control. The introduction of limited entry licensing, together with controls such as pot limits, must be coupled with the traditional use of minimum landing size. The regulations of the nineteenth century are no longer appropriate. Modern fisheries research is beginning to provide the scientific justifications for the stringent management measures we all know are necessary to provide a secure future for European lobster fisheries.

III. THE NORWAY LOBSTER

In contrast with the fisheries of the European lobster, *Homarus gammarus*, the *Nephrops* fisheries in Europe have expanded considerably in the last 25 years.

This is mainly a trawl fishery, and although other measures of control are exercised, fishery management is primarily achieved by mesh size regulation.

The *Nephrops* fishery faces two distinct management problems that could require different or opposing regulations. There is a need to ensure that the optimum sustainable yield is being obtained from the *Nephrops* fishery, which so far has apparently withstood an increasing fishing mortality. Conversely, there is considerable concern that small mesh fisheries, like *Nephrops*, are adversely affecting finfish stocks by catching large quantities of juvenile fish. Reaching a balance between these objectives could be difficult.

This section briefly reviews the development of the *Nephrops* fishery in recent years, and discusses the current conflict between *Nephrops* fishing and finfish conservation.

A. The *Nephrops* Fishery

Since 1950, *Nephrops* landings in most European countries have increased dramatically (Table III). The European *Nephrops* catch averaged nearly 11,000 t during the 1950s, over 25,000 t in the 1960s, and now averages 36,000 t. This expansion has been enjoyed by most countries with major fisheries and is attributable to the increased demand for "scampi" on the menu of European restaurants. Thus, many new fishing grounds have been exploited with the increasing fishing effort.

Little is known about the state of exploitation of the *Nephrops* stocks. Where catch per unit effort data are available, they fluctuate widely without any obvious trends (ICES, 1976). In Iceland, where data are available from the start of the fishery, there has been a decrease in catch per hour fishing.

B. Regulations

For many years, *Nephrops* was a by-catch of finfish fisheries, and little attempt was made to directly manage *Nephrops*. Some countries have minimum mesh size regulations, reinforced by minimum landing sizes. However, most of these regulations have been chosen without the use of yield assessment techniques. Trawls with a mesh size below that permitted for protected finfish are allowed in many *Nephrops* fisheries. *Nephrops* regulations have been based on those advised by the North East Atlantic Fisheries Commission (NEAFC). They have been discussed and tabulated in Chapter 8 of this volume.

C. Future Management Strategy

Although the main stocks of *Nephrops* occur within national waters, some stocks are subject to international exploitation. When *Nephrops* was a bycatch in the finfish fisheries, there was little concern for its management. However, with

TABLE III

European *Nephrops* Landings[a,b]

Year	Denmark	England and Wales	France	Iceland	Ireland	Northern Ireland[c]	Spain	Scotland	All European countries
1970	1,244	612	10,022	4,026	2,019	2,107	3,234	8,179	32,491
1971	1,233	1,044	9,025	4,657	1,775	2,190	3,231	9,029	33,148
1972	2,096	948	9,581	4,321	1,823	2,998	3,759	10,780	37,207
1973	1,339	814	12,098	2,791	2,150	2,732	4,530	9,780	37,233
1974	1,734	669	12,549	1,983	1,380	2,489	3,486	8,319	33,772
1975	2,613	1,157	12,828	2,357	1,057	3,439	4,715	8,223	37,414
1976	1,643	1,733	12,194	2,780	1,909	3,217	3,372	10,906	38,646
1977[d]	1,265	2,072	10,798	2,723	2,866	3,163	4,698	9,815	38,321
Averages									
1950–1959	1,021	215	5,197	—	214	185	1,748	740	10,883
1960–1969	1,665	625	8,554	3,117	1,025	772	2,873	5,069	25,565
1970–1977	1,646	1,131	11,137	3,205	1,872	2,792	3,878	9,379	36,029

[a] Bulletin Statistique, ICES.
[b] Data in tonnes (t).
[c] Landings corrected for wrong conversion factor from tail to total weight before 1976.
[d] Provisional

the increasing exploitation of *Nephrops*, it has become important in its own right, and there is now a need to manage it effectively. At the same time, with the pressure on fish stocks resulting from the loss to many European countries of distant water grounds and excessive catching capacity, fishery scientists have become more concerned about the effects of the small mesh fisheries on finfish recruitment. *Nephrops* fishing is among a number of fisheries in Europe using small mesh nets. There is considerable industrial fishing for reduction to fish meal of the Norway pout (*Trisopterus esmarkii*) and sandeels (*Ammodytes* sp) (Macer, 1974), and also smaller fisheries for shrimps (*Crangon crangon, Pandalus borealis*) that use small mesh nets that catch juvenile fish.

Nephrops has presented crustacean fishery scientists with similar problems to those encountered with the European lobster, *Homarus gammarus*. Discontinuous growth and the lack of aging structures have made it difficult to apply fish stock assessment techniques. In addition, since *Nephrops* is caught by trawls, selectivity is important. The rather irregular shape of *Nephrops* makes estimation of the selection factor difficult (Garrod, 1976). This has hampered attempts to use cohort analysis on length to carry out a yield assessment (ICES, 1976). The ICES *Nephrops* Working Group (ICES, 1978) has made estimates of the immediate and long-term effects of changes in mesh size on the catch per recruit. An increase in mesh size to 70 mm would result in short-term losses. However, despite uncertainties about growth, selectivity, and natural mortality estimates, it is considered that an increase in yield per recruit is possible in many fisheries.

Thus, preliminary yield assessments now exist, which provide the scientific advice necessary for the management of European *Nephrops* fisheries in their own right. However, a major concern at this time is the question of gadoid bycatches. With the long experience (compared with crustacea) of fish mesh assessments behind them, most fishery scientists agree that an increase in the minimum mesh size for *Nephrops* trawls will enhance recruitment to fish stocks. In the Irish Sea, it has been estimated that the number of juvenile whiting (*Merlangius merlangus*) discards from *Nephrops* fishing is comparable to the recruitment to the directed whiting fishery for human consumption (K. M. Brander, personal communication). Thus, an increase in *Nephrops* mesh size in the Irish Sea would significantly improve the whiting fishery. Similar arguments are being put forward to reduce industrial fishing for Norway pout to protect juvenile haddock (*Melanogrammus aeglefinus*).

An alternative or adjunct to increases in mesh size are the attempts to develop trawls that retain their efficiency to catch *Nephrops*, but are less effective in catching fish or at least selectively release them. Interest has been shown in wingless trawls that limit the ''herding'' effect on fish and in trawls with larger meshes in those parts of the net where fish are more likely to escape (Food and Agriculture Organization, 1973).

The future management strategy for *Nephrops* fisheries in Europe will depend

9. Perspectives on European Lobster Management

to a great extent on the degree of pressure applied by finfish interests to reduce discards of juvenile fish. Some increases in mesh size for *Nephrops* trawls can be expected. This may be backed up by minimum landing sizes and possibly limits on the proportion of finfish by-catches that may be landed. Another approach being considered is the possibility of defining *Nephrops* grounds and restricting the use of small mesh *Nephrops* trawls to such areas.

Nephrops fisheries have expanded in recent years and there has been little concern for their conservation. Yield assessment techniques are now being applied and scientific management advice for *Nephrops* is available. However, with most finfish fisheries being fully or over exploited, particularly those for human consumption, there is considerable pressure to reduce the effects on fish recruitment of small mesh fisheries like *Nephrops*. In the future, it is possible that the maximum yield of *Nephrops* in some areas may be sacrificed to improve finfish recruitment.

REFERENCES

Audouin, J., Campillo, A., and Leglise, M. (1971). Les cantonnements à crustaces des côtes Françaises de l'Atlantique et de la Manche. *Sci. Pêche, No.* **205,** pp. 1–9.

Brown, C. G. (1978). Trials with escape gaps in lobster and crab traps. *ICES CM 1978/***K:7,** pp. 1–6 (mimeo.).

Food and Agriculture Organization (1973). Report of the expert consultation on selective shrimp trawls, Ijmuiden, The Netherlands, 12–14 June 1973. *FAO Fish. Rep.* **139,** 1–71.

Garrod, D. J. (1976). Mesh selection of *Nephrops. Fish. Res. Tech. Rep., MAFF Direct. Fish. Res., Lowestoft No.* **26,** pp. 1–9.

ICES, (1975). Procès-verbal de la Réunion 1974. *Cons. Int. Explor. Mer: P-v.,* pp. 1–146.

ICES (1976) Report of the working group on *Nephrops* stocks. *Coop. Res. Rep., Int. Counc. Explor. Sea No.* **55,** pp. 1–42.

ICES (1978). ICES Crustacean working groups' reports 1977. *Coop. Res. Rep., Int. Counc. Explor. Sea No.* **85,** pp. 1–107.

Macer, C. T. (1974). Industrial fisheries. pp. 193–221. *In* "Sea Fisheries Research" (F. R. Harden Jones, ed.) Elek Science, London, 510 pp.

Meek, A. (1925). Experimental legislation with reference to the crab and lobster fisheries of the east coast of Britain. *J. Mar. Biol. Ass. UK.* **13(4),** pp. 755–768.

Thomas, H. J. (1964). Artificial hatching and rearing of lobsters - a review. *Scott. Fish. Bull.* **21,** pp. 6–9.

Thomas, H. J. (1965). Berried lobsters. *Scott. Fish. Bull.* **23,** pp. 10–12.

Unger, I. (1966). Artificial reefs - a review. *Am. Litt. Soc., Spec. Publ. No.* **4,** pp. 1–74.

Chapter 10

Aquaculture

JON C. VAN OLST, JAMES M. CARLBERG,
AND JOHN T. HUGHES

I.	Introduction	333
II.	Culture of Palinurid and Scyllarid Lobsters	335
III.	Culture of Nephropid Lobsters	337
	A. Biological Information	339
	B. Procedures for Artificial Rearing	360
	C. Economics	374
IV.	Conclusions	378
	References	378

I. INTRODUCTION

In 1978, the world population consumed 62% of the estimated maximum sustainable yields for finfish. Fisheries scientists of the United Nations estimate that by 1985, the world demand will outstrip maximum world supply potential in all major fisheries. Of the several possibilities for increasing the supply of seafood commodities, perhaps the most imaginative strategy is by means of aquaculture, the farming and husbandry of freshwater and marine organisms. The conventional sources of terrestrial protein produced by agriculture and animal husbandry have resulted from man's ability to domesticate and breed suitable species. Similarly, man is beginning to manage the aquatic environment by cultivating certain species of aquatic plants and animals high in food value to supplement natural production.

Production and release of juvenile aquatic organisms in order to augment the fishery is a form of aquaculture. Between 1885 and 1920, more than 22 lobster hatcheries were established in New England, the Maritime Provinces of Canada, and in Europe (Carlson, 1954). Efforts were made in the culture of both the American lobster, *Homarus americanus,* and the European lobster, *H. gammarus.* The objective was to stock the fishing grounds with hatchery-reared juveniles whose larval stages were cultured under optimal conditions, e.g., absence of predation, appropriate food, controlled temperature, and disease treatment. Bowers (1900) and Scattergood (1949) reported that larval culture of American lobsters was first attempted in the 1870s by a method known as "parking," in which lobster larvae were reared in naturally enclosed basins. By 1885, the U.S. Fish Commission had constructed a hatchery at Woods Hole, Massachusetts, where eggs were stripped from gravid females and hatched artificially (Galtsoff, 1937). Research in Canadian and in Rhode Island hatcheries revealed that eggs could be maintained and hatched more efficiently by the female than by any artificial contrivance (Carlson, 1954). In 1949, the Massachusetts State Lobster Hatchery and Research Station began culturing and releasing fourth stage larvae. Research there by John T. Hughes has concentrated on refinement of hatchery technology. It is difficult to evaluate the effects on the fishery of releasing artificially propagated larvae. Determination of the extent to which hatchery-reared lobsters contribute to the natural population and the fishery catch awaits results from tagging studies and detailed ecological investigations. As a result, the restocking programs of both Europe and North America have been reduced considerably. Currently, one hatchery is operating in the United States and one in France.

A second method of increasing yields from the lobster fishery is to extend the fishing grounds by transplanting lobsters to other geographical areas with suitable habitats. Lobsters have been transplanted for their food value, for aquarium use and public display, and accidentally. Studies have been conducted to determine the feasibility of transplanting the American lobster. In British Columbia, Canada, a small, viable population was temporarily established and the biological feasibility of relocating the American lobster was demonstrated (Ghelardi and Shoop, 1972).

Hatchery and transplantation programs are forms of aquaculture in that they involve a degree of manipulation of the natural population and some culturing methods. However, the complete, controlled culture of lobsters is still in its infancy. Since 1885, lobster culturists have been preoccupied with the rearing of larvae for subsequent release. Beyond these efforts, the source of lobster supply has been the harvesting of the sea. The lobster industry has been in the hunter stage rather than in the farmer stage.

Consumers are willing to pay more per pound for lobsters than for almost any other major foodstuff. In 1975, world consumption was approximately 394 million pounds (Bell and Fullenbaum, 1973). As lobster fisheries throughout the world encounter decreases in harvest and catch per unit effort, it is not surprising that this has evoked considerable interest in developing methods for domesticating and culturing lobsters. Unfortunately, the complex biological requirements of aquatic organisms in general, and of lobsters in particular, make this task more difficult than the domestication and rearing of terrestrial animals such as poultry and livestock. Presently there are no commercially viable lobster farming operations anywhere in the world. However, there have been many recent advances in research laboratories in the United States, Canada, Western Europe, Australia, and Japan, which have greatly increased the knowledge of the biology of lobsters and are leading to the eventual commercial culture of these species.

There are several essential requirements for the commercial farming of an aquatic organism (Cobb, 1976). These include an adequate consumer demand and profit potential for the species, the ability to reproduce in captivity, simple larval development, high food conversion efficiency, and resistance to disease. The American lobster, *H. americanus*, meets several of these requirements and has therefore been selected by the U.S. National Oceanographic and Atmospheric Administration as one of four marine animals having "high priority" for aquaculture research and development (Glude, 1977). Less biological information is available for other lobster species such as the palinurids. Consequently, commercial culture of these species does not appear to be promising at this time.

II. CULTURE OF PALINURID AND SCYLLARID LOBSTERS

The spiny lobsters and the sand or slipper lobsters have several characteristics that make them attractive for commercial cultivation. Most of these species have an intermediate stage between larval and juvenile development. These puerulus stages of the spiny lobster have been captured at sea and cultured successfully by several investigators (Kensler, 1967; Provenzano, 1968; Chittleborough, 1974a; Serfling and Ford, 1975a; Phillips *et al.*, 1977). Chittleborough succeeded in rearing *Panulirus cygnus* from the puerulus stage to reproductive maturity in approximately five years. Blecha (1972) cultured pueruli of *P. interruptus* on a diet of mussels, *Mytilus edulis,* which appeared to provide adequate nutrition. Very few disease problems have been encountered, and even under conditions of high density and crowding there is little aggression and cannibalism. However, the major obstacle to the successful culture of these species is the extremely long developmental period and the poorly understood requirements of the phyllosoma larvae. No research group has succeeded in culturing a palinurid lobster from the egg to puerulus stage. Suitable larval diets are difficult to obtain, and the need for high water quality and high temperature has further complicated culture efforts.

A number of investigators have been successful in hatching palinurid lobsters, and in many cases have raised the phyllosoma larvae through several molts in the laboratory. Efforts to culture the larvae of *P. interruptus* were first begun in California in 1911. European and Soviet scientists have had similar limited success with *Palinurus elephas* in tanks and nursery ponds (Bardach et al., 1972). At the University of Miami, the larvae of *Panulirus argus* were raised through nine molts, and Japanese workers in southern Kyushu have reared larvae of *P. japonicus* in suspended cages to seventh stage (Idyll, 1970). *Jasus lalandii* has been reared through the fifth molt in South Africa. Other palinurid species, which have been successfully hatched, but cultured through only the first few stages include *P. inflatus* in California, *P. longipes* in Japan, and *P. polyphagus* in Malaysia (Bardach et al., 1972). Researchers in the Sinai Peninsula are attempting to culture *P. penicillatus* phyllosoma in the laboratory and juvenile stages in wire cages suspended in the sea (E. Gilead, personal communication).

In Japan, brine shrimp (*Artemia salina*) nauplii have been used as food for phyllosoma. Larvae of *P. japonicus* survived for 178 days and passed through 16 molts on this diet (Saisho, 1966). Recently these phyllosomes were successfully reared in the laboratory to the last phyllosoma stage (approximately 30 mm total length) in 253 days, on a diet of nauplii and adult *Artemia*, adult *Sagitta* spp., and fish fry (Inoue, 1978). Dexter (1972) reared *P. interruptus* through six of the phyllosomal stages in 114 days, also on a diet of brine shrimp nauplii. Apparently this diet is only suitable for the early phyllosomes and is less useful in the latter stages, perhaps because of the small size of the nauplii (Provenzano, 1968).

The scyllarid lobsters are of little commercial importance to the lobster fishing industry. However, they possess some characteristics that may make them more suitable for culture than the palinurids. Often they exhibit shorter and less complex larval development. The scyllarids *Ibacus ciliatus*, *I. novemdentatus*, *Parribacus antarcticus*, and *Scyllarus bicuspidatus* have been hatched in Japan and *S. sordidus* in India (Bardach et al., 1972), although none were reared for more than a few days. One species of sand lobster, *Scyllarus americanus*, has been reared through the phyllosoma stages to metamorphosis, on a diet of brine shrimp nauplii (Robertson, 1968). One reason for success in this case probably was the short period of larval development in this species (32-40 days).

Batham (1967) and Silberbauer (1971) reported that *Jasus edwardsii* would not feed upon *Artemia* nauplii. Mitchell (1971) found that phyllosomes of *P. interruptus* preferred larger prey species such as fish larvae, hydromedusae, and ctenophores, and speculated that perhaps the spiny lobster phyllosomes are adapted to consuming large, soft-bodied organisms. Other work indicated that palinurid larvae apparently are able to grasp and extract nourishment from large marine polychaete worms suspended in the culture systems (Batham, 1967; C. Ingrahm, personal communication).

Until reliable techniques for culturing phyllosomes can be developed, the commercial feasibility of culturing palinurid and scyllarid lobsters appears to be remote. An alternative suggested by Ingle and Witham (1968), Chittleborough (1974b), and Serfling and Ford (1975b) is to capture sufficient numbers of the puerulus stage in areas where excess recruitment might permit this, without jeopardizing the natural population. For example, pueruli of *p. argus* are often found in large numbers attached to mangrove roots in embayments and can be captured in artificial habitats resembling natural clumps of algae (Ingle and Witham, 1968). The pueruli collected could then be reared extensively in coastal pond systems or in net enclosures along protected coastlines, using methods similar to those for marine and freshwater shrimp culture. However, little is known at present regarding larval recruitment of these species. Agencies concerned with regulating the fishery probably would not condone the harvest of large numbers of pueruli for aquaculture unless it could be demonstrated that this would not affect the natural population.

In recent years, the Australian Department of Agriculture, Fisheries Division (1975), has been conducting an assessment of the biological and economic potential of culturing the juvenile stages of the rock lobster, *P. cygnus,* to marketable size. Their findings indicated that this species has low biological potential, due to the delicate larval stages that require 7–11 months for development, slow growth (4 years to marketable size), and the current lack of knowledge about food and environmental requirements. Nevertheless, because of high demand and market price, they recommended a study of the potential for collection and rearing of the juvenile stages. Due to the expense associated with the capture of juveniles from the wild, the estimated production cost for cultured lobster was about twice the price received for adults by the fisherman. Therefore, they concluded that farming juveniles would not be profitable.

III. CULTURE OF NEPHROPID LOBSTERS

Commercial culture of some lobster species of the family Nephropidae now appears to be feasible from a biological standpoint. Efforts to develop techniques for the culture of lobsters in controlled environmental systems have concentrated on species of the genus *Homarus*, i.e., the American lobster (*H. americanus*) and the European lobster (*H. gammarus*). Preliminary studies by Figueiredo and Vilela (1972) in Portugal indicate that culture of *Nephrops norvegicus* may be hindered by the slow growth rate exhibited by juveniles of that species. However, too little information is now available to assess the potential for commercial culture of this organism.

Culture research on *H. gammarus* first began in France in 1865 (Herrick, 1911). In the United States, eggs from berried female *H. americanus* have been successfully hatched and cultured in the laboratory since 1885, in fisheries enhancement programs. *Homarus* spp. are extremely hardy, have a very simple and abbreviated larval period, will feed on many natural and artificial foods, have been reared from the egg to reproductive maturity, have been bred successfully, are resistant to diseases, exhibit greatly accelerated growth in warmed water, and have worldwide acceptance as a gourmet seafood. Although they are very aggressive, recent advances in techniques for individual rearing permit culture at high densities. Since *H. americanus* and *H. gammarus* are closely related, most of the culture techniques developed for one species are directly applicable to the other.

There are many problems involved in farming lobsters. A production facility would benefit from monthly or at least regular sale of market-sized animals. Therefore, methods to provide larvae on a regular and controlled basis are necessary. Although research groups have been allowed to obtain berried female lobsters from the wild, it is unlikely that commercial culturists could take large numbers of females for this purpose under existing fishery regulations. Methods to control egg deposition and hatching time are currently being developed. It also may be necessary or advisable to develop a brood stock that produces offspring with characteristics such as rapid growth or increased resistance to disease. The juveniles produced in the hatchery would be cultured to a marketable size of one pound (454 g), or perhaps a smaller size if new markets can be developed. This should be accomplished in the shortest time possible, in order to reduce labor and operating costs. Technique and systems are needed to reduce or eliminate cannibalism and disease, which can be major problems in high-density culture. Inexpensive foods will be required that will yield high growth rate, survival and conversion efficiency, that are convenient to prepare, store, and deliver to the culture systems, and that will not dissolve prematurely or contaminate the seawater environment. Culture systems must operate reliably and efficiently for long periods of time and permit convenient stocking, observation, and harvesting of the lobsters.

Data from fisheries investigations are available concerning growth, mortality, size frequency distributions, migration, and recruitment. Some of the information from these studies has direct application in determining the environmental conditions necessary for the culture of lobsters. In Europe, the fishery for *H. gammarus* is being evaluated by extensive tagging studies in England and Wales (Gundersen, 1977). Similar studies of population characteristics are being conducted in Norway, Scotland, Sweden, and North America. Information has been obtained concerning seawater temperature and other physical conditions of the natural habitat, behavior and feeding in the wild, diseases and parasites, and migration patterns and population densities. Several groups also are studying the requirements for holding lobsters in impoundments. In Scotland, lobsters caught in the fishery are being held in cages, where they are fattened prior to marketing (Milne, 1972). In Ireland, up to 85% of the lobster catch is held in floating boxes, tidal ponds, or recirculation ponds in high density storage units (Gibson, 1959). A similar practice is used by the lobster industry in Canada and was once popular along the East Coast of the United States (Prudden, (1962). The tank systems and conditions for long-term storage of live lobsters in England and

Wales have been described by Ayres and Wood (1977). For several years, the Laboratoire de l'Institut des Peches Maritimes has held adult lobsters in nursery ponds and released hatchery-reared juveniles at Roscoff and at the Ille d'Yeu along the Brittany Coast of France. The practice of using holding pounds is addressed in Chapter 8 of this volume.

More recently, several research programs have been conducted which are specifically concerned with the development of techniques for farming lobsters. A major effort in the United States has been sponsored by the National Sea Grant Program. Other major research programs are underway in Canada, France, Great Britain, and Japan. These programs have adapted the information gained in fishery assessment programs, and also have produced a large amount of new and relevant information, which is described in the following sections.

A. Biological Information

Much is known about the biology of nephropid lobsters. In addition, many technological problems have been solved that may ultimately lead to the commercial farming of these forms. Major accomplishments include (1) determination of the optimal temperature for rapid growth and efficient food conversion, (2) detailed knowledge of the nutritional requirements of lobsters, (3) techniques for controlling reproduction and improving genetic composition, (4) techniques for diagnosis, treatment, and prevention of disease, (5) techniques for communal rearing of juveniles, (6) identification and establishment of water quality standards, (7) satisfactory systems and techniques for the culture of the larval stages, (8) development of culture systems for the individual rearing of juveniles to market size, and (9) methods to assess the economic feasibility of various culture strategies. Each of these topics is discussed in the following sections.

1. Effects of Temperature on Growth Rate

The growth rate of lobsters is dependent on several physical and chemical factors. In its natural habitat, *H. americanus* may require from 5 to 9 years to grow from newly hatched larvae to a marketable size of approximately 82 mm in carapace length (CL) and a weight of one pound (454 g). This period would probably be prohibitively long for controlled farming of lobsters, considering initial capital costs, annual operating costs, and increased risk of crop losses due to equipment failure or disease during this time. However, growth rate is strongly influenced by water temperature. McLeese (1956) has shown that *H. americanus* is tolerant of a wide range of temperatures, from near freezing to 31°C (Fig. 1). Growth is most rapid at higher temperatures (20°–22°C). In the northern areas of the Atlantic, where colder seasonal seawater temperatures are encountered, growth is reduced. The rate of growth of lobsters is the combined result of the size increment per molt and the length of the intermolt period. Low temperatures

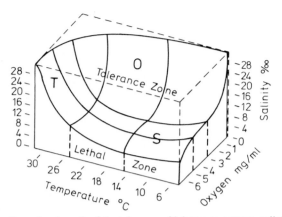

Fig. 1. Three-dimensional graph of the tolerance of lobsters to oxygen, salinity, and temperature. Any point in the graph represents a specific combination of oxygen, temperature, and salinity. Redrawn from McLeese (1956).

lengthen the intermolt period and high temperatures reduce it. The increase in size per molt is not as strongly affected by temperature (Mauchline, 1977). Under conditions of constant elevated temperature (22°C), growth can be enhanced greatly (Hughes *et al.*, 1972). Market size lobsters have been produced in as little as 2 years (Fig. 2). Such accelerated growth would mean decreased production costs in commercial lobster farming due to savings in labor and energy costs. Recent research in California (Botsford *et al.*, 1974; Van Olst,

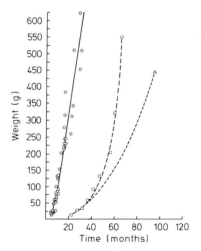

Fig. 2. Growth of *H. americanus* in warmed (circles) and ambient (squares, triangles) temperature seawater. (From Hughes *et al.*, Copyright 1972 by the American Association for the Advancement of Science.)

1975) has confirmed that the temperature promoting the most rapid growth of *H. americanus* while maintaining high survival is 21°-22°C. Since it would be expensive to heat seawater to these temperatures, research has been conducted on methods to conserve heat by recycling the seawater in a closed-system culture (Hand, 1977), and by using warmed seawater effluents from coastal steam-electric generating stations to provide an inexpensive or free source of waste heat for lobster culture (Dorband *et al.*, 1976; Van Olst *et al.*, 1976). Both methods appear encouraging. Gross food conversion efficiency was not reduced significantly at these higher temperatures (Johnson, 1977). Therefore, it appears that accelerating growth through the use of elevated temperature has considerable potential in lobster culture.

At elevated temperatures, the maximum rate of growth in carapace length observed for *H. americanus* has been approximately 0.1 mm per day. This rate represents approximately a 1 cm increase in total length of the lobster each month. In nature, lobsters molt several times during the first few years of life and then acquire a pattern of molting only once or twice a year, during the spring or summer. Older lobsters may go several years between molts. It is estimated that lobsters molt about 20 times before attaining the marketable size of approximately 80 mm CL. When lobsters are cultured at a constant elevated temperature, the seasonal timing of molting is eliminated, and the growth rate is thus increased. Under optimal conditions, each molt will contribute about a 15% increase in length and a 40% increase in weight. Food conversion efficiencies of 4:1 and 11:1 were found in warm (22°-24°C) seawater and ambient (2°-24°C) seawater, respectively (Hughes *et al.*, 1972).

2. Formulation of Artificial Diets

The first accounts of the foods selected by adult lobsters showed that most often they consumed fish, either dead or alive, and a variety of benthic invertebrates (Bowers, 1900; Herrick, 1911). More recent data has indicated that *H. americanus* is omnivorous, feeding primarily on living crabs and mollusks (Squires, 1970; Weiss, 1970). In aquaculture, natural foods such as these are probably not suitable for use, since they must be harvested periodically from the sea or cultured artificially. These approaches would be expensive. Also, natural foods are not easily stored and would have to be fed quickly before they began to decompose. Leaching of nutrients from partially eaten fragments of these foods could pollute the culture water very rapidly, and it is probable that the lobsters would not eat certain parts of some of the foods, such as bones, shells, or integument.

Thus, most research on the development of foods for lobster farming has been concerned with formulation of artificial diets. These diets consist of dry or moist pellets made from powdered, mixed feed ingredients, similar to the trout and catfish diets available for commercial culture of these species. The diets can be made in flake, pellet, or "noodle" form by large extrusion machines. Many observers believe that a noodle or spaghetti-shaped pellet would be the form most readily accepted, and due to the way lobsters masticate their food, would result in the least waste. A binding ingredient such as wheat gluten or alginate is included in these rations so that the pellet will exhibit high stability in water. Binders also reduce the rate at which water-soluble nutrients such as vitamins are leached from the diet.

Considerable research on diet formulation has been conducted in the United States and Canada. The exact nutritional ingredients of this diet appear to be critical. As yet, no artificial food pellet has been developed that will produce growth as rapidly as a diet of live adult brine shrimp, *Artemia salina*. This food is easily obtained in many areas and has been used as a standard or reference diet in nutritional studies (Conklin, 1975; Van Olst *et al.*, 1976). Growth rates as rapid as 0.1 mm/day, based on carapace length measurements, have been obtained on a diet of live brine shrimp (Hedgecock *et al.*, 1976; Van Olst *et al.*, 1976). Growth of *H. americanus* fed frozen brine shrimp has been significantly less (Fig. 3). Similarly, Forster and Beard (1973) showed that fresh mussel mantle was better than dried or processed mantle in promoting growth of the prawn, *Palaemon serratus*, and postulated the existence of a growth factor in fresh foods, perhaps a steroid, which is easily destroyed by processing. Lobsters fed poor diets often become pale blue or even white, a sign of nutritional deficiencies. Other indications of such deficiencies are a soft cuticle, loss of appendages, vulnerability to stress and disease, and a reduction in growth rate.

In an effort to discover which nutritional ingredients in brine shrimp might be responsible for the rapid growth rates, a detailed evaluation of its biochemical

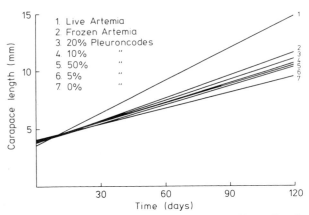

Fig. 3. Growth rates of juvenile *H. americanus*, fed seven different diets for 120 days at a constant temperature of 20°C. Diets 3–7 consisted of a commercial shrimp pellet, supplemented with the stated percentage of red crab, *Pleuroncodes planipes*. (From Van Olst *et al.*, 1975.)

composition was conducted by Gallagher and Brown (1975). In addition, Leavitt et al. (1979) analyzed the biochemical composition of food taken from stomachs of *H. americanus* captured in the wild. Hopefully this type of research will lead to the formulation of diets that compare favorably to natural foods. The problem is made more complex, since the nutritional requirements of lobsters may change with age, and therefore different diet formulations could be required. Food consumption and nutritional requirements also vary during the molt cycle. Recent observations of lobsters fed exclusively on a diet of live adult brine shrimp for a period of 2 years revealed that none of these animals were able to survive subsequent transfer to a diet of natural foods (J. C. Van Olst and J. M. Carlberg, unpublished data). This may have been due to the lack of the necessary digestive enzymes or bacteria in the gut.

In culture, the cast molt or exuvia is often left in the rearing compartment, allowing the lobster to consume much of the shed exoskeleton. This provides an additional calcium source, which facilitates rapid hardening of the new cuticle following molting.

There is a lack of agreement among workers regarding the level of protein necessary in a lobster diet. Gallagher et al. (1976) found 11% protein to be inadequate. Castell and Budson (1974) found best growth rates at protein levels of 60%, although other investigators have shown good rates of growth at 30–40% protein and most recently as low as 20–25% (Capuzzo and Lancaster, 1979). Conklin et al. (1977) have indicated the importance of water-soluble vitamins and speculated that rapid leaching of these dietary components from the artificial pellets may result in decreased growth and survival.

Conklin (see Chapter 5, Vol. I) points out that much more information is needed in order to develop an artificial lobster ration. To date, perhaps the most favorable results have been obtained by Conklin et al. (1977), who have developed an artificial diet that produces growth rates of approximately 0.07 mm/day (Table I). Similar growth has been obtained by workers at San Diego State University (Van Olst et al., 1976) on a commercially produced shrimp ration, which was supplemented with ground tissue and shell of the pelagic red crab, *Pleuroncodes planipes*, to provide an additional source of protein and pigment precursors (Fig. 3). If these linear rates of growth were maintained, lobsters would grow from larvae to marketable size in about 3 years, versus about 2 years on a diet of live brine shrimp.

Since the ingredients in a formulated diet do not function in isolation, the possible combinations and permutations indicate a number of interrelationships that are almost beyond the possibility of study. Therefore, in most diet evaluations, nutritional value is judged by the growth responses of lobsters in controlled feeding trials, where the levels of only a few ingredients are varied. However, the ideal growth function is not always the most rapid; rather it is the rate that is most efficient or economical. In feeding trials where lobsters are fed *ad libitum*

TABLE I

Composition of diet 76S[a]

Ingredient	% by Weight	Source[b]
Casein (vitamin-free)	31.0	ICN
Gluten	15.0	ICN
Yeast ("Yeaco 20")	15.0	Milbrew
Lipid mix S[c]	6.0	See below[c]
Corn starch	5.0	ICN
Albumin	4.0	ICN
Salt mix BTM[d]	3.0	ICN
Vitamin mix D[e]	2.0	ICN
Choline Cl	1.0	ICN
Thiamin HCl	1.0	ICN
Cholesterol	0.5	Sigma
Vitamin E ("Rovimix")	0.2	Hoffman-LaRoche
Cellulose	16.3	ICN

[a] From Conklin et al., 1977.

[b] ICN, Nutritional Biochemicals Corp., Cleveland, Ohio; Milbrew, Inc., Juneau, Wisconsin; Sigma Chemicals, Inc., St. Louis, Missouri; Hoffman-LaRoche, Nutley, New Jersey.

[c] Lipid mix S = corn oil, 33.8% (ICN); cod liver oil, 66.0% (ICN); and ethoxyquin ("Santoquin"), 0.2% (Montsanto, Co., St. Louis, Missouri).

[d] Salt mix BTM = Berhart–Tormelli modified salt mix.

[e] Vitamin mix D = modified from Castell and Budson (1974), thiamin (HCl), 0.32%; riboflavin, 0.72%; niacinimide, 2.56%; biotin, 0.008%; Ca-pantothenate, 1.44%; pyridoxine (HCl), 0.24%; folic acid, 0.96%; cobalamine, 0.266%; i-insitol (meso), 12.5%; ascorbic acid, 6.0%; p-aminobenzoic acid, 2.0%; vitamin D_2 (500,000 USP/gm), 0.042%; vitamin A (500 IU/gm), 0.5%; BHA 0.075%; Celite, 72.369%.

with extremely rich food mixtures, many diets frequently prove to be uneconomical for use in culture.

Ablation of the eyestalks has been employed in nutrition research and may have potential for accelerating growth in aquaculture. Lobsters that are naturally blind or whose eyestalks have been removed surgically grow at about three times the rate of normal lobsters (Mauviot and Castell, 1976; Aiken, 1977). It is known that the eyestalks are the sites for the synthesis of the molt-inhibiting hormone. Ablated animals molt more frequently. It may be possible to use eyestalk ablation on subadult ("canner") lobsters to accelerate their molting to marketable size (Castell et al., 1977). However, attempts to employ this technique on small lobsters have usually resulted in high rates of mortality, especially during molting.

Although many advances have been made in the study of lobster nutrition, much more research is necessary. The lack of a suitable artificial food ration is the greatest single deterrent to commercial lobster farming at present.

3. Reproduction and Genetics

There has been little research on the very critical problem of brood stock management, the successful maintenance of reproducing adults. Culturists now rely to a large extent on obtaining egg-bearing females from the wild. Egg extrusion and the time to hatching in *H. americanus* is determined by the culture temperature at which the female is held (Perkins, 1972). In the wild, the period from molting and mating to hatching ranges between 12 and 18 months, depending on whether the mating occurs after a spring or fall molt. For large females, this period may be greater than 2 years. The time for external egg development varies between 8 and 14 months.

Techniques must be developed to control egg extrusion, retention, and development under laboratory conditions. Workers have succeeded in reducing the egg development time from 9 months to as little as 4 months by holding egg-bearing females in warm water, but there is insufficient information about the processes involved either to predict or control larval production on a large scale (Van Olst, 1975; Hedgecock, 1977).

It is possible to capture females in the fall just after mating and allow them to extrude eggs in captivity. Most of these females will extrude eggs if given sufficient time and proper holding conditions. This is one temporary source of egg-bearing females, until techniques to control the complete life cycle of the American lobster become more refined.

Several researchers have been successful in mating *H. americanus* in the laboratory (Hughes and Matthiessen, 1962; Aiken and Waddy, 1976; Hedgecock, 1977; Van Olst and Carlberg, 1978a). Several of these matings conducted in laboratories on the east coast of the United States and Canada have resulted in successful egg extrusion and hatching. For unknown reasons, few of these matings performed on the west coast of the United States have resulted in the production of viable larvae. Hybrid matings of *H. americanus* and *H. gammarus* with both sexes of each species have been more successful. It may be that *H. americanus* requires a winter period of cold seawater temperatures that is not encountered in the Pacific at California latitudes. Why hybrid matings are not influenced similarly is not yet understood. Recent work by Nelson *et al.* (1979) indicated that providing an increasing photoperiod may be useful in promoting successful egg extrusion.

Related, controlled mating studies have been conducted to develop colormorphs of blue, red, white, gold, or mosaic-patterned lobsters, which would function as tagged individuals to aid in assessing the influence of hatchery release programs on the fishery (Hughes, 1968). Other efforts are directed toward the selection and mating of adults with rare allozymes to produce progeny with genetic markers, which would be used to evaluate the hatchery program and to follow specific family responses in culture research (Hedgecock, 1977).

4. Hybridization

In aquaculture, genetic selection and hybridization may contribute to the development of strains with desirable culture and market characteristics. Often hybrid progeny will exhibit more variability than will either of the parent species, in desirable characteristics such as rapid growth, disease resistance, efficient feed conversion, low aggression, and attractive market traits. Assessments of traits desirable for aquaculture have been made for the larval and juvenile stages of *H. americanus* and *H. gammarus*, and for the progeny from successful hybrid crosses of these species (Fig. 4). The morphological characters of the four larval stages of the hybrid showed close similarities to those of *H. americanus* described by Herrick (1911) and by Gruffydd *et al.* (1975). The large larval size of *H. gammarus* and some of the hybrids is a desirable characteristic for aquaculture. However, there were no differences in the growth or survival of juveniles of the two lobster species or the hybrids (Carlberg *et al.*, 1978). Progeny from a cross performed in France are reported to be growing faster than *H. gammarus* (J. Audouin, personal communication). Other observations suggest that *H. americanus* is less aggressive in communal rearing. This, together with evidence of higher levels of mortality for *H. gammarus*, suggests that the European species may be more susceptible to stress and less well adapted to the culture environment. These studies demonstrated successful hybridization between these

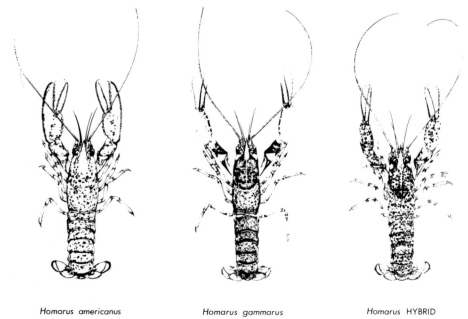

Fig. 4. Juvenile American lobster, *Homarus americanus;* European lobster, *Homarus gammarus;* and a hybrid of the two species. (From Carlberg *et al.*, 1978.)

two closely related nephropid species that are geographically isolated in nature. The fact that they produce viable F_1 hybrids, together with their relatively high degree of genetic similarity, suggests that these species might have been isolated only recently (Hedgecock et al., 1977). Whether any reproductive isolation exists between the two species must await studies of F_1 reproductive success and the variability of the F_2 generation. If reproductive barriers to gene exchange are absent, then extensive hybridization might create lobsters better suited to aquaculture. However, for the present, *H. americanus* exhibits traits that make it adequately suited for culture. There are significant differences in growth rates among different progeny of *H. americanus*, which indicates that continued research in selective breeding for rapid growth may prove useful in lobster culture (Hedgecock et al., 1976).

5. Disease

Disease is often a major problem encountered in the culture of aquatic organisms at high densities. There are two major divisions of mariculture pathogens. Primary pathogens, such as *Aerococcus viridans*, can be lethal in lobsters even when environmental factors are adequate. Secondary pathogens are opportunistic. These include the vibrios, pseudomonads, and aeromonads, which are lethal when physiological or environmental factors are marginal, e.g., when conditions of poor water quality or inadequate nutrition exist. Five diseases are of some concern in the culture of *H. americanus;* gaffkemia, filamentous bacterial disease of larvae, shell disease, and two fungal diseases, as reviewed in Chapter 6, Vol I. Although other fungi, viruses, protozoa, and parasites have been observed in cultured lobsters, none of these diseases have caused major or sustaining problems.

The bacterial disease gaffkemia, caused by the micrococcus *Aerococcus viridans* var. *homari*, has been isolated from hemolymph cultures and swabs of the cuticle of adult lobsters (Kellogg et al., 1975). This disease has caused serious losses in commercial holding pounds (see Chapter 6, Vol. I). The infection is invariably fatal within 18 days at temperatures of 20°C (Stewart et al., 1969; Dewees and Schapiro, 1974), and therefore diseased animals obtained from the wild can be eliminated during routine quarantine and temperature acclimation procedures. Gaffkemia has been treated successfully with sulfonamides and antibiotics such as penicillin or vancomycin (Fisher et al., 1978). Its spreading can be reduced by cleaning the culture systems with calcium hypochlorite or by flushing with chlorine (Dewees and Schapiro, 1974). Some indication of host immune response to this infection has been reported by pathologists (Schapiro et al., 1974). If further research is successful, prophylactic immunization may become a standard method of prevention of these systemic bacterial infections in mariculture facilities. This pathogen probably will be of minor importance to the successful culture of lobsters, if care is taken to avoid contamination of the cultured animals.

Eggs of *H. americanus* have been found to be susceptible to epibiotic growth of filamentous microorganisms, including the marine bacterium *Leucothrix mucor* and other bacteria, cyanophytes, diatoms and vorticellids that restrict metabolic exchange with the environment (Nilson *et al.*, 1975). The pathogenic fungus *Lagenidium* sp. also has been found on lobster eggs (Nilson *et al.*, 1976). This disease is easily detected microscopically and can be successfully treated by immersing the berried female in a 5 ppm solution of malachite green for 10 minutes (Fisher *et al.*, 1976a,b, 1978). This inhibits the growth of the disease agents without harm to the female or the eggs. Other similar compounds also have been found to be effective.

A variety of filamentous epiphytes are known to contaminate the larval stages of lobsters under culture conditions (Nilson *et al.*, 1975). Infestation with *L. mucor* has caused heavy mortalities in some batches of larvae. Generally, the earlier the onset of infestation, the more severe the losses. The filamentous growth on external surfaces hinders the animal as it attempts to feed or molt and may restrict respiration. The long dense filaments of the bacteria entangle larvae, food, and exuvia. Maintenance of a 1.0 mg/liter concentration of streptomycin sulfate or neomycin sulfate in the larval system has been used to control this disease and is reported to reduce mortalities (Schuur *et al.*, 1976). The use of antibiotics, however, is often undesirable, because of their incompatibility with biological filtration, the difficulty in maintaining therapeutic levels, and the possibility of enhancing the growth of fungi and resistant strains of bacteria.

A disease that also is carried over to lobster larvae from the embryo is the fungus *Lagenidium* sp. (Nilson *et al.*, 1976). Treatment with streptomycin sulfate to counteract filamentous bacteria contamination has been observed to select out competing organisms and allow the *Lagenidium* to invade the entire rearing system. The disease is capable of destroying greater than 90% of the exposed animals in less than 3 days under such conditions. Fortunately, both the epiphytic and the fungal diseases are usually restricted to larval stages. Although some persistent cases have been observed in young juveniles, the animals apparently become strong enough to ward off lethal infestation. For this reason, larvae infected near the end of the larval development will have better survivorship than those infected earlier. As with the eggs, the larvae can be successfully treated with a malachite green dip.

Chitinolytic bacteria are responsible for another disease found in lobster larvae, causing erosion and marring of the exoskeleton (Hess, 1937; Fisher *et al.*, 1976c, 1978). This is similar to shell disease of adult lobsters and may be caused by the same organisms. Young juveniles show increased susceptibility to this disease under conditions of stress, especially that associated with dietary deficiencies. Although this disease is infrequently found in larval rearing systems, it is potentially more detrimental to the thin exoskeletons of the larvae than to adult lobsters (Schuur *et al.*, 1976). At present no treatment or methods of prevention exist, other than to reduce stress caused by poor water quality or inadequate diet.

The fungus, *Haliphthoros milfordensis,* is generally found only in young juveniles, although cases have been observed in third- and fourth-stage lobster larvae (Fisher *et al.*, 1975, 1978). No treatment is yet recommended for this disease, although losses have been found to be greatly reduced with thorough cleaning of the culture system. The extent of this disease is minimal in the larval stages, but greater losses occur in the young juveniles.

Several workers have reported protrusion of the epidermis through a small circular hole in the dorsal surface of the carapace of cultured juvenile lobsters. This wartlike condition usually appears just to one side of the midline and has been observed in as many as 5% of the juveniles in culture (Van Olst and Carlberg, 1978a). Although infrequent, this condition causes the new cuticle to fuse to the old carapace, which interferes with molting and occasionally can result in mortality. Usually this irregularity in the exoskeleton disappears after a few molts. Although the cause is still unknown, it does not appear to present a serious disease problem in culture.

The most important aspect of disease control in lobster culture is the reduction of physiological stress. Lobsters should not be subjected to such stress caused by improper culture conditions. These include insufficient space, poor water quality, excessive temperature, and inadequate diet. In addition, the rearing systems should be kept free of uneaten food, exuvia, and dead animals that harbor the disease agents. If these condtions are met, most researchers believe that disease will not be a major problem in lobster culture. This is indicated by the fact that lobsters have been cultured from the egg to marketable size routinely at several laboratories, without major disease problems. However, the costs associated with monitoring, prophylaxis, and treatment of diseases in intensive culture systems are yet to be determined.

6. Water Quality

The American lobster is a relatively hardy aquatic organism. It is found in areas where the seawater temperatures may range seasonally from 1° to 25°C (Gates *et al.*, 1974). Studies by McLeese (1956) and Ford *et al.* (1979) indicated that lobsters will even tolerate brief exposure to temperatures as high as 31°C. They also can tolerate periods of low salinity and oxygen. In these respects, they are more tolerant to fluctuations in water quality conditions than most finfish considered for culture. The three-dimensional graph shown in Fig. 1 (McLeese, 1956) indicates the ranges of temperature, salinity, and oxygen tolerated by *H. americanus.* Optimal levels for several water quality parameters are shown in Table II. Of these, the compound that may be most critical in intensive culture is ammonia. Ammonia is the major nitrogenous excretion in lobsters (Logan,

TABLE II

Water Quality Requirements for the Culture of *H. americanus*[a,b]

Parameter	Optimal conditions	Natural conditions	Lethal values
Temperature (°C)	20–22 (1)	1–25 (2)	<0, >31 (3)
Salinity (0/00)	30 (2)	28–35 (6)	<8, >45 (3,4)
Dissolved oxygen (mg/liter)	6.4 (2)	4.0–7.3 (16)	<1, >saturation (3)
pH	8.0 (4)	7.5–8.3 (5)	<5, >9 (4)
Nitrogen compounds (mg/liter)			
NH_3-N			
larvae	<0.14 (8)	0–.3 (7)	1.4 (96 hr LC_{50}) (8)
juveniles	<9.4		94 (48 hr LC_{50}) (13)
NO_3-N	<50	0.07–21 (7)	>500 (9)
NO_{50}-N	<10	0.014 (7)	>100 (9)
Metals (μg/liter)[c]			
Cu	<6	3 (10)	60 (96 hr LC_{50}) (11)
Zn	<170	10 (10)	1700 (96 hr LC_{50}) (11)
Hg	<3.3	0.03 (10)	33 (96 hr LC_{50}) (12)
Pb	<100	0.03 (10)	1000 (96 hr LC_{50}) (11)
Cr	<70	0.05 (10)	700 (96 hr LC_{50}) (11)
Cd	<18	0.11 (10)	180 (96 hr LC_{50}) (11)
Co	<81	0.5 (10)	810 (96 hr LC_{50}) (11)
Chlorine residual (mg/liter)	<0.01 (14)	0	0.69 (48 hr LC_{50}) (14)
Chlorinated hydrocarbons (μg/liter)			
DDT	0	0.15 (15)	7 (96 hr LD_{50}) (15)
Aroclor 1254	0	<1 (15)	50 (408 hr LD_{50}) (15)

[a] Adapted from Wright, 1976.

[b] Numbers in parentheses are references. Key: (1) Hughes *et al.*, 1972; (2) Gates *et al.*, 1974; (3) McLeese, 1956; (4) Harriman, 1953; (5) Cornick and Stewart, 1977; (6) Anonymous, 1978; (7) Cox, 1965; (8) Delistraty *et al.*, 1977; (9) Gravitz *et al.*, 1975; (10) Horne, 1969; (11) Dorband *et al.*, 1976; (12) Connor, 1972; (13) Sastry and Laczak, 1975; (14) Capuzzo *et al.*, 1976; (15) D. Merk, unpublished data, San Diego State University; (16) Sverdrup *et al.*, 1942.

[c] Assuming an application factor of 0.1 (Sprague, 1971).

1975). Knowledge of both ammonia excretion rates and safe ammonia tolerance limits is required to optimize design of the seawater circulation and waste treatment components in an intensive lobster culture system. Ammonia toxicity could present a major problem in culture systems in which a major part of the seawater is recycled.

In an aqueous solution, NH_3, or un-ionized ammonia, exists in equilibrium with its conjugate acid, NH_4^+. The term "ammonia" usually refers to the sum (NH_3 + NH_4^+). Wuhrmann and Woker (1948) have shown that un-ionized ammonia is the toxic component of ammonia. Thus, it is necessary to know the fraction of un-ionized ammonia in solution when evaluating toxicity. Delistraty *et al.*

(1977) determined that the 192 hr incipient median lethal concentration (LC_{50}) and the tentatively safe ammonia concentration for fourth-stage larvae of *H. americanus* were 1.4 and 0.14 mg/liter of un-ionized ammonia–nitrogen (NH_3-N/liter), respectively (Fig. 5). Even lower levels were toxic when pH or temperature was elevated. Sastry and Laczak (1975) found a 24 hr LC_{50} of 8.2 mg NH_3-N/liter for fourth stage *H. americanus*, and L. Gleye (personal communication) found a 96 hr LC_{50} of 1.2 mg NH_3-N/liter for lobsters of 1.0 g wet body weight, representing approximately twelfth-stage juveniles. These estimates must be compared judiciously, since Sastry's value was for a relatively short exposure period of 24 hr, and Gleye's value was obtained for larger individuals and over an exposure period of 96 hr. Gravitz *et al.* (1975) have indicated that toxic levels of nitrate (NO_3) and nitrite (NO_2) are approximately two orders of magnitude higher than those for ammonia. If this is the case, ammonia excretion rate and tolerance limits probably will be the factors determining the design of filtration and recycling systems for lobster culture, since ammonia values will exceed the stated toxic level long before nitrate and nitrite reach their toxic limits.

Logan (1975) determined that *H. americanus* of Stage IV excrete ammonia at the rate of 0.002 mg/hr. Additional sources of nitrogenous wastes in the culture system will be uneaten food, decomposing organic matter present in the seawater, excretion from fouling organisms, and decaying lobsters that may die for various reasons. Also, Sastry and Laczak (1975) found changes in the ammonia excretion rate in relation to feeding, molt cycle, and size. Thus, the minimal flow rates for filtration and water supply needed to avoid problems with ammonia toxicity are difficult to determine at present. In a large production facility, the energy costs to provide sufficient seawater for diluting ammonia to nontoxic levels may be substantial; more research is needed regarding optimal flow rates

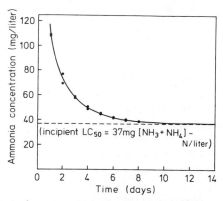

Fig. 5. Graph illustrates the concentrations of ammonia that result in 50% mortality of larval lobsters after exposure to this toxicant for varying lengths of time. (From Delistraty *et al.*, 1977.)

in culture systems. Furthermore, in the United States it will be necessary to ensure that discharges from the system are in compliance with pollutant discharge regulations as summarized by Tchobanoglous and Shleser (1974). In some cases these water quality regulations may be more stringent than the requirements of the lobsters themselves.

7. Open and Closed Seawater Systems

Open seawater systems may be used to rear lobsters successfully. However, these systems are subject to pathogens and toxicants that may appear periodically in the ocean environment and adversely affect the lobsters. Complex filtration and ultraviolet sterilization systems may be necessary. Filtration with pressure sand filters or commercial cartridge filters that remove particles 20 μm or larger greatly increase the effectiveness of ultraviolet treatment of seawater.

An open, flow-through system is the least complex mode of operation in lobster culture. In areas where a reliable source of high quality seawater is available, the need for filtration is eliminated. However, most locations, even those with the required water quality, do not have ambient water temperatures appropriate to promote rapid growth of lobsters on more than a seasonal basis.

A recirculated, closed system mode of operation has three major advantages over the open system mode (1) the capability to rear lobsters at a site without continually available seawater, (2) the capability, if artificial sea salts are used, to initially exclude pathogens from the system, and (3) reduced costs for pumping and heating replacement water. However, the buildup of metabolites is a major concern in operating closed systems. These can usually be controlled by a biological filter bed, which is populated with nitrifying bacteria that metabolize excreted ammonia and convert it to nitrate and nitrite. The animals tolerate concentrations of nitrate many times higher than that of ammonia, and the water can be reused almost indefinitely if the nitrification process remains stable (Spotte, 1970).

8. Use of Thermal Effluent

In areas where warm seawater can be obtained, open system lobster culture will be the least expensive mode of operation. Large quantities of waste heat are often available from coastal electric generating stations. The thermal effluent can be used to accelerate growth of aquatic organisms and thereby reduce production time and costs. Large-scale secondary use of cooling water as a heat source for aquatic farms may provide a means of reducing effluent temperatures, and thereby help alleviate adverse ecological effects of thermal effluent. Growth of *H. americanus* is accelerated markedly in thermal effluent at 20°–22°C (Van Olst *et al.*, 1976). Marketable lobsters (454 g) were cultured from the egg stage in as little as 24 months.

10. Aquaculture

Despite this potential, there may be a number of specific problems involved in aquaculture using thermal effluent. Gaucher (1968), Yee (1971), Yarosh et al. (1972), and Becker and Thatcher (1973) have considered many of these, including possible adverse effects caused by chemical composition of the effluent and by the periodic shutdown and temperature cycling of generating units. These are important considerations that have bearing not only on the ability to culture market species economically, but also to assure that their quality will meet standards for human consumption. Levels of toxic substances above their normal concentrations in natural seawater can cause chronic, sublethal effects, such as reductions in growth and fecundity, as well as acute, lethal effects (Doudoroff and Katz, 1953; Bowen, 1966; Sprague, 1969). Heated discharge water can be used in heat exchangers to avoid contamination, but the associated costs are increased.

Concentrations of toxic chemicals in animals harvested for human consumpttion are subject to regulation by agencies such as the U.S. Food and Drug Administration. The urgency of many water pollution problems is leading to the establishment of official water quality criteria for aquatic life (Sprague, 1969). These water quality criteria are even more critical in commercial aquaculture systems, where a very large number of animals are maintained in a single water source. The commercial aquaculturist must be confident about the consistent high quality of the water used in the culture system. Unfortunately, we are just becoming aware of the problems of chemical pollution from sources such as petrochemicals, sewage, and generating station cooling water systems. Few studies have been conducted to assess the specific effects of such chemicals on marine crustaceans. In addition, very few standardized techniques have been established to analyze such effects. Only in the last 10 years have the Food and Agriculture Organization of the United Nations and the U.S. Environmental Protection Agency attempted to evaluate methods of detecting and monitoring pollutants in the marine environment (Goldberg, 1970).

A considerable amount of research has been done at several coastal power plant sites in southern California concerning the potential problems involved with heavy metal contamination (Ford et al., 1975, 1979; Dorband et al., 1976). Juvenile and adult lobsters removed from culture systems at several generating station laboratories were utilized in comparative analyses of trace metal concentrations in the body tissues of each individual (Table III). The mean concentrations of six metals in both the whole-body and separate tissue samples of larval, juvenile, and adult *H. americanus* were similar to concentrations of the same metals found in other decapod crustaceans taken from the wild (Bryan, 1968; Eisler et al., 1972; Gale, 1973). The gills of decapod crustaceans are the main sites for absorption and loss of zinc and other heavy metals across the body surfaces. Bryan (1968) observed that *H. gammarus* probably obtains much more copper and zinc from seawater than is required, but that temporary storage by the

TABLE III

Tissue Concentrations of Six Metals in Lobsters, *H. americanus*.[a]

Tissue	Metal (μg/g dry weight)					
	Copper	Zinc	Cadmium	Lead	Cobalt	Chromium
Hepatopancreas						
Mean±	71.04±	29.29±	12.33±	12.33±	34.57±	45.78±
SD	82.27	13.11	11.33	5.25	60.94	58.33
% above detection	100	100	80	30	95	95
N = 20						
Gills						
Mean±	135.03±	76.99±	11.20±	192.27±	233.59±	315.38±
SD	129.16	72.64	14.95	173.29	139.53	315.55
% above detection	100	100	47	47	32	89
N = 19						
Claw muscle						
Mean±	60.85±	68.77±	3.70±	19.99±	18.16±	80.70±
SD	61.13	27.42	4.79	15.49	10.11	65.57
% above detection	100	100	45	50	55	86
N = 22						
Tail muscle						
Mean±	17.24±	30.17±	1.86±	8.27±	12.03±	37.54±
SD	10.87	16.42	2.49	3.39	8.37	53.30
% above detection	95	100	71	38	57	95
Exoskeleton						
Mean±	13.99±	10.79±	2.67±	17.91±	48.15±	18.54±
SD	10.37	7.79	2.65	11.77	30.86	18.54
% above detection	100	100	87	70	91	96
N = 23						
Digestive tract						
Mean±	64.69±	28.85±	6.53±	13.47±	39.63±	181.50±
SD	47.18	17.55	3.52	4.47	26.79	347.41
% above detection	100	100	53	20	60	87
N = 15						

[a] From Dorband et al., 1976.

hepatopancreas and removal across the gills and in the urine is so efficient that this species can withstand above normal concentrations in seawater. It seems likely that the regulatory mechanism of *H. gammarus* also is exhibited by *H. americanus*. High levels of six trace metals (copper, zinc, cadmium, lead, cobalt, and chromium) in the gills and hepatopancreas, and correspondingly low levels in the exoskeleton, indicated that *H. americanus* also accumulates metals from the water, primarily across the gill filaments (Dorband et al., 1976). The presence of high concentrations of the six metals in the hepatopancreas confirmed the storage function of this organ in *H. americanus*. The ability to regulate internal concentrations of trace metals may give *H. gammarus* and *H. americanus* a degree of protection in regions where metal pollution is found.

The levels of copper, zinc, cadmium, lead, cobalt, and chromium in the edible portions of the lobster (claw and tail muscle) cultured in thermal effluent were relatively low, in comparison with metal levels in vertebrate muscle tissues (Valle, 1959). The levels of all six metals in whole-body samples and in the edible tissues were well below the limits for these metals in foodstuffs established by the U.S. Food and Drug Administration (Dorband, 1975; Dorband et al., 1976).

Although "safe" levels can be established only by long-term chronic bioassays, some indication of the concentrations of a chemical that can be tolerated by *H. americanus* larvae can be drawn from the median lethal limit (LD_{50}) estimates for eight chemicals reported by Dorband et al. (1976). LD_{50}'s for *H. americanus* larvae (Table IV) presumably are conservative estimates of the minimum tolerance limits for lobsters of all life history stages, since the larvae usually are more sensitive than juveniles and adults (McLeese, 1974). The concentrations of all of the eight chemicals in several effluent sources were at least an order of magnitude lower than the median lethal limit estimates for these chemicals. In these studies, there was no evidence to suggest that any deaths or sublethal harmful effects had been caused by toxic concentrations of metals in the thermal effluent. These data provide encouraging evidence that thermal effluent from fossil fuel or nuclear generating stations may be an essentially nontoxic source of heated seawater for commercial lobster farming.

TABLE IV

Median Lethal Limits (LD_{50}) for American Lobster Larvae Subjected to Several Toxic Chemicals[a]

Metal[b]	Time intervals (hr)			
	24	48	72	96
Copper	165	75	60	60
Cobalt	1090	1090	810	810
Cadmium	900	430	220	180
Chromium	1500	700	700	700
Lead	4100	2600	2100	1000
Zinc	2500	1700	1700	1700
Chlorine (ml/liter)	0.31	0.25	0.25	0.22

[a] From Dorband et al., 1976.
[b] Concentrations in µg/liter.

9. Behavior

Herrick (1911) described the behavior of the American lobster as solitary and aggressive. Laboratory studies have shown that dominant lobsters usually molt more rapidly than subordinates, resulting in a wide size range (Cobb and Tamm, 1974). In addition, communally reared lobsters have a significantly higher rate of mortality due to cannibalism (Van Olst *et al.*, 1975; Sastry and Zeitlin-Hale, 1977). This increase in cannibalism is partly associated with molting, which leaves the soft-shelled, postmolt individuals vulnerable to hard-shelled, intermolt juveniles. Thus, when lobsters are cultured communally at high densities, low yields and large size ranges often result. In commercial production, these poor yields would present considerable problems. Some research has been conducted to develop methods of increasing yields in communal rearing.

McIntosh (1920) noted that juvenile lobsters are often found in tunnels and crevices excavated in mud bottom areas. Although Scarratt (1968) has observed that American lobsters are rarely found in nature at densities greater than $1/m^2$, it seems likely that by providing large numbers of suitable refuges in the form of caves, tunnels, and crevices, one could substantially increase the carrying capacity of a suitable lobster habitat. McBurney and Wilder (1973) found that the addition of shelters in an adult lobster impoundment reduced losses due to cannibalism. Ghelardi and Shoop (1972) reared juvenile lobsters to an age of 1 year in concrete pens with a mud substrate. They stocked these systems with stage IV individuals at $16/m^2$ and observed a carrying capacity of about $3/m^2$ after one year. Van Olst *et al.* (1975) conducted an experiment concerning the effects of substrate type on communally reared juveniles and found that shell or rock shelters had a higher carrying capacity than did sand bottoms (Fig. 6). The use of plastic tubes or large oyster shells gave similar results (Aiken and Waddy, 1977). Other research indicated that densities as high as 40 juveniles per square meter are possible when three-dimensional stacks of plastic honeycomb material are provided as vertical substrate (Carlberg *et al.*, 1979). However, a large range in size from 10 to 44 mm CL was observed after 7 months of culture (Fig. 7). The comparable range for individually reared juveniles was 20 to 25 mm.

Mass rearing of juvenile lobsters may prove useful as one phase of commercial lobster production. It appears that a certain minimum food level is necessary to reduce cannibalism, as shown by Van Olst *et al.* (1975), but that the degree of aggressive interaction is approximately the same regardless of the food dose. This suggests that two different behavioral mechanisms for cannibalism and aggression exist. It is possible that the tendency for cannibalism among lobsters is modulated by the level of hunger satiation, but the tendency to interact aggressively is a strong and constant drive, independent of food level. Under conditions of hunger, the aggression in agonistic encounters probably proceeds to cannibalism more often than under conditions of satiation. Lobsters that, for genetic

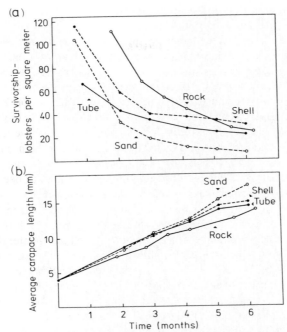

Fig. 6. (a) Survivors vs. time and (b) average carapace length vs. time for American lobsters cultured in four different substrates. The points plotted are means of three replicates. (From Van Olst et al., 1975.)

or other reasons, grow to a larger size than their siblings, win more encounters. These dominant animals have access to a greater proportion of the daily food ration.

Communal rearing of juveniles probably would result in cost savings over culturing lobsters in individual containers, because the communal rearing systems for holding, water supply, feeding, and waste removal are less complex and more economical to operate than are those incorporating single cell rearing containers. Cannibalism in communal rearing systems may serve as a process of selection, so that those individuals better adapted to culture, in terms of growth and other characteristics, will tend to survive. In addition, since the survivors will feed in part on their siblings, more rapid growth rates may result. Nutritional studies (Van Olst et al., 1975) have shown that lobster flesh used as food for lobsters promotes growth almost as well as brine shrimp.

Social communication among lobsters is partly mediated by chemical compounds secreted into seawater. Upon molting, female lobsters release sex pheromones that may suppress aggression and induce courtship behavior in

Fig. 7. Extreme variability in size of cultured lobsters produced in a communal rearing system. After seven months of culture, the carapace length of sibling lobsters ranged from 10 to 44 mm. (From Van Olst and Carlberg, 1978b.)

males (Atema and Engström, 1971). Similar pheromones may be employed for communication among juvenile lobsters. If these compounds can be demonstrated to exist; then identified, isolated, and even synthesized, they might be used to control agonistic behavior in lobsters.

Previous work has shown that lobsters are nocturnal, and exhibit most of their feeding and shelter-related behavior at night (Cobb, 1971). Current research is directed toward using this diel variation in behavior, to reduce encounters by holding lobsters in prolonged light periods. Similar studies with penaeid shrimp suggest that the wavelength of light also plays an important role (H. J. Ceccaldi, personal communication). Blue light had a calming effect, whereas orange light elicited aggression. Preliminary results of studies of day length with lobsters showed that 18 hr of light resulted in higher survival in communal rearing systems (Carlberg et al., 1979). Other studies determined that the most efficient stocking density in these systems was 100 stage IV lobsters per square meter, and that feeding of frozen brine shrimp yielded the same survival as live *Artemia*. Also, it appeared that elevated temperatures accelerated growth without adversely affecting survival. In studies by Aiken and Waddy (1977), periodic harvesting of dominant individuals increased the subsequent growth of previously stunted subordinates and resulted in increased biomass and production of more uniform size juveniles.

Immobilization of the chelipeds significantly reduced mortalities due to cannibalism among communally reared lobsters (Aiken and Waddy, 1978). Large juvenile lobsters with banded claws were stocked at a density of approximately 1.0 kg/m^2 and provided with sufficient shelter. Growth and survival of these lobsters was similar to that of juveniles held individually at the same relative density. Other methods of claw immobilization, such as surgical removal or impairment of the dactylopodite of both the ripper and the crusher claw, increased yields of individuals of uniform size (Aiken and Young-Lai, 1979; Carlberg et al., 1979). Periodic removal of regenerated claws may reduce cannibalism and allow for the harvest and sale of lobster claws.

Communal rearing of *H. americanus* also has been attempted in polyculture, with bivalve molluscs held in intensive raceway systems (Fig. 8) supplied with phytoplankton (Ryther, 1977). The feces and pseudofeces from the shellfish were available to invertebrate detritivores, and they provided a source of food for secondary crops of carnivores, such as lobsters and flounder. When juvenile lobsters were stocked in oyster trays in a segregated portion of a raceway, they attained a mean size of 25 mm CL and a mean weight of 18 g after 8 months. Although this rate of growth was comparable to those held in individual rearing systems, the carrying capacity of the raceway system was significantly lower, approximately 3 lobsters per square meter. Lobsters grown in smaller compartments of the oyster trays produced higher yields per unit area but showed lower rates of growth. Culture of lobsters in green water (high density phytoplankton) resulted in fewer mortalities and probably reduced cannibalism because of visual isolation (Hand, 1977). Investigations have begun in Norway to culture *H. gammarus* communally on the bottom of natural fiords supplemented with artificial shelter. Food will be supplied by cultures of mussels on rafts floating on the surface. In this commercial operation, the lobsters will be retained in the fiords

Fig. 8. Raceways used for culturing molluscs, algae, and lobsters in a polyculture waste treatment experiment at the Woods Hole Oceanographic Institution.

by electro-fencing devices that have been used successfully in salmonid culture (J. G. Balchen, personal communication). More research will be necessary to assess the usefulness of these new methods of communal culture.

B. Procedures for Artificial Rearing

The advances in understanding of lobster biology described in the previous sections have been encouraging, and indicate that one strategy for commercial farming of *H. americanus* might involve a three phase process. In the first phase, larvae would be reared in special larval culture units, referred to as planktonkriesels (Fig. 9). In the second phase, the small juveniles produced would be cultured in communal rearing systems for up to 6 months. The advantages of this approach are the decreased costs in construction of rearing systems and the lower operating costs related to feeding, maintenance, and water distribution for animals held in a few large communal tanks, as opposed to that of many smaller, individual containers. During the third phase, lobsters would be cultured to final market size in a system incorporating individual rearing containers, in order to eliminate losses due to cannibalism.

1. Larval Culture Techniques

A culture system for larvae of *H. americanus* should (1) provide good water circulation to maintain suspension and even distribution of both the larvae and their food; (2) have a filtration system capable of maintaining high quality water, well oxygenated and free of toxic chemicals or detritus; (3) be operable on a fully closed, recirculating basis for several weeks if problems with the seawater supply system make this necessary; (4) function with only routine surveillance; (5) incorporate temperature control; and (6) have provisions for effective disease control through ultraviolet radiation, ozone injection, or addition of antibiotics. The larval culture system used at the Massachusetts State Lobster Hatchery meets many of these requirements (Hughes *et al.*, 1974). The following paragraphs describe typical hatchery operation.

Egg-bearing females are held in large rectangular coated wood or fiberglass tanks. They can be held individually by adding partitions or communally if the claws are banded. As soon as the larvae have hatched and are free-swimming, they are removed from the brood tank by directing them with a gentle current of water into a screen basket and are then transferred to the rearing kreisel (Fig. 9). The rearing process for lobster larvae is relatively simple compared to the rearing of other Crustacea. The newly hatched animal is capable of eating live adult brine shrimp, and a variety of other foods including frozen brine shrimp, daphnia, squid, fish, finely chopped clam, abalone, live plankton, and artificial preparations. Live foods are taken more readily than frozen.

One function of the planktonkriesel is to keep the larvae separated. If the larvae merge in groups, large mortalities will result from cannibalism and fouling. When food and larvae are uniformly distributed and proper densities of both are maintained, larval survival to stage IV will average 60–75%. Hughes *et al.* (1974) described the hydrodynamic characteristics of the circulation device and the culture tank in which the larvae and food are constantly stirred in a spiral upwelling pattern. The inside of the container is white to permit convenient viewing of the small blue larvae. The maximum number of larvae that the 40-liter kriesel can successfully contain is about 3000, although higher survival is attained at lower densities. If live adult brine shrimp (*Artemia salina*) are used as food, a proportion of approximately four brine shrimp to one larva should be maintained (Carlberg and Van Olst, 1976). The constant availability of food and turbulent water flow will reduce cannibalism considerably. If other foods are used, about 10 ml of the food should be added every 3 hr. As the larvae develop

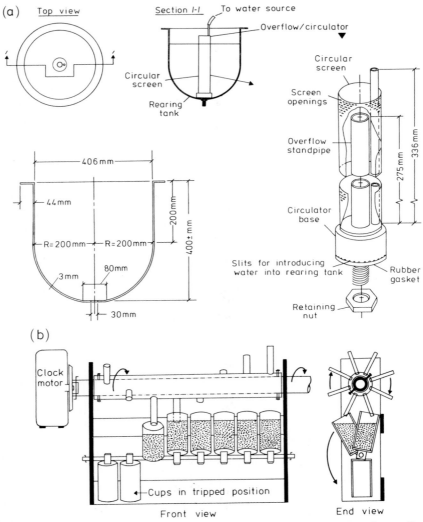

Fig. 9. (a) Larval culture container (Hughes *et al.*, 1974) used at the Massachusetts State Lobster Hatchery and Research Station at Martha's Vineyard, Massachusetts. (b) Automatic feeding device used by Serfling *et al.* (1974ab) to deliver small amounts of live food to the larval culture containers at regular intervals.

to the second, third, and fourth stages, food requirements increase and feeding levels must be adjusted to meet these needs.

Larvae may be fed heavily late at night and then not again until morning. Serfling *et al.* (1974b) devised a mechanical feeder that adds a portion of live or

Fig. 9. (c) Larval rearing container equipped with automated feeding system.

frozen brine shrimp to the kriesel at timed intervals during the night (Fig. 9). Excess food and exuviae will collect on the central cylinder filter screen, which should be removed and cleaned daily.

Warmer temperatures permit more rapid growth of larvae. At 22°C, *H.*

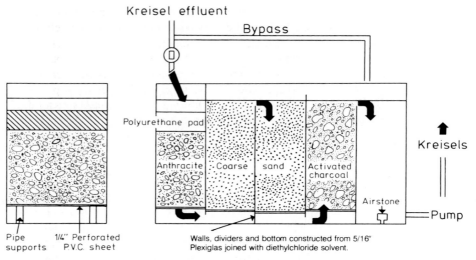

Fig. 10. Multimedia filter used in recirculating seawater culture system for larvae of *H. americanus*. (From Schuur *et al.*, 1976.)

americanus larvae reached the fourth stage in as few as 9 days, while those cultured at 15°C took as long as 35 days (Hughes and Matthiessen, 1962). Larvae also will suffer if the temperature fluctuates erratically. If a temperature change is required, a gradual adjustment of the temperature of no more than 3°C per day to the desired level is preferable to larger daily fluctuations (Schuur *et al.*, 1976).

A number of different culturing system configurations have been developed that employ the planktonkriesel developed at the Massachusetts hatchery (Serfling *et al.*, 1974a,b). Investigators in Canada and France have developed larger square or conical tanks, but the principle of operation is similar. The system described by Serfling *et al.* (1974a) is a closed system incorporating a sand substrate filter. Sand filters properly conditioned and not overly taxed by nitrogenous wastes have been used effectively to clarify seawater in larval culture. Intensive water treatment has been achieved with a multimedia filter (Fig. 10). The flow is columnar, passing through a series of media designed to remove particulate material, reduce ammonia through bacterial action, and remove organics by carbon absorption. Ultraviolet treatment may be added if pathogenic microbes become a source of animal mortality. The columnar configuration increases the total surface area of the filter and therefore allows more biological filtration per unit of floorspace than does a substrate filter.

2. Communal Rearing of Juveniles

The stage IV individuals resulting from larval culture will become benthic sometime after molting to stage IV and could be stocked in larger systems designed for juvenile culture. The second phase of culture might involve a communal rearing approach as described previously in the section on behavior. During this period, even with high levels of supplemental feeding, relatively high losses due to cannibalism can be expected (Van Olst et al., 1975), but these losses come at a time early in the growth of the lobsters and may be tolerable in terms of labor and feed expenditures at this stage in the culturing process. However, it is clear that at some point in the culturing process, further losses due to cannibalism could not be tolerated from an economic standpoint. Thus, in the third phase of the culturing process, individual or single-cell rearing would probably be mandatory.

3. Culture Systems for the Individual Rearing of Lobsters

Several different methods have been developed for culturing American lobsters individually. All of them attempt to provide a separate compartment for each animal, a constant supply of oxygenated seawater to each individual, a method of providing food and removing particulate and dissolved wastes, and in general an environment that will promote rapid, uniform growth and high survival. Once the criteria for providing high water quality and an adequate diet are met, the most important remaining requirement is the amount of space provided for each animal, in terms of cross-sectional bottom area (Shleser, 1974). Van Olst and Carlberg (1978b) showed that growth and survivorship are reduced in small containers (Figs. 11 and 12). Lobster growth is limited due to a reduction in the length and weight gains at each molt and also by a lengthening of the intermolt period. Stewart and Squires (1968) observed that holding *H. americanus* in small individual cages inhibited molting significantly. Other studies by Sastry and French (1977) and Aiken and Waddy (1977) have indicated that space limitation affects both growth components, with a greater influence on increment gain per molt than on the frequency of molting. The effects of space limitation can thus be compared to the effects of inadequate nutrition, which also influences both components of growth.

A rectangular shape for the containers would probably be advantageous due to the elongated body form of the lobster. Also, the available area in the culture systems could be used effectively without wasted space. Shleser (1974) showed no differences in lobster growth rate among circular, rectangular, and square containers.

It appears that for unrestricted growth, a container should have an area approximately three times the total body length of the lobster squared (approximately one square meter for a 454 g lobster), which would make the size of the commercial facility required to hold the lobsters extremely large (J. C. Van Olst and J. M. Carlberg, unpublished data). On the other hand, some reduction in growth rate may be tolerable if sufficient savings are realized due to the costs of constructing and operating the smaller culture facility, which would be needed if smaller containers were used. Van Olst and Carlberg (1978b) suggest that a

Fig. 11. Culture system used to rear American lobsters in eight sizes of individual containers ranging from 6 to 750 cm^2. (From Van Olst and Carlberg, 1978a.)

container with a bottom area of one body length squared will cause only a 10–15% reduction in growth and survival. To accomodate lobsters of all sizes and to adjust for limited restriction due to container size, several transfers to larger rearing compartments may be necessary in the production sequence. Perhaps as many as three sizes of containers will be required. Following transfer to larger containers, some lobsters have been observed to grow more rapidly than

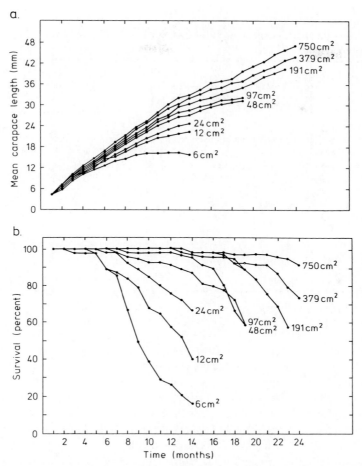

Fig. 12. (a) Growth and (b) survival of eight groups of *H. americanus* cultured in eight different sizes of individual rearing containers for a period of 24 months. (From Van Olst and Carlberg, 1978a.)

normal, once the stress due to restriction is removed. This "catch-up" phenomenon may play an important role in establishing the optimum number of transfers, and should be investigated further.

Almost all of the culture systems proposed for the individual rearing of *H. americanus* are capital-intensive, and are conservative of land and labor when compared to the typical methods utilized for rearing of less cannibalistic species in ponds. However, the complexity of the culture systems varies considerably. Table V lists some of the advantages and disadvantages inherent in these systems. Perhaps the least complex system is one in which a two-dimensional array

TABLE V

Ratings of Five Methods for the Intensive Rearing of Lobsters, in Regard to 11 Qualities Essential to the Success of the System[a]

	Culture system design				
	Offshore		Onshore		
Dimensions	2D	3D	2D	Layered 2D	3D
Yield/bottom area	Low	High	Low	Medium	High
Control of environment	Low	Low	High	High	High
Surface/volume ratio	High	Medium	High	High	Low
Aeration costs	Low	Low	Medium	Medium	High
Water treatment and pumping costs	Low	Low	Medium	High	Medium
Land costs	Low	Low	High	Medium	Low
Equipment costs	Low	Low	High	Medium	Low
Ease of feeding	Medium	Low	High	Medium	Medium
Ease of harvesting	Medium	Low	High	High	Medium
Ease of monitoring	Medium	Low	High	Medium	Low
Potential for disease transmission	High	High	Medium	Low	Low

[a] From Van Olst et al., 1977.

of individual rearing containers is placed at the surface of an existing body of water. Such an offshore, two-dimensional system is currently under consideration in the United States. A three-dimensional array of cages that would be anchored to the ocean bottom to avoid surface weather conditions has been proposed by culturists in the United States, Europe, and Japan. These underwater systems would be serviced by divers. A similar three-dimensional offshore cage system to be serviced from an access pier on the surface was patented in the United States in 1968 (Fig. 13). The principal advantages of offshore systems such as these would be lower initial capital costs and lower operating costs for water treatment and pumping. However, these may be offset by potential problems related to the lack of control over the environment offshore, the difficulty of stocking, feeding, and harvesting, the limited control over disease and predators, and the potential legal problems involving restriction of public waterways.

The majority of the proposed culture systems for the individual rearing of crustaceans has been developed as onshore systems in order to provide for increased environmental control. As Table V indicates, they can be separated into three categories. Two-dimensional systems provide an array of rearing compartments one layer deep. Layered two-dimensional systems arrange several trays in compartments vertically, but each layer is located in its own separate tray or tank of water, so that every animal has an air–water interface above it.

10. Aquaculture

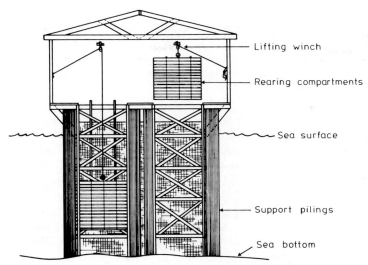

Fig. 13. Offshore three-dimensional lobster culture system proposed and patented by P. McDonald in 1968. Lobster trays would be lifted to the surface for feeding and maintenance.

Three-dimensional systems consist of stacks or layers of individual cages that are immersed in one or more deep tanks or raceways.

One example of a two-dimensional system is shown in Fig. 14. In areas where land value is low enough so that the facility can be spread out to accommodate a single layer of holding containers, this system offers the advantages of a high surface area to water volume ratio for oxygen transfer and convenient and vertically unrestricted access for monitoring, stocking, feeding, and harvesting. In this system, a matrix of perforated cages rotates around a center axis and floats

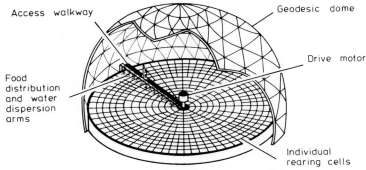

Fig. 14. Rotating "care-o-cell" lobster culture system. Several prototypes of this system have been constructed and perform satisfactorily. (From Van Olst et al., 1977.)

at the surface of a larger circular tank, similar to a primary clarifier used in sewage treatment. Each rearing compartment passes beneath radius arms suspended over the tank. The radius arms supply oxygenated water, and in a commercial system, they would be implemented with automatic food delivery hoppers and a walkway to allow human access for stocking, harvesting, and other maintenance operations.

Unfortunately, most coastal property is prohibitively expensive, and therefore it may be necessary to resort to vertical use of this land in the commercial culture of lobsters. One method of utilizing the vertical space is simply to arrange the culture trays in vertical racks. A system that is adaptable for this approach is one that consists of multiple layers of fiberglass trays supported by steel cantilever racks. Development of a multilayered tray system has been a joint effort between investigators at San Diego State University and the St. Andrews Biological Station in New Brunswick, Canada. The system consists of a fiberglass tray that is flushed peridically to remove wastes from lobster holding cells constructed of plastic (Fig. 15). Seawater for oxygenation is introduced continuously by a trough or pipe along one wall of the tray, flows through the perforated walls of each rearing container, and exits through an overflow trough or collection pipe along the opposite wall. The short partitions, oriented 90° to the long perforated partitions, are made of solid, unperforated plastic. The bottoms of the rearing

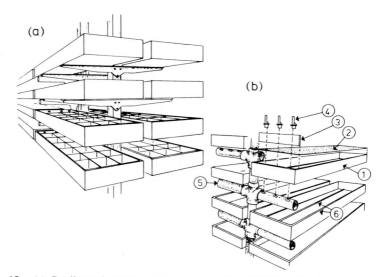

Fig. 15. (a) Cantilevered racking system used by Aiken (1977), which incorporates several flushing tray units. (b) Construction details: (1) flushing tray, (2) perforated bottom screen, (3) end partition serving as dam to water flow, (4) plastic flushing valves, (5) drain manifold, and (6) support rods to maintain space between tray and floor of rearing compartment.

10. Aquaculture

containers are covered with perforated screen and are elevated slightly off the bottom of the tray. Once per day, or more frequently if necessary, large valves are opened at the end of the flushing tray. Because the short, solid partitions act as dams to the longitudinal flow of seawater in the tray, all the water flow is forced down through the perforated bottom of each rearing container and then moves rapidly across the bottom of the tray toward the drain. This rapid flow has a scouring effect on the bottom screening and on the bottom of the tray, so that the entire system is almost completely self-cleaning. Infrequent manual cleaning is necessary to remove attached algae and bacterial slime. A similar tray system has been used at the University of Rhode Island (Sastry, 1975). Results from recent studies in which the screen container bottoms were removed and the lobsters were cultured directly on the bottom of the trays showed that even less cleaning and maintenance was required (D. E. Aiken, personal communication).

The flushing tray system appears to be superior to other two-dimensional systems in its ability to hold large numbers of animals in a minimal amount of space, deliver oxygenated seawater evenly to each rearing compartment, and

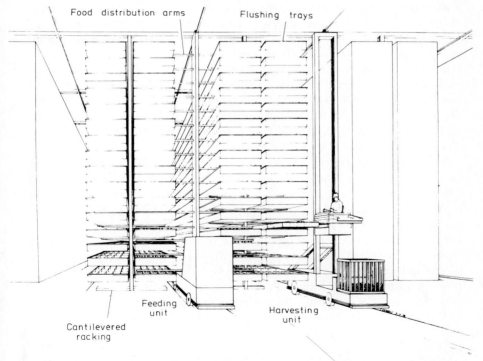

Fig. 16. Artist's conception of commercial lobster production facility incorporating flushing tray system. (From Van Olst *et al.*, 1977.)

remove wastes rapidly and economically. It is adaptable to support on cantilevered racks up to approximately 20 levels. In a commercial application, multi-armed feeding devices might pass through aisles between the tray racks to pass food dispersing arms over each tray, as shown in the artist's conception in Fig. 16. Because of the cantilever design, centralized seawater supply and drainage system, there are no obstructions to interfere with the arms of the feeding unit.

Although the layered, two-dimensional systems provide for relatively convenient access and high surface/volume ratios, they are expensive, because a separate tank is necessary for each level. These systems may be most useful in holding especially valuable animals such as brood stock, which would need to be handled and moved frequently during the culture operations.

Three-dimensional culture systems appear to offer greater savings in production costs, since they require little land area, can be placed in inexpensive holding tanks, and do not require expensive stacked trays. Disadvantages of relatively low surface area to volume ratios and inconvenient feeding and harvesting can be compensated for (at least in theory) by incorporating several recent technological developments.

For these reasons, there have been a number of three-dimensional individual culture systems designed for *H. americanus* culture. One of these (Van Olst *et al.,* 1977) consists of a large, deep holding tank in which perforated rearing container units are hung vertically, as shown in Fig. 17. The containers would be lifted vertically out of the holding tank by an overhead winch so that feeding, stocking, and harvesting could be accomplished. Food is introduced through food injection nozzles, which slip through flexible rubber slits molded into the vertical face of each container. If preliminary estimates are accurate, extremely high holding densities appear feasible, perhaps as high at 50 kg/m^2 of bottom area in concrete raceways 3 m wide and 3 m deep.

Several other three-dimensional systems have been proposed. A vertical silo culture system for *H. americanus* has been constructed in Kittery, Maine (Fig. 18). Each lobster is held on a circular disk suspended by rods in vertical acrylic tubes. Food is introduced to each level by packing it into recesses formed in the side of a dowel, which is lowered into a distribution tube which has holes at intervals matching the food recesses in the dowel. By rotating the dowel, lobsters in each level of the silo are allowed access to the food. One disadvantage of this system appears to be the high cost of the acrylic tubing used for each vertical silo. It would appear to be more cost-effective to eliminate these walls by placing the cages in a larger holding tank. This is currently being investigated.

A system of three-dimensional silo culture of *H. americanus* has been proposed by investigators at the University of California, Davis (Fig. 19). In this three-dimensional system, the lobsters are held in compartmented trays stacked vertically and immersed in large ferrocement silo tanks (Schuur *et al.,* 1974). A moveable service gantry is equipped with arms that lift the tray stack out of the tank, move each of the trays horizontally out of the stack for feeding, and then

10. Aquaculture

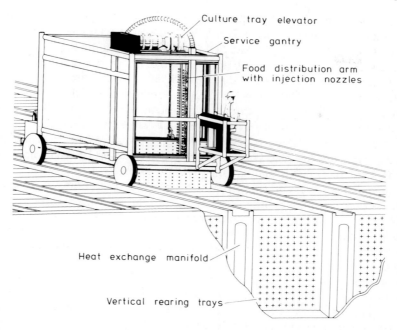

Fig. 17. Artist's conception of commercial lobster production facility incorporating a deep-tank design. (See text for details of operation.) (From Van Olst *et al.*, 1977.)

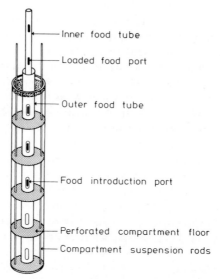

Fig. 18. Vertical tube system for culturing lobsters is being tested in a pilot lobster farm in Kittery, Maine.

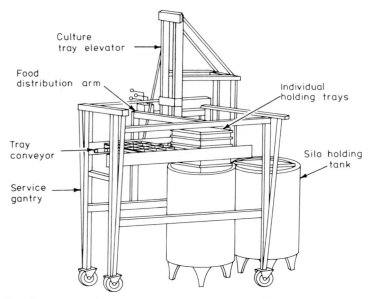

Fig. 19. Silo tank lobster culture system. A full-scale functioning prototype of the ferrocement tanks and the traveling service gantry has been constructed and evaluated (R. Garrett, personal communication).

return the tray to the stack. A full-size functioning prototype of this gantry has been constructed. This system appears to have considerable potential. However, since the perforated bottom of one tray forms the top of the tray below, it is conceivable that some animals might cling to the top of the container above and be injured or lost when that tray is shifted from one stack to another.

Workers at Brigham Young University have proposed a three-dimensional culture system for *H. americanus* in which food is delivered to the individual containers through an array of vertical tubes that reach the surface of each culture tank (R. W. Mickelsen, personal communication). High density culture also appears possible with this design.

C. Economics

The commercial production of *H. americanus* will be a very complex operation, involving expensive, specialized equipment and techniques. In order to assist private industry in deciding whether or not to construct a full-size lobster farm, investigators in California are proceeding with two types of research. At the Bodega Marine Laboratory and the University of California, Davis, a computer model of a lobster production facility has been developed (Allen and

Johnston, 1976). At San Diego State University, a prototype production system has been constructed in order to provide accurate estimates of operating costs (Van Olst and Carlberg, 1978a).

The most recent computer model was developed by Botsford (1977). The basic approach has been to model (1) the response of the biological organism to the environment, (2) the physical system which provides that environment, and (3) the associated costs as a function of the operation of the physical system. This model has been used to project culture costs and determine the accuracy of projections through sensitivity analysis. It has also been employed to determine optimal culture methods using optimal control theory. The biological model includes descriptions of growth rate, metabolic rates, and tolerance to metabolites. Growth is dependent on several factors, including temperature, amount of space available, and amount of food provided. The dependence of growth on temperature is defined as

$$dW/dt = K_0 W^\alpha (U-U_0)^m \quad \text{for} \quad U_0 < U < U_{max}$$

were W is the weight of the lobster (state variable), U is the water temperature (control variable), U_0 is the empirically determined temperature at which no growth is observed, U_{max} is the maximum practical temperature, and K_0, α, and m are empirically determined constants (Botsford et al., 1974). The effect of space is represented by

$$dW/dt = KW^\alpha S$$

where $S = 1 - (W/W_{max})^\mu$, and where K is an empirically determined growth constant, S is a space restriction factor varying between zero and one, W_{max} is the limiting weight that is a function of space, and μ and α are empirically determined constants. Table VI shows the results of a cost projection study by Johnston (1975). Approximately 3 million dollars would be required to construct this facility, which would produce 1 million pounds of lobsters annually, on a site of less than 3 hectares in area. Van Olst and Ford (1976) have presented these data in modified form (Table VII) so that the cost benefits of utilizing waste heat in lobster culture are more readily visible. The use of thermal effluent to provide a low-cost source of heated seawater reduces production costs to a level that appears to be competitive with that of the present lobster fisheries. However, interest rates and return-on-investment calculations are not included in these estimates. As the costs of lobsters from the fishery increase due to high exploitation of the resource, improvements in culture technology result in lower and lower estimates of the costs of cultured lobsters. At some point, lobster aquaculture should become an economically viable porposition. Clearly, lobster farming has evolved to the point where there is a need for experiments on a large enough scale, so that realistic estimates of production costs can be made.

In projecting the point at which lobster culture becomes profitable, it is impor-

TABLE VI

Projected Costs for Production of American Lobsters under Various Environmental Conditions and Modes of Operation[a]

System cost and physical characteristics	Alternative state variable or parameter values less than baseline model				Baseline model	Alternative state variable or parameter values greater than baseline model			
Operating temperature (°C)	13	15	17	19	21			18	20
Total cost/500 g ($)	4.47	4.22	4.34	4.88	5.65			3.76	3.13
Average ambient temperature (°C)				10	12	14	16	18	20
Total cost/500 g ($)				6.27	5.65	5.02	4.39	3.76	3.13
Time to maturity (months)			24.0	27.0	30.0	33.3	36.0		
Total cost/500 g ($)			4.91	5.29	5.65	6.05	6.46		
Plant size (monthly output):	10,000	20,000	40,000	60,000	80,000	100,000	150,000	200,000	250,000
Total cost/500 g ($)	7.02	6.18	5.81	5.70	5.65	5.61	5.57	5.55	5.54
Target weight	100	200	300	400	500	600			
monthly output (g)	296,000	166,400	122,400	94,000	80,000	68,000			
Total cost/500 g ($)	6.00	5.82	5.57	5.78	5.65	5.73			
Total space tank area/unit output (m^2)			4.58	1.92	1.24	0.96	0.80	0.70	
Total cost/500 g ($)			7.84	5.98	5.65	5.67	5.80	5.95	
Recirculation coefficient (R):					0	0.1	0.3	0.5	0.7	0.9
Total cost/500 g ($)					5.65	5.59	5.15	4.72	4.26	3.78

[a] From Johnston, 1975.

10. Aquaculture

TABLE VII

Estimated Lobster Production Costs With and Without the Use of Thermal Effluent as a Source of Heat[a]

	Heated ambient temperature seawater (12°–21°C)		Steam electric generating station thermal effluent seawater (21°C)	
	$/pound Output	% of Total Cost	$/pound Output	% of Total Cost
Space	0.68	13	0.68	30
Heat	2.90	56	0.00	0
Pumping	0.06	1	0.06	3
Waste treatment	0.08	2	0.08	4
Aeration	0.05	1	0.05	2
Food	0.96	19	0.96	43
Labor	0.37	7	0.37	16
Larval	0.04	1	0.04	2
Total cost	$5.14		$2.24	

[a] Data from W. E. Johnston, University of California, Davis.
[b] Assumptions: (1) Months to output, 30.0; (2) Plant output (thousand/month), 80.0; (3) Harvest weight (g/animal), 502.6; (4) Total capital ($100,000), 31.05; (a) Culture capital, 25.22; (b) Waste treatment capital, 5.83; (5) Tank area (1000 m^2), 98.95; (6) Water reuse (% recirculation), 0.00; (7) Intake flow (million liters/day), 43.45; (8) Land area for production facility (ha) 2.75; (9) Conversion ratio, 3.30.

tant to analyze supply and demand. The demand for lobster is influenced by population growth, by growth of per capita income, and by the income elasticity of demand. Assuming a continued 1.5% population growth rate in the United States, coupled with an average 1.5% annual growth in per capita real income, a 4.6% annual growth in aggregate supplies could be absorbed without influencing the price (Gates et al., 1974). However, supplies are expected to remain stable near the estimated maximum sustainable yield of 70.5 × 10^6 pounds for the northwest Atlantic lobster fishery (Anonymous, 1978). Therefore, lobster prices will continue to increase about 5% per year (*not* counting inflation), making lobster culture even more economically attractive in the near future.

An alternative to selling cultured, 1-pound (454 g) lobsters to the traditional market would be to enter the "prawn" market by producing smaller juvenile lobsters (100–200 g). This approach would reduce the production time and associated costs, but would require increased hatchery costs because of the greater number of larvae necessary.

Descriptive flavor analysis of both adult and juvenile cultured lobsters showed them to be indistinguishable in flavor from wild-caught adults (Van Olst and

Carlberg, 1978a). Regardless of the size of lobsters produced, the cultured product should command a higher price than those from the fishery because they are already on hand, are fresh, are available continuously, and could be marketed during periods of seasonally high prices in mid-winter when there is no supply from the fishery.

IV. CONCLUSIONS

Based on information now available, the prospect of farming palinurid and scyllarid lobsters in the immediate future is not promising. There has been a considerable amount of research conducted in recent years on the culture of the American lobster, *H. americanus*. Much of the information is encouraging and this species has been designated as one having high priority for commercial culture. Cost estimates indicate that commercial production may be economically feasible at present. However, more research is needed before full-scale production should be attempted. There are several remaining problems that need to be solved before the commercial culture of American lobsters can be expected to succeed. These include the refinement of artificial diets and reduction of diseases associated with nutritional deficiencies, development of methods for the routine production of larvae, the reduction of cannibalism, and the improvement of designs for individual rearing. Pilot or demonstration-scale programs are critically needed to provide the most reliable assessment possible concerning the ultimate potential for commercial culture of this species.

REFERENCES

Aiken, D. E. (1977). Molting and growth in decapod crustaceans, with particular reference to the lobster, *Homarus americanus*. *Circ.—CSIRO, Div. Fish. Oceanogr. (Aust.)* **7**, 41–75.

Aiken, D. E., and Waddy, S. L. (1976). Controlling growth and reproduction in the American lobster. *Proc. Annu. Meet.—World Maric. Soc.* **7**, 415–430.

Aiken, D. E., and Waddy, S. L. (1977). Effect of space and density on growth of large juvenile lobsters. *Environ. Can., Appl. Physio. Res. Rep.* No. 13, pp. 1–17.

Aiken, D. E., and Waddy, S. L. (1978). Relationship between space, density, and growth of juvenile lobsters (*Homarus americanus*) in culture tanks. *Proc. Annu. Meet.—World Maric. Soc.* **9**, 461–467.

Aiken, D. E., and Young-Lai, W. W. (1979). Cheliped ablation and immobilization as a method for improving survival and growth of juvenile American lobsters in communal culture tanks. *Proc. Annu. Meet.—World Maric. Soc.* **10** (in press).

Allen, P. G., and Johnston, W. E. (1976). Research direction and economic feasibility: An example of systems analysis for lobster aquaculture. *Aquaculture* **9**, 144–180.

Anonymous (1978). "American Fishery Management Plan" (draft). Northeast Mar. Fish. Board, Northeast Fish. Cent., Woods Hole, Massachusetts.

10. Aquaculture

Atema, J., and Engström, D. G. (1971). Sex pheromone in the lobster, *Homarus americanus*. *Nature (London)* **232**, 261–263.

Australian Department of Agriculture, Fisheries Division (1975). The potential of aquaculture in Australia. *Aus. Gov. Print Serv. Fish.* Pap. No. **21**, pp. 1–133.

Ayres, P. A., and Wood, P. C. (1977). The live storage of lobsters. *Minist. Agric. Fish. Food, Lowenstoft Lab. Leafl.* No. 37, pp. 1–10.

Bardach, J. E., Ryther, J. H., and McLarney, W. O. (1972). "Aquaculture: The Farming and Husbandry of Freshwater and Marine Organisms." Wiley (Interscience), New York.

Batham, E. J. (1967). The first three larval stages and feeding behavior of phyllosoma of the New Zealand palinurid crayfish, *Jasus edwardsii* (Hutton, 1975). *Trans. R. Soc. N.Z., Zool.* **9**, 53–64.

Becker, C. D., and Thatcher, T. O. (1973). "Toxicity of Power Plant Chemicals to Aquatic Life," WASH-1249. U.S. At. Energy Comm., Washington, D.C.

Bell, F. W., and Fullenbaum, R. F. (1973). The American lobster fishery: Economic analysis of alternative management strategies. *Mar. Fish. Rev.* **35**(8), 1–6.

Blecha, J. B. (1972). The effects of temperature on biomass production in juvenile California spiny lobster, *Panulirus interruptus* (Randall). Master's Thesis, San Diego State University, San Diego, California.

Botsford, L. W. (1977). Current economic status of lobster culture research. *Proc. Annu. Meet.—World Maric. Soc.* **8**, 723–740.

Botsford, L. W., Rauch, H. E., and Shleser, R. A. (1974). Optimal temperature control of a lobster plant. *IEEE Trans. Autom. Control* **ac-19**(5) 541–543.

Bowen, J. J. M. (1966). "Trace Elements in Biochemistry." Academic Press, New York.

Bowers, G. M. (1900). The American lobster. *In* "A Manual of Fish Culture," pp. 229–238. U.S. Comm. Fish Fish., Washington, D.C.

Bryan, G. W. (1968). Concentrations of zinc and copper in the tissues of decapod crustaceans. *J. Mar. Biol. Assoc. U.K.* **48**, 303–321.

Capuzzo, J. M., and Lancaster, B. A. (1979). The effects of dietary carbohydrate levels on protein utilization in the American lobster (*Homarus americanus*). *Proc. Annu. Meet.—World Maric. Soc.* **10** (in press).

Capuzzo, J. M., Lawrence, S. A., and Davidson, J. A. (1976). Combined toxicity of free chlorine and temperature to Stage I larvae of the American lobster, *Homarus americanus*. *Water Res.* **10**, 1093–1099.

Carlberg, J. M., and Van Olst, J. C. (1976). Brine shrimp (*Artemia salina*) consumption by the larval stages of the American lobster (*Homarus americanus*) in relation to food density and water temperature. *Proc. Annu. Meet.—World Maric. Soc.* **7**, 379–389.

Carlberg, J. M., Van Olst, J. C., and Ford, R. F. (1978). A comparison of larval and juvenile stages of the lobsters *Homarus americanus, Homarus gammarus* and their hybrid. *Proc. Annu. Meet.—World Maric. Soc.* **9**, 109–122.

Carlberg, J. M., Van Olst, J. C., and Ford, R. F. (1979). Potential for communal rearing of the nephropid lobsters (*Homarus* spp.). *Proc. Annu. Meet.—World Maric. Soc.* **10** (in press).

Carlson, F. T. (1954). The American lobster fishery and possible applications of artificial propagation. *Yale Conserv. Stud.* **3**, 3–7.

Castell, J. D., and Budson, S. D. (1974). Lobster nutrition: The effect on *Homarus americanus* of dietary protein levels. *J. Fish. Res. Board Can.* **31**, 1363–1370.

Castell, J. D., Covey, J. F., Aiken, D. E., and Waddy, S. L. (1977). The potential for ablation as a technique for accelerating growth of lobsters (*Homarus americanus*) for commercial culture. *Proc. Annu. Meet.—World Maric. Soc.* **8**, 895–914.

Chittleborough, R. G. (1974a). Western rock lobster reared to maturity. *Aust. J. Mar. Freshwater Res.* **25**, 221–225.

Chittleborough, R. G. (1974b). Review of prospects for rearing rock lobsters. *CSIRO, Div. Fish. Oceanogr., Repr.* No. 812, pp. 1–5.
Cobb, J. S. (1971). Shelter-related behavior of the lobster, *Homarus americanus*. *Ecology* **52**(1), 108–115.
Cobb, J. S. (1976). "The American Lobster: The Biology of *Homarus americanus*." University of Rhode Island Sea Grant Program, Kingston.
Cobb, J. S., and Tamm, G. R. (1974). Social conditions increase intermolt period in juvenile lobsters, *Homarus americanus*. *J. Fish. Res. Board Can.* **32**, 1941–1943.
Conklin, D. E. (1975). Nutritional studies of lobsters (*Homarus americanus*). *Proc. Int. Conf. Aquacult. Nutr., 1st, 1975* pp. 287–296.
Conklin, D. E., Devers, K., and Bordner, C. E. (1977). Development of artificial diets for the lobster, *Homarus americanus*. *Proc. Annu. Meet.—World Maric. Soc.* **8**, 841–852.
Connor, P. M. (1972). Acute toxicity of heavy metals to some marine larvae. *Mar. Pollut. Bull.* **3**(12), 190–192.
Cornick, J. W., and Stewart, J. E. (1977). Survival of American lobsters (*Homarus americanus*) stored in a recirculating, refrigerated seawater system. *J. Fish. Res. Board Can.* **34**, 688–692.
Cox, R. A. (1965). The physical properties of seawater—Part II. "Chemical Oceanography," Vol. I. pp. 75–81. Academic Press, New York.
Delistraty, D. A., Carlberg, J. M., Van Olst, J. C., and Ford, R. F. (1977). Ammonia toxicity in cultured larvae of the American lobster (*Homarus americanus*). *Proc. Annu. Meet.—World Maric. Soc.* **8**, 647–672.
Dewees, C. M., and Schapiro, H. C. (1974). Suggestions for holding live American lobsters in tanks. *Univ. Calif., Sea Grant Mar. Advis. Publ.* No. 3004, pp. 1–5.
Dexter, D. M. (1972). Molting and growth in laboratory reared phyllosomes of the California spiny lobster, *Panulirus interruptus*. *Calif. Fish Game* **58**, 107–115.
Dorband, W. R. (1975). Effects of chemicals in thermal effluent on *Homarus americanus* maintained in aquaculture systems. Master's Thesis, San Diego State University, San Diego, California.
Dorband, W. R., Van Olst, J. C., Carlberg, J. M., and Ford, R. F. (1976). Effects of chemicals in thermal effluent on *Homarus americanus* maintained in aquaculture systems. *Proc. Annu. Meet.—World Maric. Soc.* **7**, 391–414.
Doudoroff, P., and Katz, M. (1953). Critical review of literature on the toxicity of industrial wastes and their components to fish. II. The metals and their salts. *Sewage Ind. Wastes* **25**, 802–839.
Eisler, R. G. E., Zaroogian, E., and Hennekey, R. J. (1972). Cadmium uptake by marine organisms. *J. Fish. Res. Board Can.* **29**, 1367–1369.
Figueiredo, M. J., and Vilela, M. H. (1972). On the artificial culture of *Neprhops norvegicus* reared from the egg. *Aquaculture* **1**, 173–180.
Fisher, W. S., Nilson, E. H., and Shleser, R. A. (1975). Effect of the fungus *Haliphthoros milfordensis* on the juvenile stages of the American lobster, *Homarus americanus*. *J. Invertebr. Pathol.* **26**, 41–45.
Fisher, W. S., Nilson, E. H., and Shleser, R. A. (1976a). Diagnostic procedures for diseases found in egg larvae and juvenile cultured American lobsters (*Homarus americanus*). *Proc. Annu. Meet.—World Maric. Soc.* **6**, 323–335.
Fisher, W. S., Nilson, E. H., Follett, L. F., and Shleser, R. A. (1976b). Hatching and rearing lobster larvae (*Homarus americanus*) in a disease situation. *Aquaculture* **7**, 75–80.
Fisher, W. S., Rosemark, T. R., and Nilson, E. H. (1976c). The susceptibility of culture American lobsters to a chitinolytic bacterium. *Proc. Annu. Meet.—World Maric. Soc.* **7**, 511–523.
Fisher, W. S.. Nilson, E. H., Steenbergen, J. F., and Lightner, D. V. (1978). Microbial diseases of cultured lobsters: A review. *Aquaculture* **14**, 115–140.
Ford, R. F., Van Olst, J. C., Carlberg, J. M., Dorband, W. R., and Johnson, R. L. (1975). Beneficial use of thermal effluent in lobster culture. *Proc. Annu. Meet.—World Maric. Soc.* **6**, 509–522.

10. Aquaculture

Ford. R. F., Felix, J. R., Johnson, R. L., Carlberg, J. M., and Van Olst, J. C. (1979). Effects of fluctuating and constant temperatures and chemicals in thermal effluent on growth and survival of the American lobster (*Homarus americanus*). *Proc. Annu. Meet.—World Maric. Soc.* **10** (in press).

Forster, J. R. M., and Beard, T. W. (1973). Growth experiments with the prawn *Palaemon serratus* fed with fresh and compounded foods. *Fish. Invest., Ser. 2* **27**(7), 1–16.

Gale, N. L. (1973). Aquatic organisms and heavy metals in Missouri's new lead belt. *Water Res. Bull.* **9**, 673–688.

Gallagher, M. L., and Brown, W. D. (1975). Composition of San Francisco Bay brine shrimp (*Artemia salina*). *Agric. Food Chem.* **23**(4), 630–633.

Gallagher, M. L., Conklin, D. E., and Brown, W. D. (1976). The effects of pelletized protein diets on growth, molting and survival of juvenile lobsters. *Proc. Annu. Meet.—World Maric. Soc.* **7**, 363–390.

Galtsoff, P. S. (1973). Hatching and rearing larvae of the American lobster, *Homarus americanus*. *In* "Culture Methods for Invertebrate Animals" (P. S. Galtsoff, F. E. Lutz, P. S. Welch, and J. G. Needham, eds.), pp. 233–236. Dover, New York.

Gates, J. M., Mattheissen, G. C., and Griscom, C. A. (1974). Aquaculture in New England. *Univ. R. I., Mar. Tech. Rep. Ser.* No. 18, 1–77.

Gaucher, T. A. (1968). Thermal enrichment and marine aquiculture. *In* "Marine Aquiculture" (W. J. McNeil, ed.), Selected papers from the Conference on Marine Aquiculture, pp. 141–152. Newport, Oregon.

Ghelardi, R. J., and Shoop, C. T. (1972). Culturing lobsters (*Homarus americanus*) in British Columbia. *Fish. Res. Board Can., Nanaimo, B.C., Tech. Rep.* **301**, 32–42.

Gibson, F. A. (1959). Notes on lobsters storage in Ireland. *Rep. Ser. Inland Fish., Eire* Appl. No. 25, pp. 83–86.

Glude, J. B. (1977). "NOAA Aquaculture Plan." U.S. Natl. Oceanic Atmos. Admin., Washington, D.C.

Goldberg, E. D. (1970). "Seminar on Methods of Detection, Measurement. and Monitoring of Pollutants in the Marine Environment," FAO Rep. FIR:TPME/70/6 Rev. FAO, Rome.

Gravitz, N., Gleye, L., Tchobanoglous, G., and Shleser, R. (1975). Preliminary acute toxicity studies of some inorganic compounds of *Homarus americanus*. *Lobster Aquacult. Conf., Univ. R.I., 1975* Abstract only.

Gruffydd, L. D., Reiser, R. A., and Machin, D. (1975). A comparison of growth and temperature tolerance in the larvae of the lobster *Homarus gammarus* (L.) and *Homarus americanus* H. Milne Edwards (Decopoda, Nephropidae). *Crustaceana* **28**(1), 23–32.

Gundersen, K. R. (1977). Report of the working group on *Homarus* stocks. *ICES, CM* **K:11**, 1–19.

Hand, C. (1977). Development of aquaculture systems. *Univ. Calif., Sea Grant Program, Publ.* No. 58, pp. 1–97.

Harriman, D. M. (1953). Toxicities of some metals in lobster (*H. americanus*) in natural and artificial sea waters. *Maine, Dep. Sea Shore Fish., Cir.* No. 11.

Hedgecock, D. (1977). Biochemical genetic markers for broodstock identification in aquaculture. *Proc. World Maric. Soc.* **8**, 523–531.

Hedgecock, D., Nelson, K., and Shleser, R. A. (1976). Growth differences among families of the lobster. *Homarus americanus*. *Proc. Annu. Meet.—World Maric. Soc.* **7**, 347–361.

Hedgecock, D., Nelson, K., Simons, J., and Shleser, R. A. (1977). Genic similarity of American and European species of the lobster *Homarus*. *Biol. Bull. (Woods Hole, Mass)* **152**, 41–50.

Herrick, F. H. (1911). Natural history of the American lobster. *Bull. U.S. Bur. Fish.* **29**, 149–408.

Hess, E. (1937). A shell disease in lobsters (*Homarus americanus*) caused by chitinovorous bacteria. *J. Biol. Board Can.* **3**(4, pt 2), 358–362.

Horne, R. A. (1969). "Marine Chemistry," pp. 190-194. Wiley (Interscience), New York.
Hughes, J. T. (1968). Grow your own lobsters commercially. *Ocean Ind.* **3**(12), 1-4.
Hughes, J. T., and Matthiessen, G. C. (1962). Observations on the biology of the American lobster, *Homarus americanus*. *Limnol. Oceanogr.* **7**(3), 414-421.
Hughes, J. T., Sullivan, J. J., and Shleser, R. (1972). Enhancement of lobster growth. *Science* **177**, 1110-1111.
Hughes, J. T., Shleser, R. A., and Tchobanoglous, G. (1974). A rearing tank for lobster larvae and other aquatic species. *Prog. Fish Cult.* **36**. 129-132.
Idyll, C. P. (1970). Status of commercial culture of crustaceans. *Proc. Food-Drugs Sea*, pp. 55-64.
Ingle, R. M., and Witham, R. (1968). Biological considerations in spiny lobster culture. *Gulf Caribb. Fish. Inst., Univ. Miami, Proc.* **21**, 158-162.
Inoue, M. (1978) Studies on the cultured larvae of the Japanese spiny lobster, *Panulirus japonicus* - I. Morphology of the phyllosoma. *Bull. Jpn. Soc. Sci. Fish.* **44**(5), 457-475.
Johnson, R. L. (1977). Effects of temperature in using thermal effluent to culture larval and juvenile stages of the American lobster, *Homarus americanus*. Master's Thesis, San Diego State University, San Diego, California.
Johnston, W. E. (1975). Economics of aquaculture. *Univ. Calif., Sea Grant Annu. Rep., Publ.* No. 47, pp. 45-47.
Kellogg, S. T., Steenbergen, J. F., and Schapiro, H. C. (1975). Numerical taxonomy of *Pediococcus homari* and *Aerococcus viridans*. *Bacteriol. Proc.* **1**, 155.
Kensler, C. B. (1967). Notes on laboratory rearing of juvenile spiny lobsters, *Jasus edwardsii* (Hutton) (Crustacea:Decopoda:Palinuridae). *N.Z. J. Mar. Freshwater Res.* **1**, 71-75.
Leavitt, D. F., Bayer, R. C., Gallagher, M. L., and Rittenburg, J. H. (1979). Dietary intake and nutritional characteristics in wild American lobsters. *J. Fish. Res. Board Can.* **36**(8), 965-969.
Logan, D. T. (1975). A laboratory energy balance for the larvae and juveniles of the American lobster. Ph.D. Thesis, University of Delaware, Newark.
McBurney, D. R., and Wilder, D. G. (1973). Survival and condition of lobsters in a tide pond related to feeding, cover and stock density. *Fish. Res. Board Can., Biol. Stn., St. Andrews, N.B., Tech. Rep.* No. 377, pp. 1-59.
McIntosh, W. C. (1920). Tube-dwelling phase in the development of lobster. *Nature (London)* **106**, 441.
McLeese, D. W. (1956). Effects of temperature, salinity, and oxygen in the survival of the American lobster. *J. Fish. Res. Board Can.* **13**, 247-272.
McLeese, D. W. (1974). Toxicity of copper at two temperatures and three salinities to the American lobster (*Homarus americanus*). *J. Fish. Res. Board Can.* **31**, 1949-1952.
Mauchline, J. (1977). Growth of shrimps, crabs and lobsters—an assessment. *J. Cons., Cons. Int. Explor. Mer* **37**(2), 162-169.
Mauviot, J. C., and Castell, J. D. (1976). Molt-and-growth enhancing effects of bilateral eyestalk ablation on juvenile and adult American lobsters (*Homarus americanus*). *J. Fish. Res. Board Can.* **33**, 1922-1929.
Milne, P. H. (1972). "Fish and Shellfish Farming in Coastal Waters." White Friars Press, Ltd., London.
Mitchell, J. R. (1971). Food preferences, feeding mechanisms, and related behavior in phyllosoma larvae of the California spiny lobster, *Panulirus interruptus* (Randall). Master's Thesis, San Diego State University, San Diego, California.
Nelson, K., Hedgecock, D., Borgeson, W., Moffett, W., and Johnson, E. (1979). Control of lobster ovulation. *Aquaculture* (submitted for publication).
Nilson. E. H., Fisher, W. S., and Shleser, R. A. (1975). Filamentous infestations observed on eggs and larvae of cultured crustaceans. *Proc. Annu. Meet.—World Maric. Soc.* **6**, 367-378.

Nilson, E. H., Fisher, W. S., and Shleser, R. A. (1976). A new mycosis of larval lobster (*Homarus americanus*). *J. Invertebr. Pathol.* **27**, 177–183.

Perkins, H. C. (1972). Development rates at various temperatures of embryos of the northern lobster (*Homarus americanus*). *Fish. Bull.* **70**, 95–99.

Phillips, B. F., Campbell, N. A., and Rea, W. A. (1977). Laboratory growth of early juveniles of the western rock lobster. *Panulirus longipes cygnus*. *Mar. Biol.* **39**, 31–39.

Provenzano, A. J. (1968). Recent experiments on the laboratory rearing of tropical lobster larvae. *Gulf Caribb. Fish. Inst., Univ. Miami, Proc.* **21**, 152–157.

Prudden, T. M. (1962). "About Lobsters." Bond & Wheelwright, Freeport, Maine.

Robertson, P. B. (1968). The complete larval development of the sand lobster, *Scyllarus americanus* (Smith) (Decapoda, Scyllaridae) in the laboratory, with notes on larvae from the plankton. *Bull. Mar. Sci.* **18**(2), 294–342.

Ryther, J. H. (1977). Preliminary results with a pilot-plant waste recycling marine aquaculture system. *In* "Wasterwater Renovation and Reuse" (F. M. D'Itri, ed.), pp. 90–132. Dekker, New York.

Saisho, T. (1966). Studies on the phyllosoma larvae with reference to the oceanographical conditions. *Mem. Fac. Fish. Kagoshima Univ.* **15**, 177–239.

Sastry, A. N. (1975). An experimental culture-research facility for the American lobster, *Homarus americanus*. *Eur. Symp. Mar. Biol., 10th, 1975*, Vol. 1, pp. 419–435.

Sastry, A. N., and French, D. P. (1977). Growth of American lobster, *Homarus americanus* Milne-Edwards under controlled conditions. *Circ.—CSIRO, Div. Fish. Oceanogr. (Aust.)* **7** (abstr.).

Sastry, A. N., and Laczak, J. (1975). The ammonia tolerance of lobsters, *Homarus americanus Lobster Aquacult. Conf., Univ. R.I. 1975* Abstract only.

Sastry, A. N., and Zeitlin-Hale, L. (1977). Survival of communally reared larval and juvenile lobsters, *Homarus americanus*. *Mar. Biol.* **39**, 297–303.

Scarratt, D. J. (1968). An artificial reef for lobster (*Homarus americanus*). *J. Fish. Res. Board Can.* **25**, 2683–2690.

Scattergood, L. W., ed. (1949). Translations of foreign literature concerning lobster culture and the early life history of the lobster. *U.S., Fish Wild. Serv., Spec. Sci. Rep.—Fish.* **6**, 1–173.

Schapiro, H. C., Mathewson, J. H., Steenbergen, J. F., Kellogg, S., Ingram, C., Nierengarten, G., and Rabin, H. (1974). Gaffkemia in the California spiny lobster, *Panulirus interruptus*: Infection and immunization. *Aquaculture* **3**, 403–408.

Schuur, A., Fisher, W. S., Van Olst, J. C., Carlberg, J. M., Hughes, J. T., Shleser, R. A., and Ford, R. F. (1976). Hatchery methods for the production of juvenile lobsters (*Homarus americanus*). *Univ. Calif. Sea Grant Coll. Program, IMR Ref.* 76-6; *Sea Grant Publ.* No. 48, pp. 1–21.

Schuur, A. M., Allen, P. G., and Botsford, L. W. (1974). An analysis of three facilities for the commercial production of *Homarus americanus*. *Proc. Am. Soc. Agric. Eng., Pap.* No. 74-5517, pp. 1–18.

Serfling, S. A., and Ford, R. F. (1975a). Laboratory culture of juvenile stages of the California spiny lobster, *Panulirus interruptus* (Randall), at elevated temperatures. *Aquaculture* **6**, 377–387.

Serfling, S. A., and Ford, R. F. (1975b). Ecological studies of the puerulus stage of the California spiny lobster, *Panulirus interruptus*. *Fish. Bull.* **73**(2), 360–377.

Serfling, S. A., Van Olst, J. C., and Ford, R. F. (1974a). A recirculting culture system for larvae of the American lobster, *Homarus americanus*. *Aquaculture* **3**, 303–309.

Serfling, S. A., Van Olst, J. C., and Ford, R. F. (1974b). An automatic feeding device and the use of live and frozen *Artemia* for culturing larval stages of the American lobster, *Homarus americanus*. *Aquaculture* **3**, 311–314.

Shleser, R. A. (1974). Studies of the effects of feeding frequency and space on the growth of the American lobster, *Homarus americanus*. *Proc. Annu. Meet.—World Maric. soc.* **5**, 149–155.

Silberbauer, B. I. (1971). The biology of the South African rock lobster *Jasus lalandii* (H. Milne Edwards). I. Development. *Oceanogr. Res. Inst. (Durban), Invertebr. Rep.* **92**.
Spotte, S. H. (1970). "Fish and Invertebrate Culture." Wiley (Interscience), New York.
Sprague, J. B. (1969). Measurement of pollutant toxicity to fish. I. Bioassay methods for acute toxicity. *Water Res.* **3**, 793-821.
Sprague, J. B. (1971). Measurement of pollutant toxicity to fish. III. Sublethal effects and "safe" concentrations. *Water Res.* **5**, 245-266.
Squires, H. J. (1970). Lobster (*Homarus americanus*) fishery and ecology in Port Au Port Bay, Newfoundland, 1960-65. *Proc. Natl. Shellfish. Assoc.* **60**, 22-39.
Stewart, J. E., and Squires, H. J. (1968). Adverse conditions as inhibitors of ecdysis in the lobster, *Homarus americanus*. *J. Fish. Res. Board Can.* **25**(9), 1763-1774.
Stewart, J. E., Cornick, J. W., and Zwicker, B. M. (1969). Influence of temperature on gaffkemia, a bacterial disease of the lobster, *Homarus americanus*. *J. Fish Res. Board Can.* **26**(9), 2503-2510.
Sverdrup, H. U., Johnson, M. W., and Fleming, R. H. (1942). "The Oceans." Prentice-Hall, Englewood Cliffs, New Jersey.
Tchobanoglous, G., and Shleser, R. A. (1974). Waste treatment costs for saltwater aquaculture facilities. *Proc. Annu. Meet.—World Maric. Soc.* **5**, 357-369.
Valle, B. L. (1959). Biochemistry, physiology, and pathology of zinc. *Physiol. Rev.* **39**, 443-490.
Van Olst, J. C. (1975). "Beneficial Use of Thermal Effluent in Aquaculture." Annual report to Southern California Edison Company. San Diego State Univ., San Diego, California.
Van Olst, J. C., and Carlberg, J. M. (1978a). "Beneficial Use of Thermal Effluent in Aquaculture." Annual report to the Southern California Edison Company. San Diego State Univ., San Diego, California.
Van Olst, J. C., and Carlberg, J. M. (1978b). The effects of container size and transparency on growth and survival of lobsters cultured individually. *Proc. Annu. Meet.—World Maric. Soc.* **10**, 469-479.
Van Olst, J. C., and Ford, R. F. (1976). Use of thermal effluent in aquaculture. *Univ. Calif., Sea Grant Annu. Rep., Publ.* No. 57, pp. 39-44.
Van Olst, J. C., Carlberg, J. M., and Ford, R. F. (1975). Effects of substrate type and other factors on the growth survival, and cannibalism of juvenile *Homarus americanus* in mass rearing systems. *Proc. Annu. Meet.—World Maric. Soc.* **6**, 61-74.
Van Olst, J. C., Ford, R. F., Carlberg, J. M., and Dorband, W. R. (1976). Use of thermal effluent in culturing the American lobster. *In* "Power Plant Waste Heat Utilization in Aquaculture," Workshop 1, pp. 71-100. PSE&G Co., Newark, New Jersey.
Van Olst, J. C., Carlberg, J. M., and Ford, R. F. (1977). A description of intensive culture systems for the American lobster, *Homarus americanus*, and other cannibalistic crustaceans. *Proc. Annu. Meet.—World Maric. Soc.* **8**, 271-292.
Weiss, H. M. (1970). The diet and feeding behavior of the lobster, *Homarus americanus*, in Long Island Sound. Ph.D. Thesis, University of Connecticut, Storrs.
Wright, J. H. (1976). "Lobster Mariculture Project Feasibility Study," Environ. Syst. Dep. Tech. Rep. April 1976. Westinghouse Electric Corp., Pittsburgh, Pennsylvania.
Wuhrmann, K., and Woker, H. (1948). Toxicology of fish. II. Experimental investigations on the toxicity of ammonia and cyanide. *Schweiz. Z. Hydrol.* **11**, 210-244.
Yarosh, M. M., Nichols, B. L., Hirst, E. A., Michel, J. W., and Yee, W. C. (1972). "Agricultural and Aquacultural Uses of Waste Heat," ORNL-4797. Oak Ridge Natl. Lab., Oak Ridge, Tennessee.
Yee, W. C. (1971). Food values from heated waters—an overview. *Proc. 32nd Annu. Meet. Chemurgic Counc., 1970,* Chemurgic Council, New York.

Index

A

Abundance, 193, 209, 212, 229, 235–238, 295, 304
Activity rhythms, 87, 124–127, 153–155, 159–161
Aerococcus viridans var. *homari*, 347, 271
American lobster, see *Homarus americanus*
Antennules, 80
Aquaculture, 187, 333
 communal rearing, 356–357, 359–360, 364–365
 heavy metal contamination of culture system, 353–355
 individual rearing, 365–367
 water quality, 349–356
 water systems, 352–355, 364

B

Bait, 269–270
Behavioral ecology, 79, 92, 108–111, 124–132, 151–162, 356–360
Biological rhythms, 86–91
Borer, 105–107

C

Cannibalism, 356–357, 359, 361, 365
Catchability, 182, 251
Chemoreception, 80
Circadian clock, 91

Clawed lobsters, see Nephropidae
Collectors, 46
Commercial production,
 computer model, 374–375
 demand, 377–378
Communication, 357–358
Coral lobsters, see Synaxidae
Culture, larvae, 18, 33, 39
Culture systems, 367–368
 three dimensional array, 369, 372–374
 two dimensional array, 367–372

D

Defensive behavior, 83
Delayed recruitment models, 220, 235–238
Dens, 66, 67, 68, 87, 92
Diets,
 artificial, 341–344
Disease,
 bacterial, 347–348
 fungal, 347–349
Dispersal, 67, 78–79
Dynamic pool models, 185, 206–212, 227

E

Ecdysis, 25
Egg-carrying capacity, 73–74
Egg-carrying time, 76
Electrophoresis, polyacrylamide gel, 192
Endocrinology, 25

European lobster, see *Homarus gammarus*
Excretion, 349-351

F

Fecundity, 199-200, 228-229
Feeding behavior, 79, 87, 125, 127-130, 269-270
Fisheries,
  boats, 268
  shipment of lobsters, 270-271
  storage of lobsters, 270-271, 338-339, 284
  temperature effect on yield, 290-294, 300-304
Fisheries models,
  Beverton and Holt, 185, 201, 203-204, 206-208, 212, 226-227, 253
  Newton-Raphson, 203
  Ricker, 184-185, 203, 206-209
  Shaefer, 185, 254

G

Gaffkemia, 347, 284
Growth, 5-6, 48, 69-70, 72, 85, 118-122, 165, 170-173, 183, 192-198, 212, 226-228, 339-344, 365-367
Growth factor, 195

H

Hatcheries, 333-334, 361, 364
Hatching, 12, 13, 29
Homarus, 273, 337-338
  dens, 66
  larvae, 19
  larvae,
    behavior, 26
    culture, 18, 48
Homarus americanus, 5, 97, 98, 187, 220, 348-349
  aquaculture, 334-335, 337-338, 359-360, 367, 372, 374, 378
  behavior, 130, 356
  determinants of geographical distribution, 113-114
  distribution, 98, 219
  feeding behavior, 119, 122, 127-129

fishery, 221-222, 265-266, 268, 270-271, 277, 304-305, 318
fishery,
  canneries, 283-284
  live lobster industry, 283-284
  marketing, 285-287
  models, 229-233
  temperature effect on yield, 290-294, 300-304
  total catch, 287-295, 297
growth, 48, 70, 118-122, 226-227, 339-343, 352
habitat type, 99-107
hatching, 12-13
heavy metal regulation, 353-355
home range, 122-123, 224-225
hybridization, 346-347
interspecific relations, 131-132
larvae,
  biochemistry, 25
  crowding effects, 24
  culture, 18, 19, 360-361, 363-364
  distribution, 28
  ecology, 14-18
  growth, 14, 363-364
  light responses, 7, 22, 24, 26-27
  metabolism, 24-25
  pollutant effects, 28, 351
  pressure response, 7, 27
  respiration rate, 25
  salinity effects on growth and survival, 22
  substrate effects on growth and survival, 24
  swimming behavior, 27
  temperature effects on growth and survival, 19, 21, 22
management, 221-222, 294-300, 304-305
migrations, 119, 122-124
molting, 25, 118-122, 365
mortality, 228
nocturnal behavior, 124-127
population density, 113
predation, 125-127
reproduction, 345
sex ratio of populations, 115-117
shelter occupancy, 110-113
size composition of populations, 115-117
Homarus gammarus, 97-98, 225, 330
  aquaculture, 334, 337-339, 359-360
  determinants of geographical distribution, 113-114
  distribution, 99, 219

Index

feeding behavior, 130
fishery, 220, 222–223, 230, 265–266, 270–271, 274, 277, 297, 304–305, 317–318, 338–339
fishery,
  commercial value, 280–282
  management, 278–279
  temperature effect on yield, 300–304
  total catch, 279–282, 318, 320
growth, 226–227
habitat type, 107–108
hatching, 12, 13
heavy metal regulation, 353–355
hybridization, 346
larvae,
  culture, 18
  ecology, 15
  growth, 14
  light response, 27
  pollutant effects, 28
  pressure response, 27
  swimming behavior, 27
  temperature tolerances, 21
management strategies,
  closed seasons, 322
  controlled fishing effort, 324–325, 327
  habitat improvement, 326
  licensing, 322–323
  minimum landing size, 320, 323–324, 326
  protection of ovigerous females, 320–322
  recruitment enhancement, 325
metamorphosis, 14
mortality, 228
nocturnal behavior, 127
sex ratio of populations, 117–118, 229
shelter occupancy, 110–113
size composition of populations, 117–118
stocks, 318–320
Hybridization, 345–347

I

Ibacus ciliatus ciliatus, 29, 31, 39, 336
Ibacus novemdentatus, 39, 336
  naupliosoma larva, 29
Ibacus peronii, 39

J

Jasus, 29, 45–47

distribution, 61
reproduction, 199–200
Jasus edwardsii, 35, 182, 193, 336
  fishery models, 206–207
  growth, 196
  larvae,
    feeding, 39
    naupliosoma, 29
  predation, 45
  salinity tolerance, 86
  sexual dimorphism, 69
  sexual maturation, 74, 199, 245
Jasus frontalis, 190
Jasus lalandii, 29, 86, 91, 193, 203, 211, 244, 336
  biological rhythms, 87
  breeding season, 71
  defensive behavior, 83
  dens, 66–67
  feeding, 80–81
  growth, 85, 192
  larvae,
    abundance, 41
    feeding, 39
    naupliosoma, 29–30
  molting, 72, 194
  reproduction, 73–74, 199–200
  sex ratio, 70
  sexual dimorphism, 69
  social structure, 67
Jasus novaehollandiae, 70, 194–195, 200, 206, 245
Jasus paulensis, 80, 201, 204, 246
Jasus tristani, 62, 80, 190, 194–195, 245
Jasus verreauxi, 29, 39, 200, 246
Justitia, 61

L

Larvae,
  antennae, 46
  culture, 18, 33, 39, 334–338, 360–364
  diet, 361–363
  dispersal, 36–37
  distribution, 15–16, 27, 37
  metabolism, 24
  movement, 26–27
  predation, 17, 45
  salinity effects, 27, 41–42
  stage specific mortality rates, 47
  vertical migration, 37

Linuparus, 61
Lobster fisheries, 3–5, 97–98, 181–187

M

Management of fisheries, 75–76, 183–187, 221–224, 261–263
Maturation, 72–74, 165–166, 198–199
Maximum equilibrium yield, 235–238
Maximum sustained yield, 185–187
Metanephrops andamanicus, 12
Metanephrops challengeri, 276–277
Metanephrops rubellus, 276
Molting, 25, 32, 72, 77–78, 118–122, 170–173, 194–196, 199, 212, 226, 339–341
Mortality rates,
  density dependent, 209–210
  fishing, 173–174, 200, 202–203, 206–208, 227–228
  natural, 174, 200–201, 203–204, 228, 235
Mouthparts, 18

N

Natant larvae, 33
Naupliosoma larva, 29–31
Nephropidae, 3–4, 11
Nephrops norvegicus, 5, 7, 125, 182, 187, 225, 337
  behavior, 151–162
  behavior,
    endogenous rhythms, 159–161
    feeding, 155–158
    light influence on emergence, 153–155, 160–161
    local movements, 161–162
  breeding cycle, 165–169
  density of burrows, 150–151
  fecundity, 168–169, 229
  fishery, 220, 223–224, 265–266, 268–269, 271, 304, 317–318, 327–331
  fishery,
    marketing, 273–274
    temperature effect on yield, 300–302
    total catch, 272–273
  geographic distribution, 17, 143–144, 219–220
  growth, 165, 170–173, 227
  habitat, 144–151
    burrow structure, 147–150, 170–171
    sediment characteristics, 145–146
  hatching, 12–13
  larvae, 13–14, 17, 19, 21, 26
    temperature tolerances, 22
  management, 318, 328–331
    minimum size, 275–276
    minimum stretched mesh size of trawl net, 275–276, 328, 330–331
    protection of ovigerous females, 274–275
  metamorphosis, 170
  molting, 170–173
  mortality, 173–174, 228
  oxygen level influence, 146
  predation, 158–170
  recruitment, 170, 173
  sex ratio of populations, 163–166
  size composition of populations, 163–165
  temperature influence, 146
Nephropsis aculeatus, 276
Nestler, 105
Nisto, 33
Norway lobster, see *Nephrops norvegicus*

O

Optimum sustained yield, 186

P

Palinurellus, 29
Palinurellus gundlachi gundlachi, 29
Palinuridae, 3, 4, 59, 189–217
  aquaculture, 335–337
  behavioral ecology, 79–91
  fisheries, 243–246
  growth, 69–70, 85, 192–197
  habitat selection, 60–62, 64
  larvae, 6–7, 11, 29, 33
    culture, 39
    phyllosoma, 11, 47
  population dynamics, 253–254
  reproductive ecology, 69–78, 84
  sexual maturation, 72, 73
Palinurus, 29, 61
Palinurus delagoae, 61
Palinurus elephas, 199–200, 336

Index

Palinurus gilchristi, 65, 182, 245
Palinurus vulgaris, see *P. elephas*
Palinustus, 61
Panulirus, 29, 43, 45-47, 61, 200
Panulirus argus, 5, 84, 86, 91, 182, 189, 192, 211, 258
  activity rhythms, 87-91
  aquaculture, 336-337
  dens, 66, 67
  egg carrying capacity, 74
  feeding, 79-82, 87
  fishery, 78, 206-208, 211, 245
  growth, 70, 192, 194-196
  habitat, 63, 64
  larvae,
    development, 30, 31, 38
    distribution, 39
    prenaupliosoma, 31
  mating behavior, 69
  migration, 85
  molting, 194-195
  molting and mating, 72, 77, 78
  natural mortality, 83
  population dynamics,
    density, 193
    mortality rate, 204
  predation of phyllosoma larvae, 45
  puerulus, 35, 38
    settlement, 44, 62
  reproduction, 71, 73, 75, 85, 198-200
  sex ratio, 70, 198
  social structure, 68
Panulirus burgeri, 31
Panulirus cygnus, 5, 87, 91, 183
  aquaculture, 335, 337
  dens, 66
  feeding, 80
  fishery, 192, 203-204, 206-211, 245-246, 254-255
  catchability, 251
  marketing, 250
  methods of operation, 246-247
  processing, 248-249
  growth rates, 69, 72, 192, 210-211
  habitat, 63, 65, 248
  larvae,
    development, 30, 32, 33
    dispersal, 36-38
    phyllosoma, 45
    response to salinity, 42

  vertical migration, 43
  management, 255-256
    limitation of fishing effort, 256
    objectives, 251
    yield models, 253-254
  molting, 72, 194, 196
  population dynamics,
    density, 193
    mortality rate, 201, 203, 210-211
  puerulus, 35, 46
    phototactic responses, 46
    settlement, 42, 44
  reproduction, 72, 73, 84, 198-200
Panulirus dasypus, 31
Panulirus delagoae, 65
Panulirus echinatus, 211
Panulirus gracilis, 35, 36
Panulirus guttatus, 70, 71
Panulirus homarus homarus, 39, 45-46, 64, 192, 194, 196, 198-200
Panulirus homarus rubellus, 35, 64, 66, 68, 73, 75, 82
Panulirus inflatus, 32, 39, 192, 336
Panulirus interruptus, 46, 66, 80, 192, 246, 262
  aquaculture, 335-336
  growth, 85, 192
  larvae, 41
    dispersal, 36, 37
    feeding, 7, 41
  puerulus, 35, 42, 44, 46
  reproduction, 199
Panulirus japonicus, 39, 45, 46, 87, 192, 194, 199, 246, 336
Panulirus laevicauda, 189, 199, 246
Panulirus longipes longipes, 39, 64, 69, 336
Panulirus ornatus, 46, 182
  habitat, 64-66
  puerulus, 35, 46
  sex ratio, 70
  sexual dimorphism, 69
Panulirus pascuensis, 211
Panulirus penicillatus, 4, 36, 46, 64-65, 336
Panulirus polyphagus, 46, 189, 336
Panulirus rissoni, 211
Panulirus versicolor, 64-66, 74, 198
Paribacus antarcticus, 336
Phyllosoma larva, 29, 31, 36, 37
  culture, 335-336
  light responses, 43
  movement, 45

Phyllosoma larva (cont.)
  nutritional requirements, 39
Plankton kriesel, 360-361, 364
Planktonic larvae, 47
Population dynamics,
  autoregressive-moving average model, 230-231
  multiple regression on principle components, 231-232
  polynomial distributed lag model, 232-233
  simple regression and correlation models, 229-230
Population growth curve, 195-197
Postlarvae, 33
Pots, see Traps
Prenaupliosoma, 29, 31
Prephyllosoma larva, 29, 31
Projasus, 12, 61
Projasus parkeri, 35, 61, 65
Pseudibacus larva, 33
Pueblo village, 105-107
Puerulus, 33, 35, 37, 41, 46, 86, 209-210, 336-337
Puerulus, 61
Puerulus angulatus, 61, 65, 91
Puerulus carinatus, 91
Puerulus sewelli, 86

R

Recruitment, 206-211, 235-238
Reproductive ecology, 69-78, 165-169
Reproductive potential, 73
  index, 74-75
Reptant larvae, 33
Rock lobsters, see Palinuridae

S

Sampling, 4, 71, 75
  mark-recapture, 193, 203, 220, 161-162, 175
  plankton, 15-16, 47
  trawling, 152-153, 161-163, 174-175, 182, 220, 223-224

Schaefer surplus-yield model, 234-237, 185
Scyllaridae, 4
  aquaculture, 335-336
  larvae, 29, 45
  culture, 39, 41, 45
  phyllosoma, 33, 47
  vertical distribution, 43
Scyllarides aequinoctialis, 29
Scyllarides herklotsi, 30
Scyllarides squamosus, 31
Scyllarus, 45
Scyllarus americanus, 7, 31, 39, 336
Scyllarus bicuspidatus, 336
Scyllarus sordidus, 336
Sex ratio, 70, 115-118, 163-166, 197-198, 228-229
Sexual dimorphism, 69
Slipper lobsters, see Scyllaridae
Social structure, 67-68
Spawning, 71-73, 165-166, 168, 200, 212, 235-236, 345
Spiny lobsters, see Palinuridae
Statocyst, 26
Stock identification, 224-225
Stocks, 191-193, 205, 209-213, 234-235, 238, 271, 318, 320
Stridulation, 84
Surplus yield models, 205-206, 211-212, 234-238
Synaxidae, 4, 11, 29

T

Tagging, 173, 175, 193
Total allowable catch, 186, 238
Traps, 221-223, 245, 247, 262, 267-270
Trawling, 220, 271

V

Von Bertalanffy growth model, 119, 183, 195-197, 201-202, 206, 226-227